Universitext

Series editors

Universitext

Universitext is a series of textbooks that presents material from a wide variety of mathematical disciplines at master's level and beyond. The books, often well class-tested by their author, may have an informal, personal even experimental approach to their subject matter. Some of the most successful and established books in the series have evolved through several editions, always following the evolution of teaching curricula, into very polished texts.

Thus as research topics trickle down into graduate-level teaching, first textbooks written for new, cutting-edge courses may make their way into *Universitext*.

More information about this series at http://www.springer.com/series/223

Gheorghe Moroşanu

Functional Analysis
for the Applied Sciences

 Springer

Gheorghe Moroşanu
Romanian Academy of Sciences
Bucharest, Romania

Department of Mathematics
Babes-Bolyai University
Cluj-Napoca, Romania

ISSN 0172-5939   ISSN 2191-6675   (electronic)
Universitext
ISBN 978-3-030-27152-7   ISBN 978-3-030-27153-4   (eBook)
https://doi.org/10.1007/978-3-030-27153-4

Mathematics Subject Classification (2010): 32A70

This Springer imprint is published by the registered company Springer Nature Switzerland AG.
The registered company address is: Gewerbestrasse 11, 6330 Cham, Switzerland

Dedicated to my wife, Carmen

Preface

The goal of this book is to present in a friendly manner some of the main results and techniques in Functional Analysis and use them to explore various areas in mathematics and its applications. Special attention is paid to creating appropriate frameworks towards solving different problems in the field of differential and integral equations. In fact, the flavor of this book is given by the fine interplay between the tools offered by Functional Analysis and some specific problems which are of interest in the Applied Sciences.

The table of contents of the book (see below) offers a fairly good description of the material. In contrast with other books in the field, we present in Chap. 1 the real number system, describing the Cantor–Méray model which is most appropriate for our purposes here. Indeed, it is based on a completion procedure, allowing the extension from rational numbers to real numbers. This procedure involves the concepts of limit and infinity that are specific to analysis. We consider the Cantor–Méray construction as the corner stone of mathematical analysis, which is why we pay attention to this subject which is usually assumed well known.

In order to help the reader to understand the richness of ideas and methods offered by Functional Analysis, we have included a section of exercises at the end of each chapter. Some of these exercises supplement the theoretical material discussed in the corresponding chapter, while others are mathematical problems that are related to the real world. Some of the exercises are borrowed from other books, being reformulated and/or presented in a form adapted to the needs of the corresponding chapter. We do not indicate the books where individual exercises come from, but all those sources are included into the reference list of our book. In any event, we do not claim originality in such cases. Other exercises were invented by us to offer the reader enough material to understand the theoretical part of the book and gain expertise in solving practical problems. In the last chapter of the book (Chap. 12), we provide solutions to almost all exercises. This is in contrast to many other books which include exercises without solutions. For easy exercises, we provide hints or final solutions, and answers to very easy exercises are left to the reader. I encourage everybody to spend some time working on an exercise before looking at its solution.

We shall refer to an exercise by indicating the chapter and exercise numbers (and not the section number). For example, Exercise 11.3 will mean Exercise 3 in the last section of Chap. 11 (which is Sect. 11.3 in this case).

The book is addressed to graduate students and researchers in applied mathematics and neighboring fields of science.

I would like to thank the anonymous reviewers whose pertinent comments improved the initial version of the book.

Special thanks are due to a former American student of mine, Ivan Andrus, who wrote the first draft of the present book as lecture notes for my Functional Analysis lectures in 2010. He also carefully checked the final version of the book and suggested several minor changes.

I am also indebted to my former student Liviu Nicolaescu for reading the first part of the book and correcting some errors.

Last but not least, I would like to thank Mrs. Elizabeth Loew, Executive Editor at Springer, for our very kind cooperation that led to the successful completion of this book project.

Cluj-Napoca, Romania   Gheorghe Moroşanu

Contents

Chapter 1

Introduction

This chapter comprises definitions, notation, and basic results related to set theory, real and complex numbers, and linear spaces.

1.1 Sets

We assume that the reader is familiar with the basic concepts and results of set theory. However, we are going to recall or specify some concepts and symbols that will be frequently used in this book.

First of all, in this book the notation $A \subset B$ or $B \supset A$ indicates that every element (member) of the set A is also an element of the set B. In particular, $A \subset A$. The empty set, i.e., the set with no elements, will be denoted as usual by $\emptyset$. The empty set is a subset of every set A, $\emptyset \subset A$. The sets A, B are equal, $A = B$, if and only if $A \subset B$ and $B \subset A$.

We assume that the sets

$\mathbb{N} = \{1, 2, \dots\}$ (natural numbers),

$\mathbb{Z} = \{\dots, -2, -1, 0, 1, 2, \dots\}$ (integers), and

$\mathbb{Q} = \{0\} \cup \{\pm m/n; \ m, n \in \mathbb{N}, \ (m, n) = 1\}$ (rational numbers)

are well known, including their axiomatic definitions.

A set A is called *countable* if there exists an injective function from A to $\mathbb{N}$. If one can find a bijective function from A to $\mathbb{N}$ then S is called *countably infinite*. In particular, $\mathbb{N}$, $\mathbb{Z}$, and $\mathbb{Q}$ are countably infinite sets. In fact, a countable set is either finite or countably infinite.

© Springer Nature Switzerland AG 2019

G. Moroşanu, *Functional Analysis for the Applied Sciences*,

Universitext, https://doi.org/10.1007/078 3 030 27153-4_1

Ordered Sets. A *partial order* on a given set A is a binary relation $\leq$ over A satisfying the following conditions for $x, y, z \in A$: (a) $x \leq x$; (b) if $x \leq y$ and $y \leq x$, then $x = y$; (c) if $x \leq y$ and $y \leq z$, then $x \leq z$. We say that $x < y$ if $x \leq y$ and $x \neq y$. The symbols $\geq$ and $>$ have natural meanings: $x \geq y$ iff $y \leq x$, and $x > y$ iff $y < x$.

If A is endowed with a partial order, then A is called a *partially ordered set*. For example, $\mathbb{N}$ is partially ordered with respect to the divisibility relation ($m \leq n$ if m is a divisor of n); also, the set of subsets of a given set S is partially ordered by the inclusion relation. Note that in these examples there are pairs of elements which are not comparable with respect to the corresponding order, which is why the order is called partial.

If A is a set with a partial order $\leq$, then a subset $B \subset A$ is said to be *totally ordered* (or a *chain*) if any two elements $x, y \in B$ are comparable, i.e., either $x \leq y$ or $y \leq x$ (including the case $x = y$).

Let B be a subset of A. An element $z \in A$ is an *upper bound* for B if $x \leq z$ for all $x \in B$. If B has an upper bound, it is said to be *bounded above*. An element $m \in A$ is a *maximal element* of A if there is no $x \in A$, $x \neq m$, such that $m \leq x$. A maximal element of A is not necessarily an upper bound for A.

The set A is called *inductive* if any totally ordered subset of A has an upper bound.

Now, let us recall an important result which is known as Zorn's Lemma[1]:

Theorem 1.1 (Zorn's Lemma). *Every nonempty, partially ordered, inductive set has a maximal element.*

If B is a nonempty subset of a partially (possibly totally) ordered set A, the *supremum of B*, denoted $\sup B$, is defined as the *least upper bound* of B. An element $b \in A$ is the least upper bound of B if and only if

(i) $x \leq b$ for all $x \in B$;

(ii) if $a < b$ then a is not an upper bound of B, i.e., there exists an $x \in B$ such that $a < x$.

If $\sup B$ exists, then it is unique. If B has a greatest element b (i.e., $x \leq b$ for all $x \in B$), then $b = \sup B$.

---

[1]Max August Zorn, German mathematician, 1906–1993.

1.2 Sequences

A *sequence* in a nonempty set X is an ordered list of elements from X, and can be defined as a function $f : D \to X$ whose domain D is a countable, totally ordered set. The case when D is finite is not considered in this book. We shall mostly consider that $D = \mathbb{N}$ and the sequence is usually denoted $(a_n)_{n \in \mathbb{N}}$, or simply (a_n), where $a_n = f(n)$ for all $n \in \mathbb{N}$. Sometimes we consider infinite subsets of $\mathbb{N}$, for instance, $D = \{m, m + 1, \dots\}$, $m \in \mathbb{N}$, $m > 1$, and in this case the sequence is denoted $(a_n)_{n \geq m}$. A sequence can also be indicated by listing its terms: $(a_n)_{n \in \mathbb{N}} = (a_1, a_2, \dots)$. For example, $(1, 3, 5, 7, \dots)$ is the sequence of odd natural numbers. It is worth pointing out that a term (element) can appear several times in a sequence, e.g., $(a_n)_{n \in \mathbb{N}} = (0, 1, 0, 1, 0, 1, \dots)$, where $a_{2k-1} = 0$ and $a_{2k} = 1$ for all $k \in \mathbb{N}$.

A *subsequence* of a given sequence $(a_n)_{n \in \mathbb{N}} = (a_1, a_2, \dots)$ is a new sequence $(b_k)_{k \in \mathbb{N}}$, obtained by removing some terms from $(a_1, a_2, \dots)$ and preserving the order of the remaining terms, i.e.,

$$b_k = a_{n_k}, \ k \in \mathbb{N},$$

where $n_1 < n_2 < \cdots$

We close this section by noting that further details on sequences will be discussed later.

1.3 Real Numbers

While everybody feels comfortable dealing with rational numbers, in order to understand the larger set of real numbers some effort is needed.

Real numbers are needed since the set of rational numbers $\mathbb{Q}$ is not sufficiently large for many purposes. For example, the equation $p^2 = 2$ has no solution in $\mathbb{Q}$. This assertion was first proved by Euclid.[2] In fact, it was observed that the diagonal and the side of any square are incommensurable, i.e., the length p of the diagonal of the unit square is not a rational number. Indeed, p must satisfy the equation $p^2 = 2$. One needs to find a number p (which cannot be a rational one) to represent the length of that diagonal. Many other similar examples

---

[2]Greek mathematician, known as father of Geometry, born around 330 BC, presumably in Alexandria, Egypt.

appear when trying to express areas, volumes, weights, etc. So, it was really necessary to enlarge the set $\mathbb{Q}$ to obtain a set $\mathbb{R}$, called the set of real numbers, within which inconveniences as those described above do not occur. The elements of $\mathbb{R} \smallsetminus \mathbb{Q}$ will be called *irrational numbers*. In particular, the irrational number $\sqrt{2}$ will be the precise representation for the length of the diagonal of the unit square. In fact, we will see that the equation $p^2 = 2$ discussed above has two solutions in $\mathbb{R}$, $+\sqrt{2}$ and $-\sqrt{2}$.

Roughly speaking, $\mathbb{R}$ is the completion of $\mathbb{Q}$, as we will explain below. First of all, let us recall an axiomatic definition of $\mathbb{R}$: $\mathbb{R}$ *is an ordered field, containing $\mathbb{Q}$ as a subfield, and having the least upper bound property.* More precisely, $\mathbb{R}$, endowed with two internal operations, addition and multiplication, denoted "+" and "·", and a total order, denoted "$\leq$", satisfies the following axioms:

(A1) $x + y = y + x$ for all $x, y \in \mathbb{R}$;

(A2) $(x + y) + z = x + (y + z)$ for all $x, y, z \in \mathbb{R}$;

(A3) there exists an element $0 \in \mathbb{R}$ such that $x + 0 = x$ for all $x \in \mathbb{R}$;

(A4) for all $x \in \mathbb{R}$ there exists an element $-x \in \mathbb{R}$ such that $x + (-x) = 0$;

(M1) $xy = yx$ for all $x, y \in \mathbb{R}$ (note that here and in what follows $x \cdot y$ is also denoted xy);

(M2) $(xy)z = x(yz)$ for all $x, y, z \in \mathbb{R}$;

(M3) there exists an element $1 \in \mathbb{R}$, $1 \neq 0$, such that $1 \cdot x = x$ for all $x \in \mathbb{R}$;

(M4) for all $x \in \mathbb{R} \smallsetminus \{0\}$ there exists an element $x^{-1} \in \mathbb{R}$ (called the inverse of x, also denoted $\frac{1}{x}$ or $1/x$) such that $x \cdot x^{-1} = 1$;

(D) $x(y + z) = xy + xz$ for all $x, y, z \in \mathbb{R}$ (the distributive law);

(O1) if $x, y \in \mathbb{R}$ and $x \leq y$, then $x + z \leq y + z$ for all $z \in \mathbb{R}$;

(O2) if $x, y \in \mathbb{R}$ and $x \geq 0$, $y \geq 0$, then $xy \geq 0$;

(LUBP) for every nonempty subset A of $\mathbb{R}$ that is bounded above (i.e., A has an upper bound) there exists $\sup A \in \mathbb{R}$.

The axiom (LUBP) is called the least upper bound property (which is why it is so denoted) or the completeness axiom (this name will be clarified in the following).

Remark 1.2. The fact that $\mathbb{Q}$ is a subfield of $\mathbb{R}$ means that $\mathbb{Q} \subset \mathbb{R}$ and the operations of addition and multiplication in $\mathbb{R}$ are also internal operations in $\mathbb{Q}$. In fact, any ordered field $\mathbb{K}$ contains a subfield $\mathbb{Q}_\mathbb{K}$ which is isomorphic to $\mathbb{Q}$. Indeed, the function $g : \mathbb{Q} \to \mathbb{K}$, defined by $g(m/n) = (m \cdot 1_\mathbb{K}) \cdot (n \cdot 1_\mathbb{K})^{-1}$, is an injective morphism, so $g(\mathbb{Q})$ is a subfield of $\mathbb{K}$ isomorphic to $\mathbb{Q}$. Thus, the condition from the definition above, that $\mathbb{R}$ contains $\mathbb{Q}$ as a subfield is superfluous if we admit that $\mathbb{Q}$ is unique up to isomorphism. We merely wanted to make it clear that $\mathbb{R}$ is an extension of $\mathbb{Q}$.

Remark 1.3. It is worth pointing out that the extension from rational numbers to real numbers is the result of a long investigative process extended over more than 2000 years. The problem was clarified in the nineteenth century. There are several models for $\mathbb{R}$ defined by the above system of axioms, such as the **Stolz–Weierstrass model**,[3] based on decimal expansions; **Dedekind's model**,[4] based on the so-called Dedekind cuts and the **Cantor–Méray model**.[5] All these models are based on approximation (as are all models of $\mathbb{R}$). We shall describe the Cantor–Méray construction which involves Cauchy sequences of rational numbers and uses the basic properties of $\mathbb{Q}$ as an ordered field. Intuitively speaking, according to this construction, $\mathbb{R}$ will consist of all rational numbers, plus "limits" of Cauchy[6] sequences in $\mathbb{Q}$ which are not rational numbers. The most important step in this construction (completion procedure) will be to show that the completeness axiom is satisfied by this model, denoted $\mathbb{R}_{C-M}$ ($C - M$ comes from Cantor–Méray), thus ensuring that any Cauchy sequence of rational numbers is "convergent" (has a "limit") in $\mathbb{R}_{C-M}$. But such "limits" cannot be used in this construction (one cannot define real numbers by themselves!), so instead we consider as elements of $\mathbb{R}_{C-M}$ the equivalence classes of Cauchy rational sequences (two sequences being equivalent if the corresponding sequence of differences

---

[3]Otto Stolz, Austrian mathematician, 1842–1905; Karl Weierstrass, German, known as father of modern analysis, 1815–1897.

[4]Richard Dedekind, German mathematician, 1831–1916.

[5]Georg Cantor, German mathematician, 1845–1918; Charles Méray, French mathematician, 1835–1911.

[6]August-Louis Cauchy, French mathematician, engineer and physicist, 1789–1857.

approaches zero); one considers equivalence classes because the sequence which is supposed to define ("converge to") a real number is
not unique.

Finally, we will prove that any two copies of $\mathbb{R}$ are isomorphic, thus
concluding that $\mathbb{R}$ is unique up to isomorphism.

Before presenting in detail the Cantor–Méray model, we will make a
few comments and derive some abstract results regarding $\mathbb{R}$ as defined
by the axioms above.

Remark 1.4. It is easily seen that (LUBP) implies that for any nonempty set $A \subset \mathbb{R}$ which is bounded below (i.e., has a lower bound),
there exists the greatest lower bound of A, denoted $\inf A \in \mathbb{R}$. In fact,
$\inf A = -\sup\{x \in \mathbb{R};\ -x \in A\}$. The converse implication is also true,
so one may replace (LUBP) by this equivalent statement.

Remark 1.5. It is worth pointing out that the (LUBP) is precisely
what makes the difference between $\mathbb{R}$ and $\mathbb{Q}$. Indeed, $\mathbb{Q}$ is an ordered
field, but does not satisfy the (LUBP), as illustrated by the following
counterexample:

Let $A \subset \mathbb{Q}$ denote the set $\{p \in \mathbb{Q} : p > 0,\ p^2 < 3\}$. A is nonempty,
since $1 \in A$. Obviously, A is bounded above (e.g., 2 is an upper bound
of A). Assume by contradiction that there exists a number $\alpha \in \mathbb{Q}$
which is the least upper bound of A, $\alpha = \sup A$. Then $\alpha \geq 1$ and we
need to examine the following three possibilities: $\alpha^2 < 3$, $\alpha^2 = 3$, and
$\alpha^2 > 3$.

If $\alpha^2 < 3$, then $(2\alpha + 3)/(\alpha + 2) > \alpha$, and $(2\alpha + 3)/(\alpha + 2) \in A$, so α
is not even an upper bound of A.

The case $\alpha \in \mathbb{Q}$, $\alpha^2 = 3$ is impossible (prove it!).

Finally, if $\alpha^2 > 3$, then $\beta := (2\alpha + 3)/(\alpha + 2) \in \mathbb{Q}$, $\beta > 0$ (since $\alpha \in \mathbb{Q}$,
$\alpha \geq 1$), and $\alpha - \beta = (\alpha^2 - 3)/(\alpha + 2) > 0$, hence $\beta < \alpha$. On the other
hand, $3 - \beta^2 = (3 - \alpha^2)/(\alpha + 2)^2 < 0$, so $\beta^2 > 3$. It follows that β
is an upper bound for A, with $\beta < \alpha$. This contradicts the fact that
$\alpha = \sup A$.

Since none of the above cases is possible, there is no rational number
α such that $\alpha = \sup A$. Therefore $\mathbb{Q}$ does not satisfy the (LUBP).

Note that if A is considered as a subset of $\mathbb{R}$, then there exists $\sup A =
\sqrt{3} \in \mathbb{R} \setminus \mathbb{Q}$ (see below).

Now, we present a result known as the Archimedean[7] property:

---

[7]Archimedes of Syracuse, 287–212 BC.

Theorem 1.6. *If $x, y \in \mathbb{R}$ and $x > 0$, then there exists $n \in \mathbb{N}$ such that $nx > y$.*

Proof. Assume that, on the contrary, $nx \leq y$ for all $n \in \mathbb{N}$, so the set $A = \{nx; n \in \mathbb{N}\}$ is bounded above. Then the (LUBP) implies that there exists $\alpha = \sup A \in \mathbb{R}$. Since $\alpha - x < \alpha$, there exists an element of A, say mx, with $m \in \mathbb{N}$, such that $\alpha - x < mx$ which is equivalent to $\alpha < (m + 1)x \in A$. This contradicts the fact that α is an upper bound of A.   □

Theorem 1.7. $\mathbb{Q}$ *is dense in $\mathbb{R}$, i.e., between any two distinct real numbers there is a rational number.*

Proof. Let $x, y \in \mathbb{R}$, $x < y$. Since $y - x > 0$ it follows by the Archimedean property that there exists an $n \in \mathbb{N}$ such that

$$n(y - x) > 1. \tag{1.3.1}$$

By the same Archimedean property there exist $w, z \in \mathbb{N}$ such that $-w < -nx < z$. In fact, w can be replaced by $m := -\sup\{r \in \mathbb{Z} : -w \leq r < -nx\}$, so $nx < m$. Moreover,

$$nx < m \leq nx + 1. \tag{1.3.2}$$

By (1.3.1) and (1.3.2) we can conclude that $x < m/n < y$.   □

Theorem 1.8 (existence of n-th roots of positive reals). *For all $x \in \mathbb{R}$, $x > 0$, and for all $n \in \mathbb{N}$, $n \geq 2$, there exists a unique $y \in \mathbb{R}$, $y > 0$, such that $y^n = x$.*

Proof. The uniqueness of y follows from the implication $0 < y_1 < y_2 \Rightarrow y_1^n < y_2^n$. To prove the existence of y consider the set $A = \{t \in \mathbb{R}; t > 0, t^n < x\}$. A is nonempty, since it contains $t_1 = x/(1 + x)$. Indeed, $t_1^n < t_1 < x$. A is also bounded above (for example, $1 + x$ is an upper bound for A). By the (LUBP) there exists $y = \sup A \in \mathbb{R}$, $y > 0$. Let us prove that $y^n = x$. Assuming that $y^n < x$, we have for $0 < \varepsilon < 1$,

$$(y + \varepsilon)^n - y^n = \varepsilon[(y + \varepsilon)^{n-1} + y(y + \varepsilon)^{n-2} + \cdots + y^{n-1}] < \varepsilon n(y + 1)^{n-1}.$$

Hence

$$(y + \varepsilon)^n < y^n + \varepsilon z, \tag{1.3.3}$$

where $z = n(y+1)^{n-1}$. By the Archimedean property, there is a $k \in \mathbb{N}$, $k \geq 2$, such that $\varepsilon = 1/k$ satisfies

$$\varepsilon z < x - y^n. \tag{1.3.4}$$

From (1.3.3) and (1.3.4) it follows that $y + \varepsilon \in A$ which contradicts the fact that $y = \sup A$. We can also show that $y^n > x$ leads to a contradiction. Hence, $y^n = x$. $\square$

The n-th root y of the real number $x > 0$ is denoted $\sqrt[n]{x}$ ($\sqrt{x}$ if $n = 2$) or $x^{1/n}$. At this moment, we can see that in particular the equation $p^2 = 2$ can be solved in $\mathbb{R}$: $p^2 = 2 \Leftrightarrow p^2 - (\sqrt{2})^2 = 0 \Leftrightarrow (p - \sqrt{2})(p + \sqrt{2}) = 0$, so there are two solutions, $p = \sqrt{2}$ and $p = -\sqrt{2}$. The number $\sqrt{2}$, which is irrational, represents in particular the length of the diagonal of the unit square. So, the difficulty pointed out by Euclid can be handled in $\mathbb{R}$. Similarly, $\sqrt{3}$ is an irrational number representing the length of the diagonal of the unit cube.

Remark 1.9. Sometimes it is useful to represent numbers by points on a straight line. First, let us mark arbitrarily two distinct points O and A on the straight line to represent the numbers 0 and 1. The line segment OA is called the unit segment. If we choose a point P to the right of A, such that OP consists of m unit segments, $m \in \mathbb{N}$, $m \geq 2$, then P represents the natural number m. The negative integers are similarly represented by points on the left of O, following the natural order $\ldots, -3, -2, -1$. So now we have a directed straight line, called the *number line*, including the positive half-line (on the right of O) and the negative half-line. One can also associate with any rational number a point on the number line. For example, if one divides OA into 2 equal parts and choose a point R on the positive half-line, such that OR is equal to 3 such parts, then R represents 3/2. Obviously, the points corresponding to distinct rational numbers are distinct too. Note that the set of points on the number line corresponding to all the rational numbers does not cover the number line. For example, the point D corresponding to the length of the diagonal of the unit square (i.e., $\sqrt{2}$) is on the number line (D being constructible by using a ruler and compass). We will discuss later the representation of irrational numbers by points on the number line.

Sequences of Real Numbers. A sequence $(a_n)_{n \in \mathbb{N}}$ in $\mathbb{R}$ is said to be *increasing* (or *nondecreasing*) if $a_n \leq a_{n+1}$ for all $n \in \mathbb{N}$. If $a_n < a_{n+1}$ for all $n \in \mathbb{N}$, then (a_n) is called *strictly increasing*.
Similarly, if the order relations "$\leq$" and "$<$" are replaced by "$\geq$" and "$>$", we obtain the definitions for a *decreasing* (or *nonincreasing*) *sequence*, and a *strictly decreasing sequence*, respectively.

A sequence $(a_n)_{n\in\mathbb{N}}$ in $\mathbb{R}$ is said to be *bounded above* (*bounded below*) if there exists an $M \in \mathbb{R}$ such that $a_n \leq M$ ($a_n \geq M$, respectively) for all $n \in \mathbb{N}$. If (a_n) is bounded both above and below, then it is called *bounded*.

A sequence $(a_n)_{n\in\mathbb{N}}$ in $\mathbb{R}$ is said to be *convergent* if there exists a number $a \in \mathbb{R}$ (called *limit of* (a_n)) such that

$$\forall \varepsilon \in \mathbb{R}, \ \varepsilon > 0, \ \exists N = N(\varepsilon) \in \mathbb{N} \ \text{such that} \ \forall n > N, \ |a_n - a| < \varepsilon.$$

Here, $|\cdot|$ means the absolute value function, i.e., $|x| = x$ if $x \geq 0$, and $|x| = -x$ if $x < 0$. The above definition (of a convergent sequence) will be discussed again later in a more general framework. Here we are interested in some properties of sequences of real numbers.

It is easily seen that any convergent sequence is bounded, and its limit is unique.

Next, we state the so-called *Monotone Convergence Theorem*:

Theorem 1.10 (Monotone Convergence Theorem). *Any sequence* $(a_n)_{n\in\mathbb{N}}$ *in* $\mathbb{R}$ *which is increasing (decreasing) and bounded is convergent.*

Proof. We consider the case when (a_n) is increasing and bounded (the other case is similar). Since the set of all a_n's (where repetitions are eliminated) is bounded above, it follows by (LUBP) that there exists its supremum $a \in \mathbb{R}$. Thus, for all $\varepsilon \in \mathbb{R}$, $\varepsilon > 0$, there exists an $N \in \mathbb{N}$, such that $a - \varepsilon < N$. Since (a_n) is increasing, we have $a - \varepsilon < a_n$ for all $n > N$, so
$$|a_n - a| = a - a_n < \varepsilon \ \ \forall n > N. \qquad \square$$

We continue with the following result known as Bolzano–Weierstrass' Theorem.[8]

Theorem 1.11 (Bolzano–Weierstrass). *Every bounded sequence in* $\mathbb{R}$ *has a convergent subsequence.*

Proof. Let $(a_n)_{n\in\mathbb{N}}$ be a bounded sequence in $\mathbb{R}$. Let k be a natural number with the property $a_k > a_m$ for all $m > k$. Assume there are infinitely many such k's, say $k = n_j$, $n_1 < n_2 < \cdots < n_j < \cdots$. Then, the subsequence $(a_{n_j})_{j\in\mathbb{N}}$ is strictly decreasing, hence convergent since it is also bounded (cf. Theorem 1.10).

---

[8]Bernard Bolzano, Bohemian mathematician, logician, philosopher, and theologian, 1781–1848.

If the set of such k's is finite (possibly empty), we denote by K the maximum of such k's. Obviously, for $n_1 = K + 1$ there exists an $n_2 \in \mathbb{N}$, such that $a_{n_1} \leq a_{n_2}$. Now, since n_2 does not belong to the set of k's, there exists an $n_3 \in \mathbb{N}$ such that $a_{n_2} \leq a_{n_3}$. Continuing this procedure we obtain a subsequence $(a_{n_j})_{j \in \mathbb{N}}$ which is increasing and bounded, hence convergent (cf. Theorem 1.10). $\qquad\square$

A sequence $(a_n)_{n \in \mathbb{N}}$ in $\mathbb{R}$ is said to be a *Cauchy sequence* if

$$\forall \varepsilon \in \mathbb{R}, \ \varepsilon > 0, \ \exists N = N(\varepsilon) \in \mathbb{N} \ \text{such that}$$
$$\forall n, m > N, \ |a_n - a_m| < \varepsilon.$$

Theorem 1.12. *A sequence in $\mathbb{R}$ is Cauchy if and only if it is convergent.*

Proof. Let $(a_n)_{n \in \mathbb{N}}$ be a Cauchy sequence in $\mathbb{R}$. It is easily seen that (a_n) is bounded. Thus, by Theorem 1.11, there is a convergent subsequence, say $(a_{n_k})_{k \in \mathbb{N}}$. Let $a \in \mathbb{R}$ be its limit. By the triangle inequality (which obviously holds in $\mathbb{R}$), we have

$$|a_n - a| \leq |a_n - a_{n_k}| + |a_{n_k} - a|.$$

Using this inequality we easily conclude that (a_n) is convergent (with the same limit a).
The converse implication is trivial. $\qquad\square$

The facts recalled above, derived from the axiomatic definition of $\mathbb{R}$, are important in real analysis and also help us understand the Cantor–Méray model for $\mathbb{R}$.

The Cantor–Méray Construction. Assume that $\mathbb{Q}$ (the ordered field of rational numbers) is known. We want to extend $\mathbb{Q}$ to obtain a larger ordered field satisfying in addition the (LUBP). Denote by $S_{\mathbb{Q}}$ the collection of all Cauchy sequences of rational numbers. When defining a Cauchy sequence in $\mathbb{Q}$ we require $\varepsilon \in \mathbb{Q}, \ \varepsilon > 0$ (since the extension of $\mathbb{Q}$ is not yet known). Define the following equivalence relation in $S_{\mathbb{Q}}$

$$(a_n) \sim (b_n) \ \text{iff} \ \forall \varepsilon \in \mathbb{Q}, \ \varepsilon > 0, \ \exists N \in \mathbb{N} \ \text{such that}$$
$$\forall n > N, |a_n - b_n| < \varepsilon.$$

For example, the sequences (a_n), (b_n), (c_n), defined by $a_n = 1/n$, $b_n = n/(n^2 + 1)$, $c_n = 0$ for all $n \geq 1$, belong to the same equivalence class, i.e., the class of the constant sequence $(0, 0, \ldots)$, which can be identified with $0 \in \mathbb{Q}$. We identify any $r \in \mathbb{Q}$ with the equivalence class of the constant sequence $(r, r, \ldots)$.

Let us denote by $\mathbb{R}_{C-M}$ the set of all equivalence classes in $S_{\mathbb{Q}}$ (with respect to the equivalence relation defined above). Obviously, $\mathbb{Q}$ can be regarded as a subset of $\mathbb{R}_{C-M}$ (in view of the natural identification mentioned above).

Now, one defines in a natural manner the operations of addition and multiplication in $\mathbb{R}_{C-M}$. Specifically, if a, b are classes in $\mathbb{R}_{C-M}$ with representatives (a_n), $(b_n) \in S_{\mathbb{Q}}$, then $a + b$ and ab are defined as the equivalence classes of $(a_n + b_n)$ and $(a_n b_n)$, respectively. Also, $a \leq b$ if for all $\varepsilon \in \mathbb{Q}$, $\varepsilon > 0$, there exists an $N \in \mathbb{N}$ such that $b_n - a_n \geq -\varepsilon$ for all $n \geq N$. Note that the strict inequality $a < b$ (i.e., $a \leq b$ and $a \neq b$) can be equivalently expressed as follows: there exists an $\varepsilon_0 \in \mathbb{Q}$, $\varepsilon_0 > 0$, such that $b_n - a_n \geq \varepsilon_0$ for all n large enough. Likewise, these definitions do not depend on specific representatives.

It is easily seen that $\mathbb{R}_{C-M}$ is an ordered field satisfying axioms $(A1) - (A4)$, $(M1) - (M4)$, (D), and $(O1) - (O2)$.

Let us now prove that $\mathbb{R}_{C-M}$ also satisfies the (LUBP). Let Ω be a nonempty subset of $\mathbb{R}_{C-M}$ which is bounded above, with upper bound of $a \in \mathbb{R}_{C-M}$. One may assume that a is the class of a constant sequence $(u_0, u_0, \ldots)$ with $u_0 \in \mathbb{Q}$ (if this is not the case, we can use the information that a Cauchy sequence in $\mathbb{Q}$ has an upper bound in $\mathbb{Q}$, so a can be replaced by the class of a constant sequence $(u_0, u_0, \ldots)$, where u_0 is a large rational number).

Let us pick an $s_0 \in \Omega$ and a rational number l_0 such that $l_0 < s_0$, where l_0 is identified with the class of the constant sequence $(l_0, l_0, \ldots)$. Next, we construct two sequences of rational numbers (u_n) and (l_n) as follows: $u_1 = u_0$ and $l_1 = l_0$, then, successively, for $n = 1, 2, \ldots$, either $u_{n+1} = (u_n + l_n)/2$, $l_{n+1} = l_n$ if $(u_n + l_n)/2$ is an upper bound of Ω, or $u_{n+1} = u_n$ and $l_{n+1} = (u_n + l_n)/2$ if $(u_n + l_n)/2$ is not an upper bound of Ω. By induction we can see that u_n is an upper bound of Ω for all $n \in \mathbb{N}$, while l_n is not an upper bound of Ω for any $n \in \mathbb{N}$. Obviously, (u_n) and (l_n) are Cauchy sequences, so their classes $u, l \in \mathbb{R}_{C-M}$, and in fact $u = l$, since $|u_n - l_n| = u_n - l_n = (u_0 - l_0)/2^{n-1}$, $n \geq 1$. It is also obvious that u is an upper bound of Ω. Let us prove that u is the least upper bound: $u = \sup \Omega$. Assume that there exists a smaller upper bound, say $v \in \mathbb{R}_{C-M}$, $v < u = l$. Since $l_k \leq l_{k+1}$ for all $k \in \mathbb{N}$, there exists an $N \in \mathbb{N}$ such that $v < l_N$. But l_N is not an upper bound of Ω, hence $v = u$ cannot be an upper bound of Ω, leading to a contradiction. Therefore, $\mathbb{R}_{C-M}$ satisfies all the axioms and is indeed a model for $\mathbb{R}$.

Remark 1.13. Let us summarize: any element $x \in \mathbb{R}_{C-M}$ is the equivalence class of a Cauchy sequence in $\mathbb{Q}$, say (r_n) (this could be a constant sequence if $x \in \mathbb{Q}$); since $\mathbb{R}_{C-M}$ is a model for $\mathbb{R}$ (a complete ordered field), we know that (r_n) is convergent (see Theorem 1.12); its limit (which is independent of the choice of (r_n) in the class x) can be identified with x. So now we have a clear representation of $\mathbb{R}_{C-M}$, including rational and irrational numbers.

The Real Number System (Model) is Unique up to Isomorphism. Let $\hat{\mathbb{R}}$ be another model for $\mathbb{R}$. As before, we admit that $\mathbb{Q}$ is unique up to isomorphism, so $\mathbb{Q}$ is a subfield of both $\mathbb{R}_{C-M}$ and $\hat{\mathbb{R}}$. Since $\mathbb{Q}$ is dense in $\hat{\mathbb{R}}$ (see Theorem 1.7), for any $x \in \hat{\mathbb{R}}$, there exists a sequence of rational numbers (r_n) that converges to x (this sequence can be the constant sequence $(x, x, \dots)$ if $x \in \mathbb{Q}$). Of course, (r_n) is a Cauchy sequence. We associate with such an x the class of (r_n) with respect to the equivalence class "$\sim$" defined above. So we have defined a mapping $\phi : \hat{\mathbb{R}} \to \mathbb{R}_{C-M}$, $\phi(x) =$ the class of (r_n). It is easily seen that ϕ is a bijection, and

$$
\begin{aligned}
\phi(x+y) &= \phi(x) + \phi(y) \quad \forall x, y \in \hat{\mathbb{R}}, \\
\phi(x \cdot y) &= \phi(x) \cdot \phi(y) \quad \forall x, y \in \hat{\mathbb{R}}, \\
x > 0 &\implies \phi(x) > 0.
\end{aligned}
$$

Therefore, $\hat{\mathbb{R}}$ is isomorphic to $\mathbb{R}_{C-M}$, hence any two real number models are isomorphic. So in what follows we will consider that the real number system is unique and denote it by $\mathbb{R}$.

The Dedekind–Cantor Axiom on Continuity of a Straight Line. We discussed in Remark 1.9 how to represent rational numbers on a directed straight line. Now, taking into account the Cantor–Méray construction, we can complete the procedure by representing irrational numbers. We see that to every real number there corresponds a unique point of the directed straight line, and the correspondence is one-to-one. The Dedekind–Cantor axiom stipulates that there are no gaps on the line after representing all real numbers, that is *there is a one-to-one correspondence between $\mathbb{R}$ and the points of the directed straight line*. The directed straight line will be called the *real line*, and real numbers will be sometimes called *points*.

The Extended Real Number System. Sometimes it is necessary to describe mathematically what happens "beyond" real numbers. For example, $1/x$ gets closer and closer to zero when x gets larger and larger. Having in mind that the point on the real line corresponding to x goes far away to the right, we usually say that x tends to infinity, and write $x \to +\infty$. The fact that $1/x$ tends to zero as $x \to \infty$ can be written as $\frac{1}{+\infty} = 0$.

Similar situations require the introduction of the symbol $-\infty$. So we are led to the so-called *extended real number system*,

$$\overline{\mathbb{R}} := \mathbb{R} \cup \{-\infty, +\infty\}.$$

The usual order in $\mathbb{R}$ is preserved, and we define

$$-\infty < x < +\infty \quad \forall x \in \mathbb{R}.$$

Then $+\infty$ $(-\infty)$ is an upper bound (lower bound, respectively) of every nonempty subset of $\mathbb{R}$. Moreover, any nonempty subset has a least upper bound. For instance, $E = \{x + \frac{1}{x} : x \in \mathbb{R}, x \neq 0\}$ has $\sup E = +\infty$ and $\inf E = -\infty$. The symbol $+\infty$ is also denoted by ∞. In accordance with our intuition, we adopt the following conventions

$$
\begin{aligned}
x + \infty &= \infty, \; x - \infty = -\infty, \; \frac{x}{\infty} = \frac{x}{-\infty} = 0 \quad \forall x \in \mathbb{R}; \\
x \cdot \infty &= \infty, \; x \cdot (-\infty) = -\infty \quad \forall x \in \mathbb{R}, \, x > 0; \\
x \cdot \infty &= -\infty, \; x \cdot (-\infty) = +\infty \quad \forall x \in \mathbb{R}, \, x < 0; \\
\infty + \infty &= \infty, \; -\infty - \infty = -\infty, \\
\infty \cdot \infty &= \infty, \; \infty \cdot (-\infty) = -\infty, \; (-\infty) \cdot (-\infty) = +\infty.
\end{aligned}
$$

On the other hand, operations like

$$0 \cdot (\pm\infty), \; \infty - \infty, \; \frac{\infty}{\infty}$$

are not accepted. For example, $x/(1 + x^2)$ approaches 0 as $x \to \infty$, while $x/(1 + \sqrt{x})$ approaches $+\infty$ as $x \to \infty$. Thus, the quotient of two large numbers may approach either 0 or ∞. That is why we say that $\frac{\infty}{\infty}$ does not make sense.

Note that $\overline{\mathbb{R}}$ does not form a field (why?).

We assume familiarity of the reader with sequences and series of real numbers. For information see, for example, [33, 41, 42].

The Number e. Sometimes checking whether a real number is irrational is not a trivial task. The number often known as e is an example in this respect. It is defined as the sum of a series, namely

$$e = \sum_{n=0}^{\infty} \frac{1}{n!},$$

where $n! = 1 \cdot 2 \cdot 3 \cdots n$ for $n \geq 1$, and $0! = 1$. Let s_n denote the partial sum of the series, i.e., $s_n = \sum_{k=0}^{n} \frac{1}{k!}$. By the ratio test we see that the series converges, hence $e \in \mathbb{R}$. More precisely,

$$2 < e < 1 + \sum_{k=0}^{\infty} \frac{1}{2^k} = 3. \tag{1.3.5}$$

Note that

$$e - s_n < \frac{1}{(n+1)!} \sum_{k=0}^{\infty} \frac{1}{(k+1)^2},$$

hence

$$0 < e - s_n < \frac{2}{n!n}. \tag{1.3.6}$$

Let us now prove that e is irrational. Assume the contrary, that $e = p/q$, where $p, q \in \mathbb{N}$, $(p, q) = 1$. In fact, $q > 1$ (see (1.3.5)). From (1.3.6) we infer that

$$0 < q!\left(\frac{p}{q} - s_q\right) < \frac{2}{q}. \tag{1.3.7}$$

Observing that $q!s_q \in \mathbb{N}$, we have $m := q!\left(\frac{p}{q} - s_q\right) \in \mathbb{N}$. So we deduce from (1.3.7) that $0 < m < 1$ which is impossible (there is no integer between 0 and 1). Therefore e is irrational, as claimed.

Remark 1.14. By an argument from Rudin [42, p. 64] we see that e is also the limit of the sequence $(x_n)_{n \in \mathbb{N}}$ defined by $x_n = \left(1 + \frac{1}{n}\right)^n$. Using the binomial formula we can write

$$x_n = 1 + 1 + \frac{1}{2!}\left(1 - \frac{1}{n}\right) + \frac{1}{3!}\left(1 - \frac{1}{n}\right)\left(1 - \frac{2}{n}\right) + \cdots$$
$$+ \frac{1}{n!}\left(1 - \frac{1}{n}\right)\left(1 - \frac{2}{n}\right) \cdots \left(1 - \frac{n-1}{n}\right).$$

Then, for all $n, m \in \mathbb{N}$, $n \geq m \geq 2$, we have

$$1 + 1 + + \frac{1}{2!}\left(1 - \frac{1}{n}\right) + \cdots + \frac{1}{m!}\left(1 - \frac{1}{n}\right)\left(1 - \frac{2}{n}\right) \cdots$$
$$\left(1 - \frac{m-1}{n}\right) \leq x_n \leq s_n,$$

which implies

$$1 + 1 + \frac{1}{2!} + \cdots + \frac{1}{m!} \leq \liminf x_n \leq \limsup x_n \leq e.$$

Therefore, $e = \lim x_n$ exists, as claimed.

1.4   Complex Numbers

We assume that the reader is familiar with the complex field. In what follows we just recall its construction and some notation.

Let $\mathbb{C}$ denote the Cartesian[9] product $\mathbb{R} \times \mathbb{R}$ equipped with two internal operations, addition and multiplication, defined as follows:

$$(x, y) + (u, v) = (x + u, y + v), \quad (x, y) \cdot (u, v) = (xu - yv, xv + yu).$$

It is easy to check that $\mathbb{C}$ is a field, with $(0, 0)$ and $(1, 0)$ in the role of 0 and 1, respectively. In particular, for any $z = (x, y) \in \mathbb{C}$, $z \neq (0, 0)$, we have $z^{-1} = \left(\frac{x}{x^2+y^2}, \frac{-y}{x^2+y^2} \right)$.

Note that the set $\mathbb{R}_1 := \{(x, 0); x \in \mathbb{R}\}$ is a subfield of $\mathbb{C}$ with respect to the above operations that read in this case

$$(x, 0) + (u, 0) = (x + u, 0), \quad (x, 0) \cdot (u, 0) = (xu, 0).$$

Thus any $(x, 0)$ can be identified with x and $\mathbb{R}_1$ with these operations can be identified with $\mathbb{R}$ with the usual operations of addition and multiplication. So $\mathbb{R}$ can be viewed as a subfield of $\mathbb{C}$.

Any $z = (x, y) \in \mathbb{C}$ can be decomposed as $z = (x, 0) + (y, 0) \cdot (0, 1)$, so in view of the above identification, we can write $z = x + yi$, where $i := (0, 1)$. Note that $(0, 1) \cdot (0, 1) = (-1, 0)$, thus we can write $i^2 = -1$; i is called the *imaginary unit*.

Summarizing, we can write $\mathbb{C} = \{x + yi; x, y \in \mathbb{R}\}$ and observe that the two operations initially defined can be viewed as the addition and multiplication similar to those used for real numbers if we admit that $i^2 = -1$.

The elements $z = x + yi$ of $\mathbb{C}$ are called *complex numbers* and $\mathbb{C}$ is known as the *complex field* or *complex number system*. For a complex number $z = x + yi$, the real numbers x and y are called the *real part* and the *imaginary part* of z, respectively (denoted $x = \text{Re}\, z$,

---

[9]René Descartes, latinized Renatus Cartesius, French mathematician, philosopher, and scientist, 1596–1650.

$y = \operatorname{Im} z$). Complex numbers $z = x + yi$ can be represented by points (of coordinates x, y) in the complex plane which is determined by two orthogonal directed straight lines with the same unit, the x-axis (real axis) and the y-axis (imaginary axis).

Let $\bar{z} = x - yi$ be the *complex conjugate* of $z = x + yi$. Note that $z \cdot \bar{z} = x^2 + y^2 \in \mathbb{R}$. The number $|z| = \sqrt{x^2 + y^2}$ is called the *magnitude* of z, and it represents the length of the segment connecting the origin O of the complex plane and the point of coordinates x and y corresponding to z.

1.5   Linear Spaces

Recall that a nonempty set X is said to be a *linear space* (or *vector space*) over a field $\mathbb{K}$ if there exist a binary operation on X, called addition, $+ : X \times X \to X$, and an external binary operation, called scalar multiplication, $\cdot : \mathbb{K} \times X \to X$, such that the following axioms are satisfied

$(A1)$ $(x + y) + z = x + (y + z) \quad \forall x, y, z \in X$;

$(A2)$ $x + y = y + x \quad \forall x, y \in X$;

$(A3)$ $\exists 0 \in X$, called zero, such that $x + 0 = x \quad \forall x \in X$;

$(A4)$ $\forall x \in X \; \exists -x \in X$ such that $x + (-x) = 0$;

$(A5)$ $1 \cdot x = x \quad \forall x \in X$, where 1 is the unit element of the field $\mathbb{K}$;

$(A6)$ $\alpha(\beta x) = (\alpha\beta)x \quad \forall \alpha, \beta \in \mathbb{K}, \; \forall x \in X$;

$(A7)$ $(\alpha + \beta)x = \alpha x + \beta x \quad \forall \alpha, \beta \in \mathbb{K}, \; \forall x \in X$;

$(A8)$ $\alpha(x + y) = \alpha x + \alpha y \quad \forall \alpha \in \mathbb{K}, \; \forall x, y \in X$.

The first four axioms ensure that X is an Abelian[10] group with respect to addition.

In the following $\mathbb{K}$ will be either the field $\mathbb{R}$ of real numbers or the filed $\mathbb{C}$ of complex numbers, and X will be called a real or complex space, respectively.

A nonempty subset Y of X which is a linear space with respect to the same operations is called a *subspace* of X. In fact, a necessary and

---

[10]Niels Henrik Abel, Norwegian mathematician, 1802–1829.

sufficient condition for a nonempty subset Y of X to be a subspace is that Y be closed under the operations, i.e.,

$$\forall\, x, y \in Y,\ \forall\, \alpha \in \mathbb{K},\ x + y \in Y,\quad \alpha x \in Y.$$

If S is a nonempty subset of a linear space X, we denote by Span S the collection of all finite linear combinations of elements of S, i.e.,

$$\text{Span}\, S = \left\{ \sum_{i=1}^{k} \alpha_i x_i = \alpha_1 x_1 + \cdots + \alpha_k x_k;\ k \in \mathbb{N},\ \alpha_i \in \mathbb{K}, \right.$$

$$\left. x_i \in S,\ i = 1, \ldots, k \right\}.$$

Obviously, Span S is a linear subspace of X, called the *linear subspace generated by S* (and S is said to be a *system of generators*).

We recall that $x_1, x_2, \ldots, x_k \in X$ (where X is a linear space) are said to be *linearly dependent* if there exist some scalars $\alpha_1, \ldots, \alpha_k \in \mathbb{K}$, not all zero, such that $\alpha_1 x_1 + \cdots + \alpha_k x_k = 0$. Otherwise, the vectors $x_1, x_2, \ldots, x_k$ are called *linearly independent* (and $\{x_1, x_2, \ldots, x_k\}$ is said to be a *linearly independent system*). In this case, $S = \{x_1, x_2, \ldots, x_k\}$ is a *basis* of the space $Y = \text{Span}\, S$ (which could be the whole of X), and we say that Y has dimension k, $\dim Y = k$, and any vector $x \in Y$ can be uniquely expressed as a linear combination,

$$x = \sum_{i=1}^{k} \alpha_i x_i = \alpha_1 x_1 + \cdots + \alpha_k x_k,$$

where $\alpha_1, \ldots, \alpha_k \in \mathbb{K}$ are called *coordinates* of x with respect to the basis S.

A basis is not unique.

A linear space X is infinite dimensional (written as $\dim X = \infty$) if for any $k \in \mathbb{N}$ there exist k vectors in X which are linearly independent. If X contains only the null vector, then by convention we define $\dim X = 0$.

Recall that any two linear spaces X, Y are *isomorphic* if there exists a bijection $\phi : X \to Y$ which satisfies

$$\phi(\alpha x + \beta y) = \alpha \phi(x) + \beta \phi(y) \quad \forall\, \alpha,\, \beta \in \mathbb{K},\ \forall\, x, y \in X.$$

If either of the two (isomorphic) spaces is finite dimensional then the other is also finite dimensional and $\dim X = \dim Y$ (prove it!).

Scalar Product. An important concept that allows the extension of some properties of classical Euclidean geometry to general linear spaces is the scalar product.

A *scalar product* (or *inner product*) on a linear space X is a mapping from $X \times X$ to $\mathbb{K}$, denoted $(\cdot, \cdot)$, which satisfies the following axioms

$$(a_1) \quad (x, y) = \overline{(y, x)} \quad \forall\, x, y \in X\,,$$

$$(a_2) \quad (x + y, z) = (x, z) + (y, z) \quad \forall\, x, y, z \in X\,,$$

$$(a_3) \quad (\alpha x, y) = \alpha(x, y) \quad \forall\, \alpha \in \mathbb{K},\ \forall\, x, y \in X\,,$$

$$(a_4) \quad (x, x) \geq 0 \ \ \forall x \in X, \ \ \text{and} \ \ (x, x) = 0 \iff x = 0\,.$$

We have denoted by $\overline{(y, x)}$ the complex conjugate of (y, x) (obviously, $\overline{(y, x)} = (y, x)$ if $\mathbb{K} = \mathbb{R}$). A space X together with such a product is called an *inner product space*. It is easily seen that $(x, \alpha y) = \overline{\alpha}(x, y)$ for all $\alpha \in \mathbb{K}$ and all $x, y \in X$.

Two vectors $x, y \in X$ are called *orthogonal* if their scalar product is equal to zero: $(x, y) = 0$ (this is sometimes denoted $x \perp y$). One can also define the *length* of a vector $x \in X$ as $\|x\| = \sqrt{(x, x)}$. The mapping $x \to \|x\|$ satisfies the following properties:

$$(i) \quad \|x\| = 0 \iff x = 0\,;$$

$$(ii) \quad \|\alpha x\| = |\alpha| \cdot \|x\| \quad \forall \alpha \in \mathbb{K}, \forall\, x \in X\,;$$

$$(iii) \quad \|x + y\| \leq \|x\| + \|y\| \quad \forall\, x, y \in X\,.$$

The mapping $\| \cdot \| : X \to [0, \infty)$ defined above is a *norm* on X, and X is a *normed space*. In general a mapping from X to $[0, \infty)$ satisfying (i), (ii), (iii) is called a norm on X. A given space may have many different norms, but the above is a special norm, being generated by a scalar product.

While (i) and (ii) are trivial, property (iii) (called triangle inequality) follows from the Bunyakovsky–Cauchy–Schwarz[11] inequality:

$$|(x, y)| \leq \|x\| \cdot \|y\| \quad \forall x, y \in X\,, \tag{1.5.8}$$

which is valid in any normed space whose norm is generated by a scalar product. Indeed,

---

[11]Viktor Y. Bunyakovsky, Russian mathematician, 1804–1889; Karl Hermann Amandus Schwarz, German mathematician, 1843–1921.

$$\begin{aligned} \|x + y\|^2 &= \|x\|^2 + 2\operatorname{Re}(x, y) + \|y\|^2 \\ &\leq \|x\|^2 + 2|(x, y)| + \|y\|^2 \\ &\leq \|x\|^2 + 2\|x\| \cdot \|y\| + \|y\|^2 \\ &= (\|x\| + \|y\|)^2, \end{aligned}$$

which clearly implies (iii). As far as (1.5.8) is concerned, its proof is based on the inequality

$$0 \leq \|x + \alpha y\|^2 = \|x\|^2 + 2\operatorname{Re}\overline{\alpha}(x, y) + |\alpha|^2\|y\|^2, \qquad (1.5.9)$$

for all $\alpha \in \mathbb{K}$ and all $x, y \in X$. In fact, we can assume $x \neq 0$ and $y \neq 0$ (otherwise (1.5.8) is trivial). Now replacing in (1.5.9) $\alpha = -(x, y)/\|y\|^2$ yields (1.5.8).

We continue with some examples:

Example 1. For a given $n \in \mathbb{N}$, consider $X = \mathbb{R}^n$, which is the set of all ordered n-tuples (here arranged as $n \times 1$ matrices) $x = (x_1, \ldots, x_n)^T$, where $x_1, \ldots, x_n \in \mathbb{R}$. It is easily seen that $X = \mathbb{R}^n$ is a linear space over $\mathbb{R}$ with respect to the usual operations of addition and scalar multiplication:

$$x + y = (x_1 + y_1, \ldots, x_n + y_n), \quad \alpha x = (\alpha x_1, \ldots, \alpha x_n),$$

for all $x = (x_1, \ldots, x_n)^T$, $y = (y_1, \ldots, y_n)^T \in X$, $\alpha \in \mathbb{R}$. The null (zero) element of X is $(0, 0, \ldots, 0)^T$, while the inverse of any $x = (x_1, \ldots, x_n)^T \in X$ with respect to the addition is $(-x_1, \ldots, -x_n)^T$. The usual scalar product of $X = \mathbb{R}^n$ is defined by

$$(x, y) = \sum_{i=1}^{n} x_i y_i \quad \forall x = (x_1, \ldots, x_n)^T, \; y = (y_1, \ldots, y_n)^T \in X,$$

and the corresponding norm is

$$\|x\| = \sqrt{(x, x)} = \sqrt{\sum_{i=1}^{n} x_i^2}.$$

If $n = 1$ the above scalar product is the usual multiplication in $\mathbb{R}$, while the corresponding norm is the absolute value. If $n = 2$ or $n = 3$ then the above scalar product is nothing else but the scalar product (dot product) of two vectors in the Euclidean plane or space, respectively, while the corresponding norm of a vector represents its length.

Orthogonality of two vectors means the usual geometric orthogonality. For this reason $X = \mathbb{R}^n$ so equipped is called Euclidean n-space. By extension, a general normed space whose norm is generated by a scalar (inner) product is called a generalized Euclidean space (or inner product space, as previously mentioned).

Analogously, $\mathbb{C}^n$ is a linear space over $\mathbb{C}$ with respect to the usual operations of addition and scalar multiplication. Here, the usual scalar product is defined by

$$(x, y) = \sum_{i=1}^{n} x_i \overline{y_i} \quad \forall\, x = (x_1, \ldots, x_n)^T,\ y = (y_1, \ldots, y_n)^T \in \mathbb{C}^n,$$

and the corresponding (Euclidean) norm is

$$\|x\| = \sqrt{(x, x)} = \sqrt{\sum_{i=1}^{n} |x_i|^2}\,.$$

Note that any n-dimensional linear space X over $\mathbb{K}$ is isomorphic to $\mathbb{K}^n$. Indeed, an isomorphism $\phi : X \to \mathbb{K}^n$ is the mapping which associates with any $x \in X$ the vector constructed with the coordinates of x with respect to a basis in X. Thus any such space X can be equipped with a scalar product as follows:

$$(x, y)_X := (\phi(x), \phi(y)) = \sum_{i-1}^{n} \phi(x)_i \cdot \overline{\phi(y)_i} \quad \forall x, y \in X\,.$$

Therefore, any finite dimensional linear space X can be organized as a generalized Euclidean (inner product) space.

Example 2. Let X be the set of all functions from $[0, 1]$ to $\mathbb{K}$. Obviously, X is a linear space with respect to the usual operations of addition and scalar multiplication

$$(f + g)(t) = f(t) + g(t),$$
$$(\alpha f)(t) = \alpha f(t) \quad \forall t \in [0, 1],\ \forall f, g \in X,\ \forall \alpha \in \mathbb{K}.$$

Consider the set Y of all polynomial functions $f : [0, 1] \to \mathbb{K}$ (i.e., $f(t) = a_0 + a_1 t + a_2 t + \cdots + a_k t^k$, $a_0, \ldots, a_k \in \mathbb{K}$, $k \in \{0\} \cup \mathbb{N}$). Obviously Y is a (proper) subspace of X. Note that Y is infinite dimensional ($\dim Y = \infty$) and hence so is X. Indeed, for any $k \in \mathbb{N}$

the set of polynomials $\{1, t, t^2, \ldots, t^k\} \subset Y$ is an independent system. We can define on Y the scalar product

$$(f, g) = \int_0^1 f(t) \cdot \overline{g(t)} \, dt \quad \forall f, g \in Y,$$

and the corresponding norm $\|f\| = \sqrt{(f, f)} = \left(\int_0^1 |f(t)|^2 dt\right)^{1/2}$. Another norm on Y is the following

$$\|f\|_* = \sup_{t \in [0,1]} |f(t)| \quad \forall f \in Y,$$

but this one is not generated by a scalar product. Indeed, if we assume that $\|\cdot\|_*$ is generated by a scalar product, then it must satisfy the parallelogram law

$$\|f + g\|_*^2 + \|f - g\|_*^2 = 2\left(\|f\|_*^2 + \|g\|_*^2\right) \quad \forall f, g \in Y, \qquad (1.5.10)$$

which is valid in any inner product space. But, for example, the polynomial functions $f(t) = t$, $g(t) = 1 - t$ do not satisfy (1.5.10), which confirms our assertion above.

Now, let Z be the set of polynomials $f \in Y$ of degree less than or equal to $n - 1$, for a given natural number n. This is a finite dimensional subspace of Y, with basis $\{1, t, t^2, \ldots, t^{n-1}\}$ and dimension n. Therefore, Z is isomorphic to $\mathbb{K}^n$. A natural isomorphism between Z and $\mathbb{K}^n$ is the mapping which associates with any polynomial function $f(t) = a_0 + a_1 t + a_2 t^2 + \cdots + a_{n-1} t^{n-1}$ the n-dimensional vector $(a_0, a_1, a_2, \ldots, a_{n-1})^T \in \mathbb{K}^n$. Thus, besides the scalar product above, one can define on Z another scalar product

$$(f, g)_Z = \sum_{i=0}^{n-1} a_i \cdot \overline{b_i},$$

for all $f(t) = a_0 + a_1 t + \cdots + a_{n-1} t^{n-1}$, $g(t) = b_0 + b_1 t + \cdots + b_{n-1} t^{n-1}$, where $a_i, b_i \in \mathbb{K}$, $i \in \{0, 1, \ldots, n - 1\}$. This scalar product generates a new norm on Z,

$$\|f\|_Z = \sqrt{(f, f)_Z} = \left(\sum_{i=0}^{n-1} |a_i|^2\right)^{1/2}$$

$$\forall \, f(t) = a_0 + a_1 t + \cdots + a_{n-1} t^{n-1} \in Z.$$

Orthogonal Systems. Let $S \subset X \setminus \{0\}$ be a nonempty countable set, where X is an inner product space with a scalar product $(\cdot, \cdot)$ and the norm $\| \cdot \|$ generated by it. We further assume that S is a linearly independent system (otherwise we eliminate those vectors which are linear combinations of other vectors from S). Recall that an infinite set is linearly independent if all finite subsets of it are independent. Consider first the case when S is a finite independent system, $S = \{u_1, \ldots, u_n\}$. Starting from S one can construct an *orthogonal system* $S' = \{v_1, \ldots, v_n\}$, i.e., $(v_i, v_i) \neq 0$ and $(v_i, v_j) = 0$ if $i \neq j$. In what follows we present the *Gram–Schmidt*[12] *orthogonalization method.* To create S' let

$$v_1 = u_1,$$
$$v_2 = u_2 + \alpha v_1,$$

such that $v_2 \perp v_1$, i.e.,

$$
\begin{aligned}
0 = (v_2, v_1) &= (u_2 + \alpha v_1, v_1) \\
&= (u_2, v_1) + \alpha \|v_1\|^2,
\end{aligned}
$$

giving

$$\alpha = -\frac{(u_2, v_1)}{\|v_1\|^2}.$$

Note that $v_1 = u_1 \neq 0$ (by assumption) and also $v_2 \neq 0$. To see that, we suppose by contradiction that $v_2 = 0$, i.e., $u_2 + \alpha u_1 = 0$. But this is impossible since u_1, u_2 are independent vectors.
After having determined the first p members of S' define

$$v_{p+1} = u_{p+1} + \sum_{j=1}^{p} \beta_j v_j,$$

so that $v_{p+1} \perp v_k$ for all $k = 1, \ldots, p$, that is,

$$0 = (u_{p+1}, v_k) + \beta_k \|v_k\|^2,$$
$$\beta_k = -\frac{(u_{p+1}, v_k)}{\|v_k\|^2}, \quad k = 1, \ldots, p.$$

---

[12] Jorgen Pedersen Gram, Danish actuary and mathematician, 1850–1916; Erhard Schmidt, Baltic German mathematician, 1876–1959.

As before $v_{p+1} \neq 0$ since each v_k is a linear combination of u_k's, $k = 1, \ldots, p$, namely

$$v_{p+1} = u_{p+1} + \sum_{k=1}^{p} \theta_k u_k \, ,$$

and $u_1, u_2, \ldots, u_p, u_{p+1}$ are independent vectors. Continue the process until finished.

Since S' is an orthogonal system, it follows that it is independent (prove it!). S' can be simply replaced by an orthonormal (independent) system $S'' = \{w_1, \ldots, w_n\}$, by defining $w_j = \|v_j\|^{-1} v_j$, $j = 1, \ldots, n$.

In particular, any n-dimensional inner product space possesses an orthonormal basis (since any basis can be replaced by an orthonormal one).

If S is a countably infinite, independent system in X, $S = \{u_1, u_2, \ldots, u_n, \ldots\}$, then using the same Gram–Schmidt method, one can construct an orthonormal system $S'' = \{w_1, w_2, \ldots, w_n, \ldots\}$, i.e., $(w_i, w_j) = \delta_{ij}$, where δ_{ij} is the Kronecker[13] symbol, $\delta_{ii} = 1$ and $\delta_{ij} = 0$ for $i \neq j$.

Linear, Bilinear, Sesquilinear, Quadratic Forms. Let X be a linear space over $\mathbb{K}$. A function $f : X \to \mathbb{K}$ is said to be a *linear form* on X if

$$f(\alpha x + \beta y) = \alpha f(x) + \beta f(y) \quad \forall \alpha, \beta \in \mathbb{K}, \ \forall x, y \in X \, .$$

The set of all linear forms on X, denoted X^*, is a linear space with respect to the usual operations on functions and is called the *dual* of X. If X is finite dimensional, with a basis $B = \{u_1, \ldots, u_n\}$, $n \in \mathbb{N}$, then any linear form f has a specific form:

$$f(x) = \sum_{i=1}^{n} a_i \alpha_i \quad \forall x = \sum_{i=1}^{n} \alpha_i u_i \in X \, ,$$

where $a_i = f(u_i)$ are called coefficients of the linear form with respect to the basis B. X^* is isomorphic to $\mathbb{K}^n$ (hence dim $X^* = n$), since the mapping associating each $f \in X^*$ to the vector $(f(u_1), \ldots, f(u_n))^T \in \mathbb{K}^n$ is an isomorphism (prove it!).

A function $a : X \times X \to \mathbb{K}$ which is linear with respect to each variable is called a *bilinear form* on X (more precisely, $a(\cdot, y)$ is a linear form for all $y \in X$, and $a(x, \cdot)$ is also a linear form for all $x \in X$).

---

[13]Leopold Kronecker, German mathematician, 1823–1891.

If $\mathbb{K} = \mathbb{C}$ and the above condition on $a(x, \cdot)$ is replaced by the linearity of the complex conjugate function $\overline{a(x, \cdot)}$ $(x \in X)$ then a is said to be a *sesquilinear form* on X. For example, a scalar product on X is a sesquilinear form.

If X is finite dimensional, with a basis $B = \{u_1, \ldots, u_n\}$ and a is a bilinear form on X, then for all $x = \sum_{i=1}^{n} \alpha_i u_i$, $y = \sum_{j=1}^{n} \beta_j u_j \in X$ we have

$$a(x, y) = \sum_{i,j=1}^{n} c_{ij} \alpha_i \beta_j, \tag{1.5.11}$$

where $c_{ij} = a(u_i, u_j)$, $i, j = 1, \ldots, n$. Hence a is represented by the matrix $C = (c_{ij})$ (which depends on the basis of X). If a is a sesquilinear form, then instead of (1.5.11) we have

$$a(x, y) = \sum_{i,j=1}^{n} c_{ij} \alpha_i \overline{\beta_j}.$$

A bilinear form $a : X \times X \to \mathbb{R}$ is said to be *symmetric* if $a(x, y) = a(y, x)$ for all $x, y \in X$. If X is finite dimensional, then the symmetry of a bilinear form a is expressed by the symmetry of the matrix associated with a (the symmetry of that matrix being independent of the basis of space X). Any symmetric bilinear form a defines a *quadratic form* $F : X \to \mathbb{R}$ by setting $F(x) = a(x, x)$. Given a quadratic form F, one can recover the corresponding bilinear form a. Indeed, from

$$a(x + y, x + y) = a(x, x) + 2a(x, y) + a(y, y),$$

we deduce

$$
\begin{aligned}
a(x, y) &= \frac{1}{2}[a(x + y, x + y) - a(x, x) - a(y, y)] \\
&= \frac{1}{2}[F(x + y) - F(x) - F(y)].
\end{aligned}
$$

A quadratic form F is said to be *positive definite* (*positive semidefinite*) if $F(x) > 0$ for all $x \in X$, $x \neq 0$ ($F(x) \geq 0$ for all $x \in X$, respectively). F is called *negative definite* (*negative semidefinite*) if $-F$ is positive definite (positive semidefinite, respectively).

If F is a positive definite quadratic form on the real linear space X then the corresponding a is a scalar product on X.

If X is a real n-dimensional linear space, with a basis $B = \{u_1, \ldots, u_n\}$, and F is a quadratic form on X, then

$$F(x) = a(x, x) = \sum_{i,j=1}^{n} c_{ij} \alpha_i \alpha_j \,,$$

where α_j's are the coordinates of x with respect to basis B (in particular, the components of x if $X = \mathbb{R}^n$ with its usual basis). It can be shown (using the well-known Gauss[14] method), that for any such quadratic form F, there is a convenient basis of X such that F can be written as follows:

$$F(x) = \lambda_1 \beta_1^2 + \lambda_2 \beta_2^2 + \cdots + \lambda_n \beta_n^2 \,,$$

where $\beta_1, \ldots, \beta_n$ are the coordinates of x with respect to the new basis, and $\lambda_1, \ldots, \lambda_n \in \mathbb{R}$ (some of these λ's could be zero). In fact, starting from the new basis, one can simply define another basis, such that F can be written under the following canonical form:

$$F(x) = \sum_{i=1}^{n} \varepsilon_i \gamma_i^2, \quad \varepsilon_j \in \{-1, 0, +1\}, \ j = 1, \ldots, n \,,$$

where γ_i's are the coordinates of x with respect to the last basis. Obviously, F is positive definite (positive semidefinite) if and only if $\varepsilon_j = 1$, $j = 1, \ldots, n$ ($\varepsilon_j \in \{0, +1\}$, $j = 1, \ldots, n$, respectively).

Let us also recall that for a quadratic form $F : X \to \mathbb{R}$ (X being an n-dimensional real linear space), whose matrix C with respect to a basis of X has nonzero NW principal minors (i.e., $\Delta_i \neq 0$, $i = 1, \ldots, n$) there always exists a decomposition (called Jacobi's formula[15]) as follows:

$$F(x) = \sum_{i=1}^{n} \frac{\Delta_{i-1}}{\Delta_i} \beta_i^2 \,,$$

where $\Delta_0 := 1$, and $\beta_1, \ldots, \beta_n$ are the coordinates of x with respect to a new basis of X. Therefore, F is positive definite (negative definite) if and only if $\Delta_i > 0$, $i = 1, \ldots, n$ (respectively, $(-1)^i \Delta_i > 0$, $i = 1, \ldots, n$). These are known as Sylvester's conditions.[16]

---

[14]Carl Friedrich Gauss, German mathematician and physicist, 1777–1855.

[15]Carl Gustav Jacob Jacobi, German mathematician, 1804–1851.

[16]James Joseph Sylvester, English mathematician, 1814–1897.

If X is a complex linear space and $a : X \times X \to \mathbb{C}$ is a sesquilinear form, then a is called *Hermitian*[17] if $a(x,y) = \overline{a(y,x)}$ for all $x, y \in X$. Such a form a defines a quadratic form $F(x) = a(x,x)$, $x \in X$, with values in $\mathbb{R}$. If X is an n-dimensional complex linear space, then one can find a basis in X such that F takes the form

$$F(x) = \sum_{i=1}^{n} \lambda_i \beta_i \overline{\beta_i} = \sum_{i=1}^{n} \lambda_i |\beta_i|^2 \,,$$

where $\lambda_i \in \mathbb{R}$, $i = 1, \ldots, n$, and β_i, $i = 1, \ldots, n$, are the coordinates of x with respect to that basis. The Jacobi formula also works in this complex case, and Sylvester's conditions remain valid.

We close this chapter by inviting the reader to consult other books to find more information on the topics addressed in this chapter, such as [6, 16, 28, 33, 37, 41, 42, 51].

1.6   Exercises

1. Let A, B, C be some arbitrary subsets of a universe $\mathcal{U}$. Show that

   (a) $A \setminus (B \cup C) = (A \setminus B) \cap (A \setminus C) = (A \setminus B) \setminus C$;

   (b) $A \setminus (B \cap C) = (A \setminus B) \cup (A \setminus C)$;

   (c) $(A \cap B) \setminus C = A \cap (B \setminus C) = (A \setminus C) \cap B$;

   (d) $(A \cup B) \setminus C = (A \setminus C) \cup (B \setminus C)$;

2. Let A, B, C be given sets, which are subsets of a universe $\mathcal{U}$. Determine the set $X \subset \mathcal{U}$ satisfying

   $$A \cap X = B \text{ and } A \cup X = C\,.$$

   The same question if X satisfies

   $$A \setminus X = B, \text{ and } X \setminus A = C\,.$$

3. Prove that for all sets A, B, C satisfying $A \cap C = B \cap C$ and $A \cup C = B \cup C$, we have $A = B$.

---

[17]Charles Hermite, French mathematician, 1822–1901.

4. Let A, B, C, D be some arbitrary subsets of a universe $\mathcal{U}$. Which of the following statements are true?

   (a) $(A \cap B) \times (C \cap D) = (A \times C) \cap (B \times D)$;
   (b) $(A \cup B) \times (C \cup D) = (A \times C) \cup (B \times D)$;
   (c) $(A \setminus B) \times C = (A \times C) \setminus (B \times C)$,

   where "$\times$" denotes the Cartesian product?

5. Let A be a set with a partial order $\leq$. If A has a smallest element $a = \min A$, then a is the unique minimal element of A.

6. Let $A = \{a_n; a_n = \frac{1}{1 \cdot 2} + \frac{1}{2 \cdot 3} + \cdots + \frac{1}{n(n+1)}, \ n \in \mathbb{N}\}$. Find $\inf A$ and $\sup A$.

7. Define on $\mathbb{C}$ (the set of complex numbers) the binary relation $\preceq$ as follows:

   $$z_1 = x_1 + y_1 i \preceq z_2 = x_2 + y_2 i \iff x_1 \leq x_2 \text{ and } y_1 \leq y_2.$$

   Show that

   (a) $\preceq$ is a partial order on $\mathbb{C}$, but not a total order on $\mathbb{C}$;
   (b) for each $a \geq 0$, $\preceq$ is a total order on $X_a = \{z = x + yi \in \mathbb{C}; \ y = ax\}$ (i.e., X_a is a chain);
   (c) there exists a partial order on $\mathbb{C}$ such that, for each $a < 0$, X_a defined as above is a chain of $\mathbb{C}$ with respect to this partial order.

8. Show that the sequence $(a_n)_{n \geq 1}$ defined by

   $$a_1 = \sqrt{2}, \ a_n = \sqrt{2 + a_{n-1}}, \ n \geq 2,$$

   is convergent and calculate its limit.

9. Let a be a given real number and let $(a_n)_{n \geq 1}$ be a sequence in $\mathbb{R}$ such that any subsequence of it has a convergent subsequence whose limit is a. Show that $a_n \to a$.

10. Let X be a vector space. If $\{v_1, v_2, v_3\} \subset X$ is a linearly independent system, then show that $\{v_1 + v_2, v_2 + v_3, v_3 + v_1\}$ is too.

11. Let X be the real vector space of all functions $f : \mathbb{R} \to \mathbb{R}$. Show that each of the following systems of functions in X

$$S_1 = \{1, \cos x, (\cos x)^2\}, \quad S_2 = \{e^x, xe^x, \ldots, x^k e^x\}, \quad k \in \mathbb{N},$$

is linearly independent.

12. Let X be the real vector space of all continuous functions $f :$ $[0,1] \to \mathbb{R}$. Consider on X the scalar product

$$(f, g) = \int_0^1 f(t) \cdot g(t)\, dt \quad \forall f, g \in X,$$

and the induced norm.

(a) Which of the following systems of functions in X is linearly independent?

(i) $f_1(t) = 1$, $f_2(t) = t$, $f_3(t) = t^2$;

(ii) $f_1(t) = 1 - t$, $f_2(t) = t(1 - t)$, $f_3(t) = 1 - t^2$;

(iii) $f_1(t) = 1$, $f_2(t) = e^t$, $f_3(t) = 2e^{-t}$;

(iv) $f_1(t) = 3t$, $f_2(t) = t + 5$, $f_3(t) = -2t^2$, $f_4(t) = (t+1)^2$;

(v) $f_1(t) = (t+1)^2$, $f_2(t) = t^2 - 1$, $f_3(t) = 2t^2 + 2t - 3$;

(vi) $f_1(t) = 1$, $f_2(t) = 1 + t$, $f_3(t) = 1 + t + t^2, \ldots, f_k(t) =$ $1 + t + t^2 + \cdots + t^{k-1}$, where k is a given natural number.

(b) Let Y be the vector subspace of X generated by $B =$ $\{f_1, f_2, f_3\}$, where $f_1(t) = 1$, $f_2(t) = t$, $f_3(t) = t^2$ for $t \in [0,1]$. By using the Gram–Schmidt method, construct an orthonormal basis in Y with respect to the above scalar product.

13. Show that the system $B = \{1, t - 1, (t - 1)^2, (t - 1)^3\}$ is a basis of the real vector space X of all polynomials of degree ≤ 3 with real coefficients, and find the coordinates of a polynomial $p = p(t) \in X$ with respect to this basis.

14. Let X be a linear space equipped with a scalar product $(\cdot, \cdot)$. Show that a system $S = \{x_1, x_2, \ldots, x_k\} \subset X$ is linearly independent if and only if the following determinant (called the Gram determinant)

$$\det \left((x_i, x_j)_{1 \leq i,j \leq k} \right) \neq 0.$$

15. Show that the mapping $(\cdot, \cdot) : \mathbb{R}^3 \times \mathbb{R}^3 \longrightarrow \mathbb{R}$ defined by

$$(x, y) = \frac{1}{2} \sum_{i=1}^{3} x_i y_i - \frac{1}{4}(x_1 y_2 + x_1 y_3 + x_2 y_3)$$

is a scalar product. Construct a basis of $\mathbb{R}^3$ which is orthonormal with respect to this scalar product.

16. Let X be the real vector space of polynomials of degree $\leq m$ with real coefficients, where m is a given natural number. Find the expression of the linear form $f : X \to \mathbb{R}$ defined by

$$f(p) = \int_0^1 p(t)\, dt \quad \forall p(t) = a_0 + a_1 t + a_2 t^2 + \cdots a_m t^m \in X$$

with respect to each of the bases

$$B = \{1,\, t,\, t^2, \ldots, t^m\}, \quad B' = \{1,\, 1+t,\, 1+t+t^2, \ldots,$$
$$1 + t + t^2 + \cdots + t^m\}.$$

17. Let X be a vector space over $\mathbb{R}$. A bilinear form $a : X \times X \to \mathbb{R}$ is said to be *antisymmetric* if

$$a(x, y) = -a(y, x) \quad \forall x, y \in X.$$

Show that

   (i) a bilinear form $a : X \times X \to \mathbb{R}$ is antisymmetric $\Longleftrightarrow$ $a(x, x) = 0 \ \forall x \in X$.

   (ii) any bilinear form on X is the sum of a symmetric bilinear form and an antisymmetric one.

18. Let A be an $n \times n$ matrix with real entries, and let $B = aI_n + A^T A$, where A^T denotes the transpose of A, I_n denotes the $n \times n$ identity matrix, and $a > 0$. Show that the quadratic form $F : \mathbb{R}^n \to \mathbb{R}$ whose matrix with respect to the canonical basis of $\mathbb{R}^n$ is B is positive definite. What about the case $a = 0$?

19. Consider the quadratic form $F : \mathbb{R}^3 \to \mathbb{R}$,

$$F(x) = x_1^2 + x_2^2 + 3x_3^2 + 4x_1 x_2 + 2x_1 x_3 + 2x_2 x_3 \quad \forall x \in \mathbb{R}^3.$$

Determine a basis of $\mathbb{R}^3$ such that $F(x) = \sum_{i=1}^{n} \varepsilon_i \xi_i^2$, where $\xi_1, \ldots, \xi_n$ are the coordinates of x with respect to this basis, and $\varepsilon_j \in \{-1, 0, +1\}$, $j = 1, \ldots, n$. Check whether F is positive definite, negative definite, or neither.

20. Use Sylvester's conditions to show that the quadratic form $F : \mathbb{R}^4 \to \mathbb{R}$,

$$F(x) = 2x_1^2 + 2x_2^2 + x_3^2 + 4x_4^2 - x_1x_2 + x_1x_3 + x_2x_4 - x_3x_4 \quad \forall x \in \mathbb{R}^4 \,,$$

is positive definite. Determine a basis of $\mathbb{R}^4$ such that F can be written as a sum of squares with respect to this basis.

Chapter 2

Metric Spaces

Metric spaces offer a sufficiently large framework for most of the problems we discuss in this book.

2.1 Definitions

Definition 2.1. *A* **metric** *(or a* **distance function***) on a nonempty set X is a function $d : X \times X \to [0, \infty)$ satisfying*

$$(M1) \qquad d(x,y) = 0 \iff x = y\,;$$
$$(M2) \qquad d(x,y) = d(y,x) \quad \forall x,y \in X\,;$$
$$(M3) \qquad d(x,y) \le d(x,z) + d(z,y) \quad \forall x,y,z \in X\,.$$

A set X equipped with a metric d is called a **metric space** *and is sometimes denoted (X, d).*

Any set $X \ne \emptyset$ can be equipped with a metric. The "simplest" metric is the so-called **discrete metric** which is defined by

$$d(x,y) = 1 \quad \text{if } x,y \in X,\ x \ne y \quad \text{and} \quad d(x,x) = 0 \ \forall x \in X\,.$$

Note that this metric is not very useful in practice, but is suitable for counterexamples.

Now let $(X, \|\cdot\|)$ be a normed (linear) space. Then X can be equipped with the metric

$$d(x,y) = \|x - y\|, \quad x,y \in X\,. \tag{2.1.1}$$

© Springer Nature Switzerland AG 2019
G. Moroșanu, *Functional Analysis for the Applied Sciences*,
Universitext, https://doi.org/10.1007/978-3-030-27153-4_2

Note also that any finite dimensional linear space can be equipped with a norm (e.g., with the Euclidean norm—see the previous chapter), and hence with the metric generated by that norm (cf. (2.1.1)).

If (X, d) is a metric space and $\emptyset \neq Y \subset X$, then Y is also a metric space with respect to d restricted to $Y \times Y$.

Definition 2.2. *Let (X, d) be a metric space. For $x_0 \in X$ and $r > 0$ define*

$$B(x_0, r) := \{x \in X; \, d(x, x_0) < r\},$$

which is called the **open ball** *centered at x_0 with radius r.*

Definition 2.3. *A nonempty set $A \subset (X, d)$ is said to be* **open** *if for each $x \in A$ there exists an $\varepsilon > 0$ such that $B(x, \varepsilon) \subset A$. By convention the empty set is considered open.*

Obviously, the collection τ of all open sets forms a **topology**:

(a) $\emptyset, X \in \tau$;

(b) the union of any sub-collection of τ is in τ;

(c) the intersection of any finite sub-collection of τ is in τ.

Note that the intersection of an infinite collection of open sets may not be open. For example, in $X = \mathbb{R}$, with $d(x, y) = |x - y|$, we have for a fixed $x_0 \in \mathbb{R}$

$$\bigcap_{n=1}^{\infty} \left(x_0 - \frac{1}{n}, \, x_0 + \frac{1}{n}\right) = \{x_0\},$$

and obviously $\{x_0\}$ does not belong to the (usual) topology of $\mathbb{R}$ defined by $|\cdot|$.

In what follows X denotes a metric space endowed with the topology τ generated by its metric d (see above), called **metric topology**. If d is defined by a norm, i.e., $d(x, y) = \|x - y\|$ $(x, y \in X)$, then τ is called a **norm topology**.

A set $V \subset X$ is said to be a **neighborhood of a point** $p \in X$ if there is an $r > 0$ such that $B(p, r) \subset V$. In particular, any open set D is a neighborhood of any $p \in D$.

A set $C \subset X$ is said to be **closed** (with respect to topology τ) if $X \setminus C$ is open (i.e., $X \setminus C \in \tau$). In particular, for any $x_0 \in X$ and $r > 0$, we have $B(x_0, r) \in \tau$, and

$$\overline{B(x_0, r)} := \{x \in X;\ d(x, x_0) \leq r\}$$

is closed, i.e., $X \setminus \overline{B(x_0, r)} \in \tau$ (prove these assertions!).

A subset A of a metric space (X, d) is said to be **bounded** (with respect to d) if it is contained in a closed ball (equivalently, in an open ball). Otherwise, A is called **unbounded** (with respect to d). For example, $\mathbb{N} \subset \mathbb{R}$ is bounded with respect to the discrete metric on $\mathbb{R}$, but is unbounded with respect to the usual norm topology (the norm being the absolute value function $|\cdot|$).

A sequence $(a_n)_{n \in \mathbb{N}}$ in (X, d) is said to be **convergent** if there exists $a \in X$ such that $d(a_n, a) \to 0$. This is denoted $a_n \to a$, or $\lim_{n \to \infty} a_n = a$, or $\lim a_n = a$, and we say that (a_n) converges to a, or that a is the limit of (a_n). It is easily seen that the limit is unique.

Let S be a nonempty subset of a metric space (X, d). S is closed if and only if the limit of any convergent sequence of points in S is also a point of S (prove it!).

A point $p \in (X, d)$ is called an **accumulation point** (or **limit point**) of a set $S \subset X$ if $(V \cap S) \setminus \{p\} \neq \emptyset$ for every neighborhood V of p. Note that p is not necessarily an element of S. If $q \in S$ and q is not an accumulation point of S, then q is an **isolated point** of S.
Obviously, p is an accumulation point of S if and only if there exists a sequence (p_n) in S such that $p_n \to p$. By the above assertion, S is closed if and only if S contains all its accumulation points.

Let $(a_n)_{n \in \mathbb{N}}$ a sequence in (X, d). A point $p \in X$ is called a **cluster point** of (a_n) if for every $\varepsilon > 0$ there are infinitely many a_n such that $d(a_n, p) < \varepsilon$ (in other words, (a_n) has a subsequence converging to p).

A point $p \in S$ is called an **interior point** of S if there is an $r > 0$ such that $B(p, r) \subset S$. The set of all interior points of S is called the **interior** of S, and is denoted Int S.
Obviously,
• Int S is the union of all open subsets of S, and hence Int S is an open set (possibly $\emptyset$);
• S is open if and only if $S = $ Int S.

For a set $S \subset (X, d)$ the **closure** of S, denoted Cl S or $\overline{S}$, is the intersection of all closed sets containing S.

Clearly, Cl S is a closed set, and

- S is closed if and only if $S = $ Cl S;
- Cl $S = S \cup \{$accumulation points of $S\}$.

A metric space (X, d) is called **separable** if it has a countable, dense subset S, i.e., Cl $S = X$ (the closure being related to the metric topology generated by d).

For example, $\mathbb{R}$ is separable with respect to its usual topology (since $\mathbb{Q}$ is dense in $\mathbb{R}$ with respect to this topology), but is not separable with respect to the discrete topology, i.e., the topology associated with the discrete metric on $\mathbb{R}$. This is because any subset of $\mathbb{R}$ is closed with respect to the discrete topology, so there is no dense countable subset of $\mathbb{R}$.

The **boundary** of a set $S \subset (X, d)$ is defined to be the set

$$\partial S := \text{Cl } S \cap \text{Cl}\,(X \setminus S).$$

Obviously, $\partial S = \partial(X \setminus S)$, and $p \in \partial S$ if and only if $B(p, \varepsilon) \cap S \neq \emptyset$ and $B(p, \varepsilon) \cap (X \setminus S) \neq \emptyset$ for all $\varepsilon > 0$.

2.2 Completeness

We start this section with the definition of a Cauchy sequence which is essential in what follows.

Definition 2.4. *A sequence $(a_n)_{n \in \mathbb{N}}$ in a metric space (X, d) is called a* **Cauchy sequence** *if for all $\varepsilon > 0$ there exists an $N = N(\varepsilon) \in \mathbb{N}$ such that $d(a_n, a_m) < \varepsilon$ for all $m, n > N$.*

It is easy to see that any convergent sequence in a metric space is a Cauchy sequence. The converse implication is not true in general.

Definition 2.5. *A metric space (X, d) is called* **complete** *if every Cauchy sequence $(a_n)_{n \in \mathbb{N}}$ in X converges (i.e., there exists a point $a \in X$ such that $d(a_n, a) \to 0$).*

For example, $\mathbb{R}$ with its usual topology (defined by $|\cdot|$) is a complete metric space (as shown in the previous chapter, see Theorem 1.12). More generally, for any $n \in \mathbb{N}$, $\mathbb{R}^n$, equipped with the Euclidean metric (generated by the Euclidean norm), is complete, because a Cauchy sequence in $\mathbb{R}^n$ is Cauchy in each coordinate. In fact, we will see later that $\mathbb{R}^n$ endowed with any norm is complete.

On the other hand, the metric space $(\mathbb{Q}, d)$, where $d(x, y) = |x - y|$ $(x, y \in X)$ is *not* complete. For example, the sequence in $\mathbb{Q}$, defined by

$$a_1 = 2, \quad a_{n+1} = \frac{1}{2}\left(a_n + \frac{2}{a_n}\right), \quad n = 1, 2, \ldots$$

is convergent in $(\mathbb{R}, |\cdot|)$ (since $a_n \geq \sqrt{2}$ and $a_{n+1}/a_n \leq 1$, $n = 1, 2, \ldots$), hence Cauchy with respect to $|\cdot|$, but its limit is $\sqrt{2} \notin \mathbb{Q}$.

Now, let us examine another example. Let S be a nonempty set. Define

$$B(S; \mathbb{R}) = \{f : S \to \mathbb{R}; \; f(S) \text{ is bounded}\},$$

where the boundedness condition on $f(S)$ means: $\exists M > 0$ such that $|f(s)| \leq M$ for all $s \in S$. Obviously, $X = B(S; \mathbb{R})$ is a real linear space with respect to the usual operations (addition and scalar multiplication). It can be equipped with a norm $\|\cdot\|$ defined by

$$\|f\| := \sup_{s \in S} |f(s)| \quad \forall f \in X,$$

which gives a metric $d : X \times X \to [0, \infty)$, $d(f, g) = \|f - g\|$ $(f, g \in X)$. Moreover, it is easily seen that (X, d) is a complete metric space. The key condition ensuring the completeness of X is the completeness of $\mathbb{R}$ with respect to its usual metric.

Convergence in $X = B(S; \mathbb{R})$ is called **uniform convergence** on S. It is stronger than the **pointwise convergence**. In particular,

$$d(f_n, f) = \|f_n - f\| \to 0 \implies \lim_{n \to \infty} f_n(s) = f(s) \quad \forall s \in S,$$

but the converse implication is not true in general.

Definition 2.6. *A normed space* $(X, \|\cdot\|)$ *which is complete (i.e.,* (X, d) *is complete for* $d(x, y) = \|x - y\|$, $x, y \in X$) *is called a* **Banach space**.

In particular, $B(S; \mathbb{R})$ with the norm above (called uniform convergence norm) is a Banach space. The subset $X_K = \{f \in B(S; \mathbb{R}) : |f(s)| \leq K \; \forall s \in S\}$, where K is a given positive constant, is a complete metric space with respect to the same metric (generated by the uniform convergence norm), since X_K is closed in $B(S; \mathbb{R})$. Note that X_K is not a Banach space because it is not a linear space.

In general, if (X, d) is a complete metric space, then any nonempty closed set $Y \subset X$ is also a complete metric space with the metric d restricted to $Y \times Y$.

Definition 2.7. *Two metric spaces* (X_1, d_1), (X_2, d_2) *are* **isometric** *if there exists a bijection* $\phi : X_1 \to X_2$ *such that* $d_2(\phi(x), \phi(y)) = d_1(x, y)$ *for all* $x, y \in X$.

An important result, due to Hausdorff,[1] says that any metric space can be extended (uniquely up to isometry) to a complete metric space (see [44, Chapter 2]). More precisely we have

Theorem 2.8. *For any metric space* (X, d) *there exists a complete metric space* $(\bar{X}, \bar{d})$ *such that*

(j) *there exists* $X_1 \subset \bar{X}$ *such that* (X, d) *is isometric to* $(X_1, \bar{d})$;

(jj) X_1 *is dense in* $(\bar{X}, \bar{d})$.

$(\bar{X}, \bar{d})$ *with the above properties is unique up to isometry.*

Proof. One can construct an extension (completion) of (X, d) by a procedure similar to that used in the previous chapter to construct the Cantor–Méray model for $\mathbb{R}$ starting from rational numbers. Specifically, let E denote the set of all Cauchy sequences in (X, d). E is nonempty as it contains constant sequences $(c, c, \dots)$, $c \in X$. We define an equivalence relation in E as follows: (a_n), $(b_n) \in E$ are equivalent iff $d(a_n, b_n) \to 0$. In other words, two sequences convergent in (X, d) with the same limit are equivalent. It is easily seen that the relation defined above is indeed an equivalence relation. Let $\bar{X}$ be the collection of all equivalence classes in E with respect to this equivalence relation. Denote by A, B, $C, \dots$ the classes of sequences (a_n), (b_n), $(c_n), \dots$ Now, define $\bar{d} : \bar{X} \times \bar{X} \to [0, \infty)$ by

$$\bar{d}(A, B) = \lim_{n \to \infty} d(a_n, b_n) \quad \forall A, B \in \bar{X}. \tag{2.2.2}$$

The limit above exists since

$$|d(a_{n+p}, b_{n+p}) - d(a_n, b_n)| \leq d(a_{n+p}, a_n) + d(b_{n+p}, b_n),$$

---

[1]Felix Hausdorff, German mathematician, 1868–1942.

which says that $(d(a_n, b_n))$ is a Cauchy sequence in $(\mathbb{R}, |\cdot|)$.

Note also that the limit in (2.2.2) does not depend on representatives, as the following inequality shows

$$|d(a_n, b_n) - d(a'_n, b'_n)| \leq d(a_n, a'_n) + d(b_n, b'_n) \to 0.$$

Thus $\bar{d}$ is well defined. It is easy to check that $\bar{d}$ is a metric.

Now, let $\psi : X \to \bar{X}$ be the mapping which associates with every $a \in X$ the class A of the constant sequence $(a, a, \dots)$: $\psi(a) = A$. Obviously, ψ is injective, so if we denote $X_1 = \psi(X)$, then ψ is a bijection between X and X_1. Moreover, for any $A, B \in X_1$ we have

$$\bar{d}(\psi(a), \psi(b)) = \bar{d}(A, B) = \lim d(a, b) = d(a, b).$$

Hence (X, d) and $(X_1, \bar{d})$ are isomorphic, i.e., (j) holds true.

Let us now prove (jj). To this purpose, let A be an arbitrary element of $\bar{X}$ and let (a_n) be a representative of A. For each $k \in \mathbb{N}$ denote by A_k the class of the constant sequence $(a_k, a_k, \dots)$. Since (a_n) is a Cauchy sequence in (X, d), we can write

$$\forall \varepsilon > 0, \ \exists N \in \mathbb{N} : \ d(a_{k+p}, a_k) < \varepsilon \ \forall k > N, \, p \in \mathbb{N}.$$

Therefore,

$$\bar{d}(A, A_k) = \lim_{m \to \infty} d(a_m, a_k) \leq \varepsilon \ \ \forall k > N.$$

This shows that A is approximated by $A_k \in X_1$, hence (jj) holds true. In order to prove that $(\bar{X}, \bar{d})$ is complete, let (A_n) be a Cauchy sequence in $(\bar{X}, \bar{d})$. For each class A_k there is a class $B_k \in X_1$ such that $\bar{d}(A_k, B_k) < 1/k$ (see (jj)). Notice that B_k is the class of some constant sequence $(b_k, b_k, \dots)$ with $b_k \in X$. We can show that (b_k) is a Cauchy sequence in (X, d):

$$
\begin{aligned}
d(b_k, b_m) &= \bar{d}(B_k, B_m) \\
&\leq \bar{d}(B_k, A_k) + \bar{d}(A_k, A_m) + \bar{d}(A_m, B_m) \\
&\leq \frac{1}{k} + \frac{1}{m} + \bar{d}(A_m, B_m) \to 0,
\end{aligned}
$$

as $k, m \to \infty$, so the class B of the sequence (b_k) belongs to $\bar{X}$. We claim that B is the limit of (A_k) with respect to $\bar{d}$. Indeed, given $\varepsilon > 0$,

$$
\begin{aligned}
\bar{d}(B, A_k) &\leq \bar{d}(B, B_k) + \bar{d}(B_k, A_k) \\
&= \lim_{m \to \infty} d(b_m, b_k) + \frac{1}{k} \\
&< \varepsilon
\end{aligned}
$$

for large enough k, so $\lim_{k\to\infty}\bar{d}(A_k,B)=0$, as claimed. Therefore $(\bar{X},\bar{d})$ is complete.

Finally, we need to show that any two complete metric spaces $(\bar{X},\bar{d})$ and $(\hat{X},\hat{d})$ satisfying (j) and (jj) are isometric. Let $X_1\subset\bar{X}$ and $X_2\subset\hat{X}$ such that each of these spaces is isometric to (X,d). Let $g:(X,d)\to(X_1,\bar{d})$ and $h:(X,d)\to(X_2,\hat{d})$ be the corresponding isometries. Then $(X_1,\bar{d})$ and $(X_2,\hat{d})$ are isometric, and $\theta=h\circ g^{-1}$ is an isometry between these spaces.

Let A be an arbitrary element of $\bar{X}$. By (jj) there exists a sequence (A_n) in X_1 such that $\bar{d}(A_n,A)\to 0$. Obviously, $B_n=\theta(A_n)\in X_2$ and (B_n) is a Cauchy sequence in $(\hat{X},\hat{d})$, so it is convergent since $(\hat{X},\hat{d})$ is complete. Let $B\in\hat{X}$ be its limit: $\hat{d}(B_n,B)\to 0$. Denote by $\tilde{\theta}$ the mapping that takes A to B. Note that B does not depend on the choice of (A_n) so it is unique for each A, i.e., $\tilde{\theta}$ is well defined. In fact, $\tilde{\theta}$ is an extension of θ to the whole $\bar{X}$. It is easily seen that $\tilde{\theta}$ is a bijection between $\bar{X}$ and $\hat{X}$.

It remains to prove that $\tilde{\theta}$ is an isometry. Let A, $A'\in\bar{X}$ and let (A_n), (A'_n) sequences in X_1 which converge, respectively, to A and A' with respect to $\bar{d}$. Let B, B' be the limits of $B_n=\theta(A_n)$ and $B'_n=\theta(A'_n)$ in $(\hat{X},\hat{d})$. By letting n tend to infinity in the equation

$$\hat{d}(B_n,B'_n)=\bar{d}(A_n,A'_n),$$

we obtain

$$\hat{d}(\tilde{\theta}(A),\tilde{\theta}(A'))=\hat{d}(B,B')=\bar{d}(A,A'),$$

by using the inequality

$$|\bar{d}(A_n,A'_n)-\bar{d}(A,A')|\le\bar{d}(A_n,A)+\bar{d}(A'_n,A'),$$

and the similar one for B_n, B'_n, B, B'. Therefore, $(\bar{X},\bar{d})$ and $(\hat{X},\hat{d})$ are indeed isometric. $\square$

Remark 2.9. Let X be a nonempty subset of a given complete metric space (Z,d). Then $(\mathrm{Cl}\,X,d)$ is also a complete metric space, where $\mathrm{Cl}\,X$ is the closure of X in (Z,d), also denoted $\bar{X}$. Clearly, $(\mathrm{Cl}\,X,d)$ plays the role of $(\bar{X},\bar{d})$ in Theorem 2.8, so $(\mathrm{Cl}\,X,d)$ can be regarded as the completion of X with respect to d.

To illustrate this case, consider $X=(0,1]$ and $Z=\mathbb{R}$ with $d(x,y)=|x-y|$. Then, $\mathrm{Cl}\,X=[0,1]$ and so $([0,1],d)$ is the completion of $((0,1],d)$ (which is not itself complete). Further examples will be discussed later, including examples involving function spaces.

Note that in Theorem 2.8 we had to construct $\bar{X}$ because it was not a priori known.

We continue with Baire's Theorem[2] that is used to derive some important principles of Functional Analysis: the Uniform Boundedness Principle, the Open Mapping Theorem, and the Closed Graph Theorem (see Theorems 4.7, 4.8, and 4.10).

Theorem 2.10 (Baire). *Let (X, d) be a complete metric space and let $X_n \subset X$, $n \in \mathbb{N}$, be closed sets satisfying*

$$\text{Int } X_n = \emptyset \quad \forall n \in \mathbb{N}. \tag{2.2.3}$$

Then,

$$\text{Int}\left(\bigcup_{n=1}^{\infty} X_n \right) = \emptyset. \tag{2.2.4}$$

Proof. Notice first that for all $F \subset X$ we have

$$\text{Cl}(X \setminus F) =: \overline{X \setminus F} = X \iff \text{Int } F = \emptyset.$$

So by (2.2.3) $D_n = X \setminus X_n$ is dense in X and is also open for all $n \in \mathbb{N}$. We have to show that (2.2.4) holds or, equivalently, that $M := \cap_{n=1}^{\infty} D_n$ is dense in X, i.e., for every open set $D \subset X$ we have $D \cap M \neq \emptyset$. Fix such an open set D and choose some $x_0 \in D$ and $r_0 > 0$ such that the closed ball $\overline{B(x_0, r_0)} \subset D$. Since D_1 is open and dense in X there exist $x_1 \in B(x_0, r_0) \cap D_1$ and $r_1 > 0$ such that

$$\overline{B(x_1, r_1)} \subset B(x_0, r_0) \cap D_1, \quad 0 < r_1 < \frac{r_0}{2}.$$

By induction one can find sequences (x_n) and (r_n) such that

$$\overline{B(x_{n+1}, r_{n+1})} \subset B(x_n, r_n) \cap D_{n+1}, \quad 0 < r_{n+1} < \frac{r_n}{2},$$

for $n = 0, 1, 2, \dots$ It is easily seen that (x_n) is Cauchy, hence convergent (since (X, d) is complete). If a denotes its limit then $a \in D \cap M$, hence $D \cap M \neq \emptyset$, as claimed. $\square$

---

[2]René-Louis Baire, French mathematician, 1874–1932.

2.3   Compact Sets

Let A be a subset of a metric space (X, d). A **cover** of A is a collection
of sets $\{D_i\}_{i \in I}$ whose union contains A:

$$A \subset \bigcup_{i \in I} D_i \,,$$

where I is a finite or infinite index set. If all D_i are open sets then
$\{D_i\}_{i \in I}$ is called an **open cover**.

Definition 2.11. *A subset A of (X, d) is called* **compact** *if every
open cover of A has a finite subcover.*

The next result is a characterization of compact sets in metric spaces.

Theorem 2.12. *A subset A of a metric space (X, d) is compact if and
only if every sequence in A has a subsequence that converges to a point
of A (in other words A is* **sequentially compact***).*

Proof. is divided into several steps.

Step 1: If A is compact then A is closed.
We need to show that $X \setminus A$ is open. Let $x \in X \setminus A$. If $y \in A$ we have
$d(y, x) > 0$ and so y belongs to

$$D_n = \{z \in X; \, d(z, x) > 1/n\}$$

for $n \in \mathbb{N}$ large enough. Thus $\{D_n\}_{n \in \mathbb{N}}$ is an open cover of A. Since A
is compact, there is a finite subcover of A. In fact, this subcover can be
reduced to one set D_N with N large. By construction $B(x, \frac{1}{N}) \subset X \setminus A$,
and hence $X \setminus A$ is open, therefore A is closed, as claimed.

Step 2: If A is compact and $B \subset A$ is closed, then B is compact.
If $\{D_i\}_{i \in I}$ is an open cover of B, then $\{D_i\}_{i \in I} \cup \{X \setminus B\}$ is an open
cover of A. Since A is compact, we can extract a finite subcover of A,
say $\{D_{i_1}, D_{i_2}, \ldots, D_{i_m}, X \setminus B\}$. Thus $\{D_{i_1}, D_{i_2}, \ldots, D_{i_m}\}$ is a finite
subcover of B extracted from $\{D_i\}_{i \in I}$.

Step 3: A being compact implies A is sequentially compact.
Assume, by contradiction, that there is a sequence (x_n) in A that
has no convergent subsequence. So (x_n) has infinitely many distinct
points $y_1 = x_{n_1}, y_2 = x_{n_2}, \ldots$ such that for each y_m there is an open

ball $B(y_m, r_m)$ containing no other y_i (otherwise y_m is a cluster point of the sequence (y_i)). The set $C = \{y_1, y_2, \ldots\}$ is closed since all its points are isolated. By Step 2, C is compact. On the other hand, $\{B(y_m, r_m)\}_{m \in \mathbb{N}}$ is an open cover of C which has no finite subcover. Hence (x_n) must have a convergent subsequence. Its limit belongs to A, since A is closed (see Step 1).

Step 4: If A is sequentially compact, then for every open cover $\{D_i\}_{i \in I}$ of A, there exists an $r > 0$ such that $\forall y \in A$, $B(y, r)$ is contained in some D_i.

Assume to the contrary that this is not the case. Thus there exists an open cover $\{D_i\}$ of A such that $\forall n \in \mathbb{N}$ there is some $y_n \in A$ so that $B(y_n, \frac{1}{n})$ is not contained in any D_i. By hypothesis (y_n) has a subsequence $(z_1 = y_{n_1}, z_2 = y_{n_2}, \ldots)$ converging to some $z \in A$. Obviously, z belongs to some D_{i_0} and since D_{i_0} is open and $z_n \to z$, we can choose some large N such that $B(z_N, \frac{1}{N}) \subset D_{i_0}$, which is a contradiction.

Step 5: A being sequentially compact implies that for all $\varepsilon > 0$ there is a finite number of open balls of radius ε covering A (i.e., A is **totally bounded**).

We need to analyze the case when A is not finite, otherwise the conclusion is obvious. Assume that A is not totally bounded, i.e., for some $\varepsilon > 0$ we cannot cover A with finitely many open balls of radius ε. Choose $y_1 \in A$ and $y_2 \in A \setminus B(y_1, \varepsilon)$. By the same assumption there exists a point $y_3 \in A \setminus [B(y_1, \varepsilon) \cup B(y_2, \varepsilon)]$. Repeating this process we obtain a sequence

$$y_n \in A \setminus [\cup_{i=1}^{n-1} B(y_i, \varepsilon)],$$

which satisfies $d(y_n, y_m) \geq \varepsilon$ for all $n, m \in \mathbb{N}$, $n \neq m$. In other words, (y_n) has no Cauchy subsequence and hence has no convergent subsequence, thus contradicting sequential compactness.

Step 6: If A is sequentially compact then A is compact.

Let $\{D_i\}$ be an open cover of A. Associate with this cover a positive r given by Step 4. By Step 5 (see also its proof) there is a finite number of points, say $y_1, y_2, \ldots, y_p \in A$, such that

$$A \subset \cup_{j=1}^{p} B(y_j, r).$$

By Step 4, each ball $B(y_j, r)$ is contained in some D_{i_j}. Hence $\{D_{i_1}, D_{i_2}, \ldots, D_{i_p}\}$ is a finite (open) subcover of A. $\qquad\square$

Definition 2.13. *A set $A \subset (X, d)$ is called* **relatively compact** *if* Cl A *is compact.*

Corollary 2.14. *A set $A \subset (X, d)$ is relatively compact if and only if every sequence in A has a convergent subsequence.*

Proof. Assume that any sequence in A has a convergent subsequence (its limit being a point of Cl A). Then Cl A is sequentially compact (hence compact) in (X, d). Indeed, if (x_n) is a sequence in Cl A, then there exists a sequence (y_n) in A such that $d(x_n, y_n) < 1/n$ for all $n \in \mathbb{N}$. As (y_n) has a convergent subsequence (y_{n_k}), it follows that (x_{n_k}) is also convergent. So the statement of the corollary holds true by Theorem 2.12.   □

Now let us recall a result due to Bolzano and Weierstrass.

Theorem 2.15 (Bolzano–Weierstrass). *Every bounded sequence in $\mathbb{R}^k$ endowed with the Euclidean norm has a convergent subsequence.*

Proof. This theorem is known for $k = 1$ (see Theorem 1.11) and extends easily to $\mathbb{R}^k$: a bounded sequence in $\mathbb{R}^k$ is bounded in each coordinate.   □

From the proof of Theorem 2.12 we see that every compact set in a metric space is closed and bounded. The converse implication is not true in general. However, we have the following result attributed to Heine and Borel.[3]

Corollary 2.16 (Heine–Borel). *Let $\emptyset \neq A \subset \mathbb{R}^k$ endowed with the usual Euclidean metric. A is compact if and only if A is closed and bounded (with respect to the same metric).*

Proof. The forward implication is valid in every metric space, as observed above. Conversely, assume that A is closed and bounded. Then any sequence in A is bounded so it has a convergent subsequence (cf. Theorem 2.15). Its limit belongs to A because A is closed. Thus A is sequentially compact, hence compact by Theorem 2.12.   □

Remark 2.17. The Heine–Borel Theorem extends to any finite dimensional space with Euclidean metric but may not be true for other

---

[3]Heinrich Eduard Heine, German mathematician, 1821–1881; Émile Borel, French mathematician, 1871–1956.

metrics. As an example, consider $(\mathbb{R}, d_0)$, where d_0 is the discrete metric

$$d_0(x, y) = \begin{cases} 0 & x = y, \\ 1 & x \neq y. \end{cases}$$

Let $A = \mathbb{N} \subset \mathbb{R}$. A is bounded with respect to d_0 because $A \subset B(0, 2)$. $A = \mathbb{N}$ is closed with respect to d_0, but it is not compact because the open cover $\{B(n, 1/2)\}_{n \in \mathbb{N}}$ has no finite subcover.

A collection of subsets of (X, d) is said to have the **finite intersection property** if the intersection of every finite sub-collection of the family is nonempty.

Theorem 2.18. *If a collection of compact subsets of a metric space (X, d), say $\{K_i\}_{i \in I}$, has the finite intersection property, then $\cap_{i \in I} K_i \neq \emptyset$.*

Proof. The statement is trivial if I is a finite set, so let us assume that I is infinite. Assume to the contrary that $\cap_{i \in I} K_i = \emptyset$. Hence

$$\begin{aligned} X &= \cup_{i \in I}(X \setminus K_i) \\ &= (X \setminus K_{i_0}) \bigcup \left[\cup_{i \in I_1}(X \setminus K_i) \right], \end{aligned} \quad (2.3.5)$$

where $i_0 \in I$ is an arbitrary but fixed index, and $I_1 = I \setminus \{i_0\}$. It follows that

$$K_{i_0} \subset \cup_{i \in I_1}(X \setminus K_i).$$

As K_{i_0} is compact and $\{X \setminus K_i\}_{i \in I_1}$ is an open cover of K_{i_0}, there is a finite set $J \subset I_1$ such that

$$K_{i_0} \subset \cup_{i \in J}(X \setminus K_i).$$

Therefore (see (2.3.5)),

$$X = \cup_{i \in J_1}(X \setminus K_i),$$

where $J_1 = J \cup \{i_0\}$, or equivalently

$$\emptyset = \cap_{i \in J_1} K_i,$$

which contradicts our assumption because J_1 is finite. $\square$

2.4   Continuous Functions on Compact Sets

Let (X, d) and (X_1, d_1) be two metric spaces. A function

$$f : D \subset (X, d) \to (X_1, d_1)$$

is said to be **continuous at some point** $x_0 \in D$ if for every neighborhood $V \subset (X_1, d_1)$ of $f(x_0)$ there exists a neighborhood $U \subset (X, d)$ of x_0 such that $f(U \cap D) \subset V$, or equivalently

$$\forall \varepsilon > 0, \ \exists \delta > 0 : \ \forall x \in D, \ d(x, x_0) < \delta \ \Rightarrow \ d_1(f(x), f(x_0)) < \varepsilon.$$
$$\text{(2.4.6)}$$

U and δ depend on ε and x_0. The continuity of f at $x_0 \in D$ can also be equivalently expressed by using sequences

$$x_n \in D, \ \ d(x_n, x_0) \to 0 \implies d_1(f(x_n), f(x_0)) \to 0.$$

If f is continuous at all $x_0 \in D$ then we say that f is **continuous on** D (or simply **continuous**). The function f is called **uniformly continuous on** D if δ can be the same for all $x_0 \in D$, i.e., δ is independent of $x_0 \in D$ (it depends only on ε).

Theorem 2.19. *If $D \subset (X, d)$ is a nonempty compact set and $f : D \to (X_1, d_1)$ is continuous (on D), then the following hold:*

- *f is uniformly continuous on D;*

- *the set $f(D) := \{f(x); \ x \in D\}$ is compact in (X_1, d_1);*

- *$C(D; X_1) := \{f : D \to (X_1, d_1); \ f \ continuous \ on \ D\}$ is a metric space with respect to the metric $\tilde{d}(f, g) = \sup_{x \in D} d_1(f(x), g(x))$.*

If in addition (X_1, d_1) is complete, then $(C(D; X_1), \tilde{d})$ is also complete.

The proof is left to the reader as an exercise.

Theorem 2.20 (Weierstrass). *If $D \subset (X, d)$ is a nonempty compact set and $f : D \to \mathbb{R}$ ($\mathbb{R}$ being equipped with the usual metric), then $f(D)$ is closed and bounded, and there exist $x_0, y_0 \in D$ such that $f(x_0) = \inf f(D)$ and $f(y_0) = \sup f(D)$.*

Proof. The first part of the theorem follows from Theorem 2.19 which in particular says that $f(D)$ is compact (in $\mathbb{R}$), hence closed and bounded. So the infimum and supremum of $f(D)$, denoted m and M, are finite numbers. Now, for all $n \in \mathbb{N}$ there exists an $x_n \in D$ such that

$$m \le f(x_n) < m + \frac{1}{n}. \tag{2.4.7}$$

As D is a compact set, (x_n) has a subsequence which converges to some $x_0 \in D$. This fact combined with (2.4.7) implies $m = f(x_0)$. Similarly, there is a point $y_0 \in D$ such that $M = f(y_0)$. $\qquad \square$

Equivalent Norms. Let X be a linear space over $\mathbb{K}$ (as usual $\mathbb{K}$ is either $\mathbb{R}$ or $\mathbb{C}$). Two norms on X, say $\|\cdot\|$ and $\|\cdot\|_*$, are said to be equivalent if there exist two positive constants C_1, C_2 such that

$$C_1\|x\| \le \|x\|_* \le C_2\|x\| \quad \forall x \in X. \tag{2.4.8}$$

Obviously, two equivalent norms on X generate the same topology on X.

If X is a k-dimensional linear space, $k \in \mathbb{N}$, with a basis $B = \{u_1, \ldots, u_k\}$, then X can be equipped with different norms, such as

$$\|x\|_{\max} = \max_{1 \le i \le n} |\alpha_i|,$$

$$\|x\|_p = \Big(\sum_{i=1}^{n} |\alpha_i|^p \Big)^{1/p}, \quad p \in [1, \infty),$$

for all $x = \sum_{i=1}^{k} \alpha_i u_i \in X$. Note that $\|\cdot\|_2$ is precisely the Euclidean norm of X introduced before.

Theorem 2.21. *If X is a k-dimensional linear space, $k \in \mathbb{N}$, then any two norms on X are equivalent.*

Proof. It is enough to show that any norm $\|\cdot\|$ on X is equivalent to the Euclidean norm $\|\cdot\|_2$. On the one hand, for any $x = \sum_{i=1}^{k} \alpha_i u_i \in X$, we have

$$
\begin{aligned}
\|x\| &\le \sum_{i=1}^{k} |\alpha_i| \cdot \|u_i\| \\
&\le \max_{1 \le i \le k} \|u_i\| \cdot \sum_{i=1}^{k} |\alpha_i| \\
&\le \sqrt{k} \max_{1 \le i \le k} \|u_i\| \cdot \|x\|_2. \tag{2.4.9}
\end{aligned}
$$

We have used the triangle inequality and the Bunyakovsky–Cauchy–Schwarz inequality. Denoting $C := \sqrt{k}\max_{1 \le i \le k}\|u_i\|$, we can derive from (2.4.9)

$$\|x\| \le C\|x\|_2 \quad \forall x \in X. \tag{2.4.10}$$

In order to get the other inequality we use Theorem 2.20. Observe that $\|\cdot\|$ is a continuous function on $(X, \|\cdot\|_2)$:

$$\big|\|x\| - \|x_0\|\big| \le \|x - x_0\| \le C\|x - x_0\|_2,$$

so $\|\cdot\|$ is bounded and attains its infimum, say C_1, on the unit sphere $S_2(0,1) = \{x \in X;\ \|x\|_2 = 1\}$ (which is compact in $(X, \|\cdot\|_2)$), i.e.,

$$\|x\| \ge C_1 \quad \forall x \in S_2(0,1). \tag{2.4.11}$$

C_1 cannot be zero since it is the value of $\|\cdot\|$ at a point in $S_2(0,1)$. From (2.4.11) we easily derive

$$C_1\|x\|_2 \le \|x\| \quad \forall x \in X. \tag{2.4.12}$$

According to (2.4.10) and (2.4.12), the two norms are equivalent, as claimed. $\square$

Remark 2.22. In infinite dimensional linear spaces there exist norms which are not equivalent. For instance, let us consider the following two norms on the real linear space $X = C[a,b] := C([a,b];\mathbb{R})$, $-\infty < a < b < +\infty$,

$$\|f\| = \sup\{|f(t)|;\ a \le t \le b\}, \quad \|f\|_1 = \int_a^b |f(t)|\,dt.$$

We have

$$\|f\|_1 \le (b-a)\|f\| \quad \forall f \in X,$$

i.e., the sup-norm $\|\cdot\|$ is stronger than $\|\cdot\|_1$. But the two norms are not equivalent. Indeed, let (f_n) be the sequence in X defined by

$$f_n(t) = \begin{cases} 0, & a \le t \le b - \frac{1}{n}, \\ nt + 1 - nb, & b - \frac{1}{n} < t \le b, \end{cases}$$

where $n \in \mathbb{N}$, $n > 1/(b-a)$. Clearly $\|f_n\| = 1$, but

$$\|f_n\|_1 = \int_{b-\frac{1}{n}}^{b} |nt + 1 - nb|\,dt = \frac{1}{2n},$$

so there does not exist C such that $\|f_n\| \le C\|f_n\|_1$ because $\|f_n\|_1 \to 0$ as $n \to \infty$.

Remark 2.23. It follows from Theorem 2.21 that any norm on a finite dimensional linear space generates the same topology as that defined by the Euclidean norm, so any topological result involving the Euclidean norm is also valid with respect to any other norm. In particular, the Heine–Borel Theorem is valid in any finite dimensional linear space equipped with any norm. Throughout the rest of this book, $\mathbb{R}^k$ and any other finite dimensional linear space is always considered as a normed space, equipped with the norm topology (generated by any convenient norm), unless otherwise specified. The next result is a characterization (due to Riesz[4]) of the finite dimensionality of normed spaces clarifying the Heine–Borel Theorem.

Theorem 2.24 (Riesz). *Let $(X, \|\cdot\|)$ be a normed linear space. X is finite dimensional if and only if every closed bounded subset of X is compact.*

In order to prove Theorem 2.24, we need the following lemma.

Lemma 2.25. *Let $(X, \|\cdot\|)$ be a normed space. Let $X_1 \subset X$ be a proper, closed linear subspace of X. Then there exists $x_0 \in X \setminus X_1$ such that*

$$\|x_0\| = 1 ,$$
$$\|x - x_0\| \geq \frac{1}{2} \quad \forall x \in X_1 .$$

Proof. Choose $x_1 \in X \setminus X_1$ and let $\rho = d(x_1, X_1) := \inf\{\|x_1 - z\|; z \in X_1\}$. We first prove $\rho > 0$. Suppose $\rho = 0$. Then there exists a sequence $z_n \in X_1$ such that $\|x_1 - z_n\| < 1/n$, hence $z_n \to x_1$. As X_1 is closed, this implies $x_1 \in X_1$, which is a contradiction.
By the definition of ρ there exists $x_2 \in X_1$ such that $\|x_1 - x_2\| < 2\rho$. Let

$$x_0 = \frac{1}{\|x_1 - x_2\|}(x_1 - x_2) .$$

---

[4]Frigyes Riesz, Hungarian mathematician, 1880–1956.

Clearly $x_0 \in X \setminus X_1$ and $\|x_0\| = 1$. For $x \in X_1$ we have

$$
\begin{aligned}
\|x - x_0\| &= \|x - \|x_1 - x_2\|^{-1}(x_1 - x_2)\| \\
&= \frac{1}{\|x_1 - x_2\|} \|x_1 - v\| \\
&\geq \frac{1}{2\rho} \|x_1 - v\| \\
&\geq \frac{\rho}{2\rho} = \frac{1}{2},
\end{aligned}
$$

where $v = x_2 + \|x_1 - x_2\| x \in X_1$. $\square$

Proof of Theorem 2.24. The necessity part follows from the Heine–Borel Theorem extended to finite dimensional linear spaces (see Remark 2.23).

To prove sufficiency, assume by way of contradiction that X is not finite dimensional, i.e., there exist infinitely many distinct points in X, say $x_1, x_2, \ldots$, such that for all $n \in \mathbb{N}$, $B_n = \{x_1, x_2, \ldots, x_n\}$ is a linearly independent system. Let $X_n = \operatorname{Span} B_n$. Now, $(X_n, \|\cdot\|)$ is a closed space and $X_n \subset X_{n+1}$ (proper inclusion) for all $n \in \mathbb{N}$. By Lemma 2.25, there exists $y_n \in X_{n+1} \setminus X_n$ for $n \in \mathbb{N}$ such that $\|y_n\| = 1$ and

$$
\|y_n - x\| \geq 1/2 \quad \forall x \in X_n.
$$

In particular $\|y_n - y_m\| \geq 1/2$ for all $m, n \in \mathbb{N}$, $m \neq n$. So (y_n) has no Cauchy subsequence, hence no convergent subsequence. On the other hand, $y_n \in \operatorname{Cl} B(0,1) \ \forall n \in \mathbb{N}$, so (y_n) should have a convergent subsequence (since $\operatorname{Cl} B(0,1)$ is compact by assumption). This contradiction completes the proof.

Arzelà–Ascoli Criterion[5]

Let (X, d) and (X_1, d_1) be metric spaces and let $\emptyset \neq A \subset X$. Denote as usual by $C(A; X_1)$ the set of all continuous functions from $A \subset (X, d)$ to (X_1, d_1).

Definition 2.26. *A family of functions* $\mathcal{F} \subset C(A; X_1)$ *is called* **equicontinuous** *if for all* $\varepsilon > 0$ *and all* $x \in A$ *there exists* $\delta > 0$ *such that* $y \in A$ *and* $d(x, y) < \delta$ *implies* $d_1(f(x), f(y)) < \varepsilon$ *for all* $f \in \mathcal{F}$, *i.e.,* $\delta = \delta(\varepsilon, x)$ *is independent of* f.

---

[5]Cesare Arzelà, Italian mathematician, 1847–1912; Giulio Ascoli, Italian mathematician, 1843–1896.

Definition 2.27. *If in addition $\delta = \delta(\varepsilon)$ (i.e., δ is independent of x and f), then $\mathcal{F}$ is* **uniformly equicontinuous**, *i.e., $\forall \varepsilon > 0$, $\forall x, y \in A$, $d(x, y) < \delta$ implies $d_1(f(x), f(y)) < \varepsilon$ for all $f \in \mathcal{F}$.*

Remark 2.28. If $A \subset (X, d)$ is compact and $\mathcal{F} \subset C(A; X_1)$ is equicontinuous, then $\mathcal{F}$ is uniformly equicontinuous (see Exercise 2.22 below). Note also that if A is compact then $C(A; X_1)$ is a metric space with respect to the metric $\tilde{d}(f, g) = \sup_{x \in A} d_1(f(x), g(x))$; if in addition (X_1, d_1) is complete then $(C(A; X_1), \tilde{d})$ is complete too, and in particular $C(A; \mathbb{R}^k)$, $k \in \mathbb{N}$, is a Banach space with respect to the sup-norm $\|f\|_{C(A; \mathbb{R}^k)} = \sup_{x \in A} \|f(x)\|$, where $\| \cdot \|$ is a norm of $\mathbb{R}^k$.

Theorem 2.29 (Arzelà–Ascoli Criterion). *Let $\emptyset \neq A \subset (X, d)$ be compact. Assume that $\mathcal{F} \subset C(A, \mathbb{R}^k)$ is equicontinuous and bounded in $C(A; \mathbb{R}^k)$ (i.e., $\exists M > 0$ such that $\|f(x)\| \leq M$, $\forall x \in A$, $\forall f \in \mathcal{F}$). Then $\mathcal{F}$ is relatively compact in $C(A; \mathbb{R}^k)$ equipped with the sup-norm.*

Proof. For any $\delta > 0$ we have $A \subset \cup_{x \in A} B(x, \delta)$ and since A is compact, there exists a finite subcover, so that $A \subset \cup_{j=1}^{p} B(y_j, \delta)$. Let $C_\delta = \{y_1, y_2, \ldots, y_p\}$ and consider $C = \cup_{i \in \mathbb{N}} C_{1/i}$. C is dense in A and countable so $C = \{x_1, x_2, \ldots\}$.

In order to prove that $\mathcal{F}$ is relatively compact in $C(A; \mathbb{R}^k)$ it suffices to show that any sequence in $\mathcal{F}$ has a convergent subsequence in this space (cf. Corollary 2.14). So let $(f_n)_{n \in \mathbb{N}}$ be a sequence in $\mathcal{F}$. Since $\mathcal{F}$ is bounded in $C(A; \mathbb{R}^k)$, then $(f_n(x_1)_\mathbb{N})$ is bounded in $\mathbb{R}^k$ so there exists a subsequence of (f_n),

$$f_{11}, f_{12}, \ldots, f_{1n}, \ldots$$

which is convergent at $x = x_1$. By the same assumption this subsequence has a subsequence

$$f_{21}, f_{22}, \ldots, f_{2n}, \ldots$$

which is convergent at $x = x_2$ (and at $x = x_1$ as well). Continuing the process we obtain successive subsequences

$$f_{m1}, f_{m2}, \ldots, f_{mn}, \ldots$$

$$\vdots$$

Think of it as an infinite matrix and consider the diagonal sequence $(g_n) = (f_{11}, f_{22}, \ldots, f_{nn}, \ldots)$ which converges at any point of C. On the other hand, as $\mathcal{F}$ is equicontinuous and A is compact, $\mathcal{F}$ is in fact uniformly equicontinuous, i.e., for every $\varepsilon > 0$ there exists a $\delta = \delta(\varepsilon) > 0$ such that

$$\forall z, w \in A,\ d(z,w) < \delta \text{ implies } \|g_n(z) - g_n(w)\| < \varepsilon \ \forall n \in \mathbb{N}. \quad (2.4.13)$$

We can choose $\delta = 1/i$, with $i \in \mathbb{N}$ sufficiently large.

Now, for a given ε fix a $\delta = 1/i$, so $C_\delta = C_{1/i}$ is a finite set $C_\delta = \{y_1, \ldots, y_p\} \subset C$. If $x \in A$ then it belongs to a ball $B(y_j, \delta)$ for some $j \in \{1, \ldots, p\}$ and we have, by (2.4.13) and the convergence of $(g_n(y_j))$,

$$\|g_n(x) - g_m(x)\| \leq \|g_n(x) - g_n(y_j)\| + \|g_n(y_j) - g_m(y_j)\| + \|g_m(y_j)$$
$$- g_m(x)\| < \varepsilon + \varepsilon + \varepsilon = 3\varepsilon \ \forall n,m > N(\varepsilon, j).$$

Therefore,

$$\|g_n - g_m\|_{C(A;\mathbb{R}^k)} \leq 3\varepsilon \ \forall n,m > N(\varepsilon) := \max_{j \in \{1,\ldots,p\}} N(\varepsilon, j).$$

As $C(A; \mathbb{R}^k)$ is a Banach space it follows that (g_n) converges in this space. $\qquad\square$

Notice that in the above proof we have used two essential arguments: the completeness of the space $(\mathbb{R}^k, \|\cdot\|)$ (implying that $C(A; \mathbb{R}^k)$ is a Banach space) and the fact that the set $\{f(x);\ f \in \mathcal{F}, x \in X_1\}$ is bounded in $\mathbb{R}^k$ (equivalently, relatively compact in this space) for all $x \in A$. So the following generalization holds true:

Theorem 2.30. *Let $\mathcal{F} \subset C(A; X_1)$ where $A \neq \emptyset$ is a compact subset of (X, d) and (X_1, d_1) is a complete metric space. Assume that $\mathcal{F}$ is equicontinuous and $\{f(x);\ f \in \mathcal{F}\}$ is relatively compact in (X_1, d_1) for all $x \in A$. Then $\mathcal{F}$ is relatively compact in $C(A; X_1)$.*

Peano's Existence Theorem[6]

In what follows we illustrate the Arzelà–Ascoli Criterion with Peano's Existence Theorem which is a fundamental result in the theory of ordinary differential equations.

Theorem 2.31 (Peano). *Let $a, b \in (0, \infty)$, $t_0 \in \mathbb{R}$, $x_0 \in \mathbb{R}^k$, and $\mathbb{R}^k$ be equipped with the norm $\|v\| = \max_{1 \leq i \leq k} |v_i|$. Let D be the set*

$$D = \{(t,v) \in \mathbb{R} \times \mathbb{R}^k;\ |t - t_0| \leq a,\ \|v - x_0\| \leq b\} \subset \mathbb{R}^{k+1}$$

---

[6]Giuseppe Peano, Italian mathematician, 1858–1932.

and let $f : D \to \mathbb{R}^k$ be a continuous function. Then there exists a continuously differentiable function $x : [t_0 - \delta, t_0 + \delta] \to \mathbb{R}^k$ satisfying the equation

$$x'(t) = f(t, x(t)) \quad \forall t \in [t_0 - \delta, t_0 + \delta], \tag{2.4.14}$$

and the initial (Cauchy) condition

$$x(t_0) = x_0, \tag{2.4.15}$$

where $\delta = \min(a, b/M)$ with $M = \sup\{\|f(t,v)\|; (t,v) \in D\}$. M is assumed to be a positive number, because the case $M = 0 \iff f \equiv 0$ is trivial.

Proof. We shall use Euler's method of polygonal lines.[7]
Since $f \in C(D; \mathbb{R}^k)$ and D is compact, f is uniformly continuous, i.e., $\forall \varepsilon > 0, \exists \delta = \delta_1(\varepsilon) > 0$ such that

$$(t, v_1), (s, v_2) \in D, |t - s| < \delta_1, \|v_1 - v_2\| < \delta_1 \Rightarrow \|f(t, v_1)$$
$$- f(s, v_2)\| < \varepsilon.$$

We shall only prove existence on $I := [t_0, \delta]$. By symmetry we get the other side, however we have to check that the solution is differentiable at $t = t_0$. Given

$$x(t) = \begin{cases} x_r(t), & t \in [t_0, t_0 + \delta], \\ x_l(t), & t \in [t_0 - \delta, t_0], \end{cases}$$

we have

$$x'_-(t_0) = \frac{dx_l}{dt}(t_0) = f(t_0, x_0) = \frac{dx_r}{dt}(t_0) = x'_+(t_0).$$

Consider the uniform subdivision

$$\Delta : t_0 < t_1 < \cdots < t_N = t_0 + \delta,$$

i.e., $t_j = t_0 + jh_\varepsilon$, $j = 0, 1, \ldots, N$, with $h_\varepsilon \leq \min\{\delta_1(\varepsilon), \frac{\delta_1(\varepsilon)}{M}\}$.
Now for a given $\varepsilon > 0$ construct $\phi_\varepsilon : I \to \mathbb{R}^k$ as

$$\phi_\varepsilon(t) = \begin{cases} \phi_\varepsilon(t_j) + (t - t_j)f(t_j, \phi_\varepsilon(t_j)), & t_j < t \leq t_{j+1}, \\ x_0, & t = t_0. \end{cases}$$

---

[7]Leonhard Euler, Swiss mathematician, physicist, astronomer, logician, and engineer, 1707–1783.

The graph of ϕ_ε is a polygonal line called Euler's polygonal line and we shall see that it approximates for ε small the trajectory of the solution of problem (2.4.14) and (2.4.15). For $k = 1$ Euler's polygonal line can be visualized in the (t, x) coordinate plane.

Consider the family $\mathcal{F} = \{\phi_\varepsilon; \varepsilon > 0\}$. Let us first show that ϕ_ε is well defined on I for all $\varepsilon > 0$. On the interval $[t_0, t_1]$, $\phi_\varepsilon(t) = x_0 + (t - t_0)f(t_0, x_0)$ and

$$\|\phi_\varepsilon(t) - x_0\| \leq M(t - t_0) \leq M\delta \leq b,$$

so $(t, \phi_\varepsilon(t)) \in D$. In particular $(t_1, \phi_\varepsilon(t_1)) \in D$.

So on $[t_1, t_2]$, $\phi_\varepsilon(t) = \phi_\varepsilon(t_1) + (t - t_1)f(t_1, \phi_\varepsilon(t_1))$ is well defined and

$$
\begin{aligned}
\|\phi_\varepsilon(t) - x_0\| &\leq \|\phi_\varepsilon(t) - \phi_\varepsilon(t_1)\| + \|\phi_\varepsilon(t_1) - x_0\| \\
&\leq (t - t_1)M + (t_1 - t_0)M \\
&\leq (t - t_0)M \\
&\leq M\delta \\
&\leq b,
\end{aligned}
$$

so by induction $\phi_\varepsilon(t)$ is well defined and continuous on I and

$$\|\phi_\varepsilon(t) - x_0\| \leq M(t - t_0) \leq M\delta \leq b \qquad (2.4.16)$$

for all $t \in I$. Thus

$$\|\phi_\varepsilon(t)\| \leq \|\phi_\varepsilon(t) - x_0\| + \|x_0\| \leq b + \|x_0\|.$$

Therefore, $\mathcal{F}$ is a bounded subset of $C(I; \mathbb{R}^k)$. In order to apply the Arzelà–Ascoli Theorem, we need to show that $\mathcal{F}$ is equicontinuous. If $t, s \in [t_j, t_{j+1}]$ then $\|\phi_\varepsilon(t) - \phi_\varepsilon(s)\| \leq M|t - s|$. If t, s are in different intervals, say $t \in [t_p, t_{p+1}]$, $s \in [t_q, t_{q+1}]$ with $p < q$, then

$$
\begin{aligned}
\|\phi_\varepsilon(t) - \phi_\varepsilon(s)\| &\leq \|\phi_\varepsilon(s) - \phi_\varepsilon(t_q)\| + \|\phi_\varepsilon(t_q) - \phi_\varepsilon(t_{q-1})\| + \cdots \\
&\quad + \|\phi_\varepsilon(t_{p+1}) - \phi_\varepsilon(t)\| \\
&\leq M(s - t_q) + M(t_q - t_{q-1}) + \cdots + M(t_{p+1} - t) \\
&\leq M(s - t) \\
&= M|t - s|,
\end{aligned}
$$

so $\mathcal{F}$ is equicontinuous, in fact it is Lipschitz equicontinuous. Thus by the Arzelà–Ascoli Criterion there is a sequence $\varepsilon_n \to 0^+$ such that

ϕ_{ε_n} converges in $C(I; \mathbb{R}^k)$ to some $\phi \in C(I; \mathbb{R}^k)$ as $n \to \infty$. Also (see (2.4.16))

$$\|\phi(t) - x_0\| \leq b,$$

so $(t, \phi(t)) \in D$ for all $t \in I$.

Now it simply remains to prove that $x = \phi(t)$ is a solution of problem (2.4.14) and (2.4.15). Define

$$g_{\varepsilon_n}(t) = \begin{cases} \phi'_{\varepsilon_n}(t) - f(t, \phi_{\varepsilon_n}(t)) & t \neq t_j^n, \\ 0 & \text{otherwise}, \end{cases}$$

where $\{t_j^n\}$ is the subdivision of I corresponding to ε_n. If $t_j^n < t < t_{j+1}^n$ then $\phi'_{\varepsilon_n}(t) = 0 + f(t_j^n, \phi_{\varepsilon_n}(t_j^n))$. For $t \in (t_j^n, t_{j+1}^n)$, we have $|t - t_j^n| \leq h_{\varepsilon_n} \leq \delta_1(\varepsilon_n)$, and

$$\|\phi_{\varepsilon_n}(t) - \phi_{\varepsilon_n}(t_j^n)\| \leq M|t - t_j^n| \leq Mh_{\varepsilon_n} \leq \delta_1(\varepsilon_n).$$

The final inequality holds by the definition of h_ε. Because f is uniformly continuous

$$\|g_{\varepsilon_n}(t)\| \leq \varepsilon_n \ \forall n, \ \forall t \in I,$$

so g_{ε_n} converges uniformly to 0.

On the other hand, for all $t \in I$

$$\int_{t_0}^t g_{\varepsilon_n}(s)\, ds = \phi_{\varepsilon_n}(t) - x_0 - \int_{t_0}^t f(s, \phi_{\varepsilon_n}(s))\, ds. \qquad (2.4.17)$$

Now since ϕ_{ε_n} converges uniformly to ϕ on I and f is continuous on D, $f(s, \phi_{\varepsilon_n}(s)) \to f(s, \phi(s))$ uniformly on I as $n \to \infty$. Therefore, passing to the limit in (2.4.17), we get

$$\phi(t) = x_0 + \int_{t_0}^t f(s, \phi(s))\, ds, \quad t \in I,$$

so $x = \phi(t)$ is a solution to the given Cauchy problem (2.4.14) and (2.4.15). $\qquad \square$

Remark 2.32. There is no guarantee of uniqueness. For example the Cauchy problem

$$\begin{cases} x'(t) = 2\sqrt{|x(t)|}, \\ x(0) = 0, \end{cases}$$

with $a = b = 1$, $D = [-1, 1] \times [-1, 1]$, $f(t, v) = 2\sqrt{|v|}$, has the following solutions:

$$x_1(t) = 0, \quad -1 \leq t \leq 1,$$

$$x_2(t) = \begin{cases} t^2, & -1 \leq t \leq 0, \\ 0, & 0 < t \leq 1, \end{cases}$$

$$x_3(t) = \begin{cases} 0, & -1 \leq t \leq 0, \\ -t^2, & 0 < t \leq 1, \end{cases}$$

$$x_4(t) = \begin{cases} t^2, & -1 \leq t \leq 0, \\ -t^2, & 0 < t \leq 1. \end{cases}$$

Note that all these solutions are defined on the whole interval $[-1, 1]$, even if the existence interval given by Peano's Theorem is smaller: $\delta = \min\{a, b/M\} = \min\{1, 1/2\} = 1/2$. A solution which is defined on the whole initial interval $[t_0 - a, t_0 + a]$ in the case of problem (2.4.14) and (2.4.15)) is called a **global solution**. In particular, the above four solutions are global solutions. In fact, there are infinitely many solutions of the above Cauchy problem (see Exercise 2.28 below).

Peano's Theorem provides only a **local solution**, i.e., a solution defined on an interval around t_0 which in general is smaller than the initial interval. If f in (2.4.14) is defined on an open set $\Omega \subset \mathbb{R}^{k+1}$ then one can associate with each pair $(t_0, x_0) \in \Omega$ a box $D \subset \Omega$ so that Peano's Theorem gives a local solution to problem (2.4.14) and (2.4.15) defined on an interval which depends on (t_0, x_0).

By requiring additional conditions one can guarantee uniqueness. For example, we get uniqueness if, in addition, f satisfies a Lipschitz condition: $\exists L > 0$ such that

$$\|f(t, v_1) - f(t, v_2)\| \leq L\|v_1 - v_2\| \qquad (2.4.18)$$

for all $(t, v_1), (t, v_2) \in D$. Let $x = \phi(t)$, $y = \psi(t)$ for $t \in I = [t_0, t_0 + \delta]$ be two solutions of problem (2.4.14) and (2.4.15). Then

$$\|\phi(t) - \psi(t)\| \leq L \int_{t_0}^{t} \|\phi(s) - \psi(s)\| \, ds,$$

or, equivalently,

$$\frac{d}{dt}\left(e^{-Lt} \int_{t_0}^{t} \|\phi(s) - \psi(s)\| \, ds\right) \leq 0,$$

for all $t \in I$. It follows easily that $\phi(t) = \psi(t)$ for all $t \in I$. Uniqueness on $[t_0 - \delta, t_0]$ follows by converting problem (2.4.14) and (2.4.15) on $[t_0 - \delta, t_0]$ into a similar Cauchy problem on $[0, \delta]$ by using the change $\tau = t_0 - t$. Therefore, we can state the following result.

Theorem 2.33. *Under the assumptions of Peano's Theorem (Theorem 2.31), plus (2.4.18), there exists a unique function $x \in C^1([t_0 - \delta, t_0 + \delta]; \mathbb{R})$ satisfying (2.4.14) and (2.4.15), where δ is the same as in Theorem 2.31.*

Remark 2.34. Peano's Theorem is no longer valid in infinite dimensions, i.e., if $\mathbb{R}^k$ is replaced by an infinite dimensional Banach space (see [18]).

Euler's Difference Scheme.
If $x = \phi(t)$ is unique, then $\phi_\varepsilon \to \phi$ in $C(I; \mathbb{R}^k)$ as $\varepsilon \to 0^+$ so the polygonal line corresponding to ϕ_ε approximates the graph of ϕ. Let $\Delta : t_0 < t_1 < \cdots < t_N = t_0 + \delta$ with $t_j = t_0 + jh$ and $h = \frac{\delta}{N}$. The points $(t_j, \phi_\varepsilon(t_j))$ give us the polygonal line approximation. Denoting $\phi_j := \phi_\varepsilon(t_j)$ we have

$$\begin{cases} \phi_{j+1} = \phi_j + hf(t_j, \phi_j), & j = 0, 1, \ldots, N-1, \\ \phi_0 = x_0. \end{cases}$$

This is an explicit difference scheme, called Euler's scheme. Its solution provides the vertices of a polygonal line approximation, so Euler's scheme is important for the numerical analysis of the solutions of differential equations.

2.5 The Banach Contraction Principle

We saw in the previous section that under the assumptions of Peano's Existence Theorem (Theorem 2.31) plus the Lipschitz condition (2.4.18) the Cauchy problem

$$x'(t) = f(t, x(t)), \quad x(t_0) = x_0 \tag{2.5.19}$$

has a unique solution $x \in C^1(I; \mathbb{R}^k)$, where $I = [t_0 - \delta, t_0 + \delta]$, with δ as defined in the statement of Theorem 2.31. This (existence and uniqueness) result can also be derived by applying the general Banach[8]

---

[8]Stefan Banach, Polish mathematician, 1892–1945.

Contraction Principle (also known as the Banach Fixed Point Theorem) we present below. Before stating this principle let us explain how problem (2.5.19) can be reduced to a fixed point problem. Note that problem (2.5.19) is equivalent to the integral equation

$$x(t) = x_0 + \int_{t_0}^{t} f(s, x(s)) \, ds \, . \tag{2.5.20}$$

Denote $X = \{v \in C(I; \mathbb{R}^k); \|v(t) - x_0\| \leq b, \, t \in I\}$. This is a complete metric space since it is a closed subset of the Banach space $C(I; \mathbb{R}^k)$ equipped with the sup-norm, denoted $\| \cdot \|_C$, which gives the metric $d(u, v) = \|u - v\|_C$. Define on X the map (operator) T by

$$(Tv)(t) = x_0 + \int_{t_0}^{t} f(s, v(s)) \, ds \, , \quad \forall v \in X \, .$$

We prefer the notation Tv instead of $T(v)$. It is easily seen that under the assumptions above $Tv \in X$ for all $v \in X$, i.e., $T : X \to X$. Equation (2.5.20) can be simply written as

$$x = Tx \, , \tag{2.5.21}$$

so the above Cauchy problem (or Eq. (2.5.20)) reduces to solving Eq. (2.5.21) in X. In other words, the Cauchy problem (2.5.19) has a unique solution x defined on I if and only if T has a unique fixed point x: $x = Tx$. We do not go into further details concerning the above Cauchy problem, or Eq. (2.5.20), since later on we will address Volterra equations which are more general. We simply wanted to motivate the Banach Contraction Principle which is applicable to many other problems.

Theorem 2.35 (Banach Contraction Principle). *Let (X, d) be a complete metric space, and assume $T : X \to X$ is a* **contraction**, *i.e., $\exists \alpha \in (0, 1)$ such that $d(Tx, Ty) \leq \alpha d(x, y)$ for all $x, y \in X$. Then T has a unique fixed point (i.e., $\exists! \, x^* \in X$ such that $Tx^* = x^*$).*

Proof. We will use the **method of successive approximations**. Define a sequence $x_n = Tx_{n-1}$ for $n \in \mathbb{N}$ with $x_0 \in X$ arbitrary. We have by induction

$$d(x_{n+1}, x_n) \leq \alpha^n d(x_1, x_0) = \alpha^n d(Tx_0, x_0) \, , \quad \forall n \in \mathbb{N} \, . \tag{2.5.22}$$

We now prove that (x_n) is Cauchy in (X, d):

$$d(x_{n+p}, x_n) \leq d(x_{n+p}, x_{n+p-1}) + d(x_{n+p-1}, x_{n+p-2}) + \cdots$$
$$+ d(x_{n+1}, x_n)$$

which by (2.5.22) is

$$\leq \alpha^n (1 + \alpha + \cdots + \alpha^{p-1}) d(Tx_0, x_0)$$
$$= \alpha^n \frac{1 - \alpha^p}{1 - \alpha} d(Tx_0, x_0)$$
$$\leq \frac{\alpha^n}{1 - \alpha} d(Tx_0, x_0).$$

So it is Cauchy in (X, d) (as $\alpha^n \to 0$), and since (X, d) is complete, x_n converges to some $x^* \in X$: $d(x_n, x^*) \to 0$. Now,

$$d(x^*, Tx^*) \leq d(Tx^*, x_n) + d(x_n, x^*)$$
$$= d(Tx^*, Tx_{n-1}) + d(x_n, x^*)$$
$$\leq \alpha d(x^*, x_{n-1}) + d(x_n, x^*),$$

which converges to 0 as $n \to \infty$, so $d(x^*, Tx^*) \leq 0$ and thus x^* is a fixed point of T.

We now wish to show that x^* is unique. Suppose that y^* is also a fixed point of T, then $d(x^*, y^*) = d(Tx^*, Ty^*) \leq \alpha d(x^*, y^*)$, so $(1 - \alpha)d(x^*, y^*) \leq 0$ which implies $x^* = y^*$. $\qquad\square$

Remark 2.36. The assumption $\alpha < 1$ in Theorem 2.35 is essential as the following counterexample from Natanson[9] [38, p. 571] shows. If $X = \mathbb{R}$, and $T : \mathbb{R} \to \mathbb{R}$ is given by $Tx = x + \frac{\pi}{2} - \arctan x$, then T has no fixed point because $\frac{\pi}{2} - \arctan x > 0 \ \forall x \in \mathbb{R}$. On the other hand, by the Mean Value Theorem, we have for all $x, y \in \mathbb{R}$, $x \neq y$,

$$|Tx - Ty| \leq |x - y - \arctan x + \arctan y|$$
$$= \left| x - y - \frac{x - y}{1 + z^2} \right|$$

for some z between x and y

$$= |x - y| \cdot \left(1 - \frac{1}{1 + z^2} \right)$$
$$< |x - y|,$$

---

[9]Isidor P. Natanson, Russian mathematician, 1906–1963.

so, even though the inequality is strict, $\alpha = 1$ and hence T is not a contraction. Thus, the fact that this T has no fixed point is not surprising.

Remark 2.37. From the above proof we see that

$$d(x_n, x^*) \leq \frac{\alpha^n}{1-\alpha} d(Tx_0, x_0),$$

which gives us an approximation of x^*.

Remark 2.38. Suppose that $T^k = \underbrace{T \circ \cdots \circ T}_{k \ factors}$, $k \geq 2$, is a contraction (even though T may not be), then there is a unique fixed point for T.

Proof. A fixed point of T is obviously a fixed point of T^k. Conversely if x^* is a fixed point of T^k (which exists and is unique by Theorem 2.35) then $Tx^* = T^{k+1}x^* = T^k(Tx^*)$, so both x^* and Tx^* are fixed points of T^k, and consequently $Tx^* = x^*$. $\qquad\qquad\square$

2.6   Exercises

1. Let $A_1, A_2, \ldots$ be subsets of a metric space. Prove that $\mathrm{Cl}\left(\bigcup_{i=1}^{n} A_i\right) = \bigcup_{i=1}^{n} \mathrm{Cl}\, A_i$ for all $n \in \mathbb{N}$ and $\mathrm{Cl}\left(\bigcup_{i=1}^{\infty} A_i\right) \supset \bigcup_{i=1}^{\infty} \mathrm{Cl}\, A_i$. Show by an example that the latter inclusion can be proper.

2. Let A be a subset of a metric space. Do A and $\mathrm{Cl}\, A$ always have the same interior? Do A and $\mathrm{Int}\, A$ always have the same closure?

3. Prove that if $X \neq \emptyset$ and d_0 is the discrete metric on X, then any subset of (X, d_0) is open.

4. Let $\emptyset \neq A \subset (X, d)$. Prove that

$$p \in \mathrm{Cl}\, A \iff \inf\{d(p, x) : x \in A\} = 0.$$

5. Let (X, d) be a metric space, $\emptyset \neq A \subset (X, d)$, and let $(Y, \|\cdot\|)$ be a Banach space. Denote

$$BC(A; Y) := \{f : (A, d) \to (Y, \|\cdot\|);$$
$$f \text{ continuous and bounded}\}.$$

Prove that $BC(A; Y)$ is a Banach space with respect to the sup-norm: $\|f\|_{\sup} = \sup_{x \in A} \|f(x)\|$.

6. Find the accumulation points of the following subsets of $\mathbb{R}^2$ (equipped with the Euclidean metric):

   (a) $\mathbb{Z} \times \mathbb{Z}$;

   (b) $\mathbb{Q} \times \mathbb{Q}$;

   (c) $\{(\frac{m}{n}, \frac{1}{n}); \ m, n \in \mathbb{Z}, n \neq 0\}$;

   (d) $\{(\frac{1}{n} + \frac{1}{m}, 0); \ m, n \in \mathbb{Z} \setminus \{0\}\}$.

7. Find the boundaries of the following sets:

   (a) $A = [0, 1] \cap \mathbb{Q}$;

   (b) $B = \{\frac{1}{n}; \ n \in \mathbb{N}\}$;

   (c) $C = \{(x, y) \in \mathbb{R}^2; \ x^2 - y^2 > 1\}$.

8. Let (X, d) be a linear, metric space with d defined by a norm $\| \cdot \|$ (i.e., $d(x, y) = \|x - y\|$, $\forall x, y \in X$). Prove that the closure of any open ball $B(x, r) := \{v \in X; \ d(v, x) < r\}$ in (X, d) is the closed ball $\overline{B(x, r)} := \{v \in X; \ d(v, x) \leq r\}$. Show that this property fails to be true if X is equipped with the discrete metric d_0.

9. Show that any Cauchy sequence in a metric space can have at most one cluster point.

10. Find the cluster points of the following sequences:

    (a) $x_n = \sin\left(2\pi\sqrt{n^2 + 3n}\right)$, $n = 1, 2, \ldots$;

    (b) $y_n = \sin\left(\pi\sqrt{n^2 + n}\right)$, $n = 1, 2, \ldots$

11. Show that $B := \{f \in C([0, 1]; \mathbb{R}) : \ f(x) > 0 \text{ for all } x \in [0, 1]\}$ is open in $C([0, 1]; \mathbb{R})$ equipped with the metric generated by the sup-norm. What is the closure of B in this metric (in fact Banach) space?

12. Denote

$$BC(\mathbb{R}; \mathbb{R}) := \{f : \mathbb{R} \to \mathbb{R}; \ f \text{ is continuous and}$$

$$f(\mathbb{R}) \text{ is bounded}\}.$$

Let $D := \{f \in BC(\mathbb{R}; \mathbb{R}); \ f(x) > 0 \text{ for all } x \in \mathbb{R}\}$. Is D open in $BC(\mathbb{R}; \mathbb{R})$ equipped with the sup-norm? If not, what is Int D? What is Cl D?

13. Find an open cover of $(0,1] \subset (\mathbb{R}, |\cdot|)$ which has no finite sub-cover.

14. Find a necessary and sufficient condition for a discrete subset of a metric space (X,d) to be compact. [Recall that $S \subset (X,d)$ is discrete if all its elements are isolated].

15. If A is a nonempty compact subset of a metric space (X,d), then A is separable (i.e., there exists a countable subset S of A, such that $A = \operatorname{Cl} S$).

16. Let A, B be nonempty subsets of a normed space $(X, \|\cdot\|)$ equipped with the topology given by the metric d defined by $d(x,y) = \|x - y\|$, $x, y \in X$. We have the following:

   (a) If A, B are both compact sets, then $A + B := \{u + v;\ u \in A,\ v \in A\}$ is compact, too;

   (b) If A is closed and B is compact, then $A + B$ is closed, but not necessarily compact (give a counterexample).

17. Let $f : [0, \infty) \to \mathbb{R}$,

$$f(x) = \begin{cases} \sin(\pi(2x - 1)), & x \in [\frac{1}{2}, 1], \\ 0, & \text{otherwise.} \end{cases}$$

Let $f_n(x) = f(2^n x)$ for $x \in [0,1]$, $n \in \mathbb{N}$. Show that $\mathcal{F} = \{f_n;\ n \in \mathbb{N}\}$ is closed and bounded in $C[0,1] := C([0,1], \mathbb{R})$ equipped with the sup-norm, but not compact.

18. Let (X,d) be a complete metric space. If $\emptyset \neq A \subset X$ is a totally bounded set, show that A is relatively compact.

19. Let l^1 be the set of all sequences of real numbers $a = (a_n)_{n \in \mathbb{N}}$ satisfying $\sum_{n=1}^{\infty} |a_n| < \infty$. Show that

   (a) l^1 is a Banach space over $\mathbb{R}$ with respect to the norm $\|a\| = \sum_{n=1}^{\infty} |a_n|$, $a \in l^1$.

   (b) the set $A = \{a = (a_n)_{n \in \mathbb{N}} \in l^1;\ \sum_{n=1}^{\infty} n|a_n| \leq 1\}$ is compact in $(l^1, \|\cdot\|)$ (i.e., in (X,d), where d is the metric generated by $\|\cdot\|$: $d(a,b) = \|a - b\|$, $a, b \in l^1$).

20. Let $\mathcal{F}$ be the set of all functions $f : D = [0, 1] \to \mathbb{R}$,

$$f(x) = \sum_{n=1}^{\infty} a_n \sin{(n\pi x)},$$

where $a = (a_n)_{n \in \mathbb{N}}$ is a sequence in $\mathbb{R}$ satisfying $\sum_{n=1}^{\infty} n|a_n| \leq 1$. Show that $\mathcal{F}$ is a compact subset of $C[0, 1] := C([0, 1]; \mathbb{R})$ equipped with the sup-norm. Does the result hold if the domain of the f's is $D = \mathbb{R}$?

21. Let $-\infty < a < b < \infty$, $u_n \in C^1([a, b]; \mathbb{R})$, $n = 1, 2, \ldots$, such that $(u_n)_{n \in \mathbb{N}}$ and $(u_n')_{n \in \mathbb{N}}$ are bounded in $L^p([a, b], \mathbb{R})$, $p \in (1, \infty)$, equipped with the usual norm. Show that (u_n) has a subsequence which is convergent in $C([a, b]; \mathbb{R})$ with respect to the sup-norm. (Information on L^p spaces is available in Chap. 3 below.)

22. Let (X, d), (Y, ρ) be metric spaces, and let $\mathcal{F} \subset C(A; Y)$, where $\emptyset \neq A \subset X$. If A is compact (with respect to d) and $\mathcal{F}$ is an equicontinuous family, then $\mathcal{F}$ is uniformly equicontinuous.

23. For $a \in \mathbb{R}$ consider $f_a : [0, 1] \to \mathbb{R}$, $f_a(x) = \frac{x}{1+a^2x^2}$. Show that $\mathcal{F} = \{f_a; \ a \in \mathbb{R}\}$ is relatively compact in $C[0, 1] := C([0, 1]; \mathbb{R})$ equipped with the sup-norm, but not compact.

24. (a) Prove Gronwall's lemma, namely given

$$u(t) \leq a(t) + \int_{t_0}^{t} b(s)u(s)\, ds, \ \ t \in I = [t_0, T],$$

where $u, a, b : I \to \mathbb{R}$ are all continuous functions and $b \geq 0$, then

$$u(t) \leq a(t) + \int_{t_0}^{t} a(s)b(s)e^{\int_s^t b(\tau)d\tau}\, ds \ \ \forall t \in I.$$

In particular, prove Bellman's lemma, which states that in the case a is a constant function, i.e., $a(t) = C \ \forall t \in I$, then

$$u(t) \leq Ce^{\int_{t_0}^{t} b(s)ds}, \ \ t \in I.$$

(b) Let $x = x(t) : [t_0 - \delta, t_0 + \delta] \to \mathbb{R}^k$ be a solution given by Theorem 2.31 (Peano's Theorem). Assume (in addition to continuity on D) that f satisfies the Lipschitz condition (2.4.18). Use Bellman's lemma to prove that x is the unique solution of the corresponding Cauchy problem.

25. Prove that if

$$\frac{1}{2}x(t)^2 \le \frac{1}{2}c^2 + \int_{t_0}^t f(s)x(s)\,ds \quad \forall t \in I = [t_0, T]\,,$$

where $c \in \mathbb{R}$, $f, x \in C(I) := C(I; \mathbb{R})$, $f \ge 0$ for all $t \in I$, then

$$|x(t)| \le |c| + \int_{t_0}^t f(s)\,ds \quad \forall t \in I\,.$$

26. Show that the following Cauchy problem in $\mathbb{R}$

$$x'(t) = 1 + t^2 + \frac{x(t)^2}{1 + x(t)^2}; \quad x(0) = 0\,,$$

has a unique solution defined on $\mathbb{R}$.

27. Do the same for the Cauchy problem

$$x'(t) = 2e^{-t^2} + \ln\left(1 + x(t)^2\right); \quad x(0) = 0\,.$$

28. Show that the Cauchy problem

$$x'(t) = 2\sqrt{|x(t)|}, \ t \in \mathbb{R}; \quad x(0) = 0,$$

has infinitely many solutions defined on $\mathbb{R}$.

29. Show that for every $x_0 \in \mathbb{R}$ the Cauchy problem

$$x'(t) = 1 + t\left(1 + x(t)^2\right), \ t \ge 0; \quad x(0) = x_0\,,$$

has a unique solution defined on a bounded interval.

30. Show that the Cauchy problem

$$x'(t) = t^2 + x(t)^2, \ x(0) = 0\,,$$

has a solution whose maximal interval is $(-T, T)$, with $\sqrt{2}/2 < T < \infty$.

31. Let $\emptyset \ne \Omega \subset \mathbb{R}^{k+1}$, $k \ge 2$, be an open set, and let $f : \Omega \to \mathbb{R}^k$ be a continuous function. Then, for any $(t_0, x_0) \in \Omega$, the Cauchy problem

$$(CP) \qquad\qquad x'(t) = f(t, x(t)), \ x(t_0) = x_0\,,$$

has at least one solution defined on an interval around t_0. If, in addition, f satisfies the condition: $\forall$ compact $K \subset \Omega$, $\exists L_K > 0$ such that $\forall (t, u), (t, v) \in K$,

$$\|f(t, u) - f(t, v)\| \le L_K \|u - v\|,$$

where $\| \cdot \|$ is a norm of $\mathbb{R}^k$, then the (local) solution of (CP) is unique.

32. Consider in an interval $I \subset \mathbb{R}$ the Cauchy problem

$$x'(t) = A(t)x(t) + b(t), \ t \in I,$$
$$x(t_0) = x_0,$$

where $t_0 \in I$, $x_0 = (x_{01}, x_{02}, \ldots, x_{0k})^T \in \mathbb{R}^k$, $A(t) = (a_{ij}(t))$ is a $k \times k$-matrix, and $b(t) = (b_1(t), \ldots, b_k(t))^T$ with $a_{ij}, b_j \in C(I) := C(I; \mathbb{R})$, $i, j = 1, 2, \ldots, k$. Show that the above Cauchy problem has a unique solution on the **whole** interval I.

33. Let $T : \overline{B(0,1)} \to \overline{B(0,1)}$ be a map satisfying

$$\forall x, y \in \overline{B(0,1)}, \ d_2(Tx, Ty) \le d_2(x, y),$$

where $\overline{B(0,1)}$ is the closed unit ball of $(\mathbb{R}^k, d_2)$, and d_2 is the Euclidean metric. Show that T has at least one fixed point.

34. Prove that for every $f \in C[0, 1] := C([0, 1]; \mathbb{R})$ and $\alpha \in (0, 1)$ the integral equation

$$x(t) = f(t) + \int_0^1 e^{-ts} \cos\left(\alpha x(s)\right) ds, \ \ t \in [0, 1]$$

has a unique solution $x \in C[0, 1]$.

35. Let $(X, \| \cdot \|)$ be a Banach space and let $f : [0, \infty) \times X \to X$ be a continuous function satisfying

$$\|f(t, x_1) - f(t, x_2)\| \le a(t)\|x_1 - x_2\|, \ \ t \in [0, \infty), \ x_1, x_2 \in X,$$

where $a \in C([0, \infty); \mathbb{R})$. Show that the Cauchy problem

$$x'(t) = f(t, x(t)), \ t \ge 0; \ x(0) = x_0,$$

has a unique solution $x \in C^1([0, \infty); X)$.

Chapter 3

The Lebesgue Integral and L^p Spaces

In this chapter we discuss Lebesgue[1] measurable sets, Lebesgue measurable functions, Lebesgue integration, and L^p spaces. These spaces, equipped with appropriate norms, are significant examples of Banach spaces.

3.1 Measurable Sets in $\mathbb{R}^k$

Here we essentially follow [46]. First of all, for any closed cube $C \subset \mathbb{R}^k$,

$$C = [a_1, b_1] \times [a_2, b_2] \times \cdots \times [a_k, b_k],$$

where $b_i - a_i = c > 0$, $i = 1, 2, \ldots, k$, we denote $v(C) := c^k$ (which is called the volume of C).

A collection of cubes in $\mathbb{R}^k$ is said to be *almost disjoint* if the interiors of the cubes are disjoint.

It is easily seen that every open set $D \subset \mathbb{R}^k$ (equipped with the usual norm topology) can be written as a countable union of almost disjoint closed cubes: $D = \cup_{j=1}^{\infty} C_j$. To prove this, consider a grid in $\mathbb{R}^k$ of closed cubes of side length $1/n$, with n sufficiently large, retaining the cubes of the grid that are completely contained in D. Then, we bisect each cube of the above grid into 2^k cubes with side length $1/(2n)$ and

---

[1] Henri Léon Lebesgue, French mathematician, 1875–1941.

© Springer Nature Switzerland AG 2019

G. Moroşanu, *Functional Analysis for the Applied Sciences*, Universitext, https://doi.org/10.1007/978-3-030-27153-4_3

retain those new cubes that are contained in D. Thus, repeating indefinitely the procedure, we construct a countable collection of almost disjoint closed cubes whose union equals D, as claimed.

Now, for any set $M \subset \mathbb{R}^k$, we define the *exterior measure* of M by

$$m_e(M) = \inf \sum_{j=1}^{\infty} v(C_j),$$

where the infimum is taken over all countable covers of M, $\cup_{j=1}^{\infty} C_j \supset M$ with closed cubes C_j.

Some Remarks on the Exterior Measure

(a) Obviously, *the exterior measure of a singleton is zero, and* $m_e(\emptyset) = 0$.

(b) *If $M_1 \subset M_2 \subset \mathbb{R}^k$, then $m_e(M_1) \leq m_e(M_2)$.*

(c) *If C is a closed cube in $\mathbb{R}^k$, then $m_e(C) = v(C)$.*

Indeed, we clearly have $m_e(C) \leq v(C)$, and in order to prove the converse inequality it suffices to show that for any cover by closed cubes $\cup_{j=1}^{\infty} C_j \supset C$, we have

$$v(C) \leq \sum_{j=1}^{\infty} v(C_j). \tag{3.1.1}$$

Let $\varepsilon > 0$ be arbitrary but fixed. Choose for each j an open cube $C_j' \supset C_j$ such that $v(\overline{C_j'}) \leq (1 + \varepsilon)v(C_j)$. Since $\{C_j'\}_{j=1}^{\infty}$ is an open cover of the compact set C, there exists a finite subcover $\{C_{j_1}', \ldots, C_{j_m}'\}$, $C \subset \cup_{i=1}^{m} C_{j_i}'$. It follows that

$$v(C) \leq (1 + \varepsilon) \sum_{i=1}^{m} v(C_{j_i}) \leq (1 + \varepsilon) \sum_{j=1}^{\infty} v(C_j).$$

As ε was arbitrarily chosen, this implies (3.1.1).

(d) *If C is an open cube in $\mathbb{R}^k$, then $m_e(C) = v(\bar{C})$.*

(e) *If $M = \cup_{j=1}^{\infty} M_j$, then*

$$m_e(M) \leq \sum_{j=1}^{\infty} m_e(M_j). \tag{3.1.2}$$

We can assume $m_e(M_j) < \infty$ for all $j \in \mathbb{N}$, otherwise the inequality is trivially satisfied. For arbitrary $\varepsilon > 0$ we can choose for each j a cover by closed cubes $M_j \subset \cup_{q=1}^{\infty} C_{j,q}$ such that

$$\sum_{q=1}^{\infty} v(C_{j,q}) < m_e(M_j) + \frac{\varepsilon}{2^j}.$$

Then, $M \subset \cup_{j,q=1}^{\infty} C_{j,q}$, hence

$$
\begin{aligned}
m_e(M) &\leq \sum_{j,q=1}^{\infty} v(C_{j,q}) \\
&\leq \sum_{j=1}^{\infty} \left(m_e(M_j) + \frac{\varepsilon}{2^j} \right) \\
&= \sum_{j=1}^{\infty} m_e(M_j) + \varepsilon,
\end{aligned}
$$

which implies (3.1.2).

(f) *For every $M \subset \mathbb{R}^k$, we have*

$$m_e(M) = \inf\{m_e(D); \ D \text{ open}, \ D \supset M\}.$$

Clearly,

$$m_e(M) \leq \inf\{m_e(D); \ D \text{ open}, \ D \supset M\}.$$

For the converse inequality, let $\varepsilon > 0$ and choose a cover of M by closed cubes, $M \subset \cup_{j=1}^{\infty} C_j$, such that

$$\sum_{j=1}^{\infty} v(C_j) < m_e(M) + \frac{\varepsilon}{2}.$$

Choose for every j an open cube C_j', such that $C_j \subset C_j'$ and

$$v(\overline{C_j'}) \leq v(C_j) + \frac{\varepsilon}{2^{j+1}}.$$

Then, denoting $D' = \cup_{j=1}^{\infty} C'_j$, we have that D' is an open set and by (e)

$$
\begin{aligned}
m_e(D') \leq \sum_{j=1}^{\infty} m_e\big(\overline{C'_j}\big) &= \sum_{j=1}^{\infty} v\big(\overline{C'_j}\big) \\
&\leq \sum_{j=1}^{\infty} \Big(v(C_j) + \frac{\varepsilon}{2^{j+1}}\Big) \\
&= \sum_{j=1}^{\infty} v(C_j) + \frac{\varepsilon}{2} \\
&< m_e(M) + \varepsilon.
\end{aligned}
$$

Hence, $\inf\{m_e(D);\ D\ \text{open},\ D \supset M\} \leq m_e(M)$, as claimed.

(g) *If M is a countable union of almost disjoint closed cubes, $M = \cup_{j=1}^{\infty} C_j$, then $m_e(M) = \sum_{j=1}^{\infty} v(C_j)$.*

Indeed, by (c) and (e), $m_e(M) \leq \sum_{j=1}^{\infty} v(C_j)$, and for the converse inequality we consider, for a fixed $m \in \mathbb{N}$ and an arbitrary but fixed ε, closed cubes $\tilde{C}_j \subset \mathrm{Int}(C_j)$, $j = 1, \ldots, m$, such that

$$
v(C_j) < v(\tilde{C}_j) + \frac{\varepsilon}{2^j}, \quad j = 1, \ldots, m.
$$

Then,

$$
m_e(M) \geq m_e(\cup_{j=1}^{m}\tilde{C}_j) = \sum_{j=1}^{m} v(\tilde{C}_j) \geq \sum_{j=1}^{m} v(C_j) - \varepsilon,
$$

which implies $m_e(M) \geq \sum_{j=1}^{\infty} v(C_j)$.

Definition 3.1. *A set $M \subset \mathbb{R}^k$ is Lebesgue measurable (or simply measurable) if for every $\varepsilon > 0$ there exists an open set D such that $D \supset M$ and $m_e(D \setminus M) < \varepsilon$. If M is measurable, we define the Lebesgue measure (or measure) of M by $m(M) := m_e(M)$.*

Some Properties of Measurable Sets

(A) It follows from the above definition that *every open set is measurable.*

(B) *If $m_e(M) = 0$, then M is measurable and $m(M) = 0$.*

Indeed, we know (see (f) above) that

$$0 = m_e(M) = \inf\{m_e(D); D \text{ open}, D \supset M\},$$

so for any $\varepsilon > 0$ there exists an open set D_ε such that $D_\varepsilon \supset M$ and $m_e(D_\varepsilon) < \varepsilon$. As $D_\varepsilon \setminus M \subset D_\varepsilon$, we have $m_e(D_\varepsilon \setminus M) < \varepsilon$.

(C) *If $M = \cup_{j=1}^{\infty} M_j$, where each M_j is measurable, then M is measurable.*

Indeed, for a given $\varepsilon > 0$, we can choose for each j an open set D_j, $D_j \supset M_j$, such that $m_e(D_j \setminus M_j) < \varepsilon/2^j$. Hence $D = \cup_{j=1}^{\infty} D_j$ is open, $D \supset M$ and $D \setminus M \subset \cup_{j=1}^{\infty}(D_j \setminus M_j) \Longrightarrow m_e(D \setminus M) \leq \sum_{j=1}^{\infty} m_e(D_j \setminus M_j) < \varepsilon$.

(D) *If $K \subset \mathbb{R}^k$ is a compact set, then K is measurable.*

Since K is compact, hence bounded, we have $m_e(K) < \infty$. For any $\varepsilon > 0$ there exists an open set D, $D \supset K$, such that $m_e(D) < m_e(K) + \varepsilon/2$ (cf. (f)). The open set $D \setminus K$ can be written as a countable union of almost disjoint closed cubes: $D \setminus K = \cup_{j=1}^{\infty} C_j$.

Now, for a given $p \in \mathbb{N}$, $K_1 = \cup_{j=1}^{p} C_j$ is a compact set with $K_1 \cap K = \emptyset$, $K \cup K_1 \subset D$, and

$$
\begin{aligned}
m_e(D) &\geq m_e(K \cup K_1) \\
&= m_e(K) + m_e(K_1) \\
&= m_e(K) + \sum_{j=1}^{p} v(C_j),
\end{aligned}
$$

which implies that

$$\sum_{j=1}^{p} v(C_j) \leq m_e(D) - m_e(K) < \frac{\varepsilon}{2},$$

hence

$$m_e(D \setminus K) \leq m_e(\cup_{j=1}^{\infty} C_j) \leq \sum_{j=1}^{\infty} m_e(C_j)$$

$$= \sum_{j=1}^{\infty} v(C_j) \leq \frac{\varepsilon}{2} < \varepsilon,$$

so K is indeed measurable. It follows that

(D1) *any closed set $F \subset \mathbb{R}^k$ is measurable.*

Indeed, F can be written as a countable union of compact sets,

$$F = \cup_{n=1}^{\infty} F \cap \overline{B(0,n)},$$

so the assertion follows from (C) and (D).

(E) *If $M \subset \mathbb{R}^k$ is measurable, then $\mathbb{R}^k \setminus M$ is also measurable.*

To prove this, observe first that for all $n \in \mathbb{N}$ there exists an open set D_n such that $M \subset D_n$ and $m_e(D_n \setminus M) < 1/n$. Since $\mathbb{R}^k \setminus D_n$ is a closed set, it is measurable, hence $E := \cup_{n=1}^{\infty}(\mathbb{R}^k \setminus D_n)$ is also measurable (cf. (C)). We have $E \subset \mathbb{R}^k \setminus M$ and $\mathbb{R}^k \setminus (M \cup E) \subset D_n \setminus M$, hence $m_e(\mathbb{R}^k \setminus (M \cup E)) < 1/n$. Therefore $m_e(\mathbb{R}^k \setminus (M \cup E)) = 0$, so $\mathbb{R}^k \setminus (M \cup E)$ is measurable (cf. (B)). Since

$$\mathbb{R}^k \setminus M = [\mathbb{R}^k \setminus (M \cup E)] \cup E,$$

we conclude by (C) that $\mathbb{R}^k \setminus M$ is measurable, as claimed.

(F) *Any countable intersection of measurable sets is also a measurable set.*

This follows easily from $\cap_{j=1}^{\infty} M_j = \mathbb{R}^k \setminus [\cup_{j=1}^{\infty}(\mathbb{R}^k \setminus M_j)]$ (see also (C) and (E)).

Now let us state an important result related to measurable sets:

Theorem 3.2. *If $\{M_n\}_{n=1}^{\infty}$ is any collection of disjoint measurable sets, then $m(\cup_{n=1}^{\infty} M_n) = \sum_{n=1}^{\infty} m(M_n)$.*

Proof. In a first stage, we assume that each M_n is bounded. Let $\varepsilon > 0$ be arbitrary but fixed. Since $\mathbb{R}^k \setminus M_n$ is measurable, for any $n \in \mathbb{N}$ there exists a closed set $F_n \subset M_n$ such that $m_e(M_n \setminus F_n) < \varepsilon/2^n$. For each fixed $p \in \mathbb{N}$, $F_1, \ldots, F_p$ are compact and disjoint, and, denoting $M = \cup_{n=1}^{\infty} M_n$, we have

$$m(M) \geq m(\cup_{n=1}^{p} F_n) = \sum_{n=1}^{p} m(F_n) \geq \sum_{n=1}^{p} m(M_n) - \varepsilon,$$

which implies $m(M) \geq \sum_{n=1}^{p} m(F_n) \geq \sum_{n=1}^{p} m(M_n)$. This concludes the proof in the case when each M_n is bounded, since the converse inequality is also satisfied. In the general case, we consider the closed

cubes C_i centered at the origin with side length $i \in \mathbb{N}$ and define
$M_{n,1} = M_n \cap C_1$, $M_{n,i} = M_n \cap (C_i \setminus C_{i-1})$, $i = 2, 3, \ldots$ Then

$$M_n = \cup_i M_{n,i}, \ M = \cup_{n,i} M_{n,i},$$

so, as each $M_{n,i}$ is bounded, we can use what we obtained above to
write

$$m(M) = \sum_{n,i} m(M_{n,i}) = \sum_n \left[\sum_i m(M_{n,i}) \right] = \sum_n m(M_n). \qquad \Box$$

Remark 3.3. There are subsets of $\mathbb{R}^k$ which are not Lebesgue measurable. See, for example, [46, p. 24].

Remark 3.4. Denote by $\mathcal{A}$ the collection of all measurable subsets of $\mathbb{R}^k$. According to the usual terminology, as $\emptyset \in \mathcal{A}$ and (E) and (C) hold, the pair $(\mathbb{R}^k, \mathcal{A})$ is a *σ-algebra*. As the Lebesgue measure m is a nonnegative function on $\mathcal{A}$ satisfying $m(\emptyset) = 0$ and Theorem 3.2, the triple $(\mathbb{R}^k, \mathcal{A}, m)$ is a *measure space*. This definition of a measure space can be also used for sets other than $\mathbb{R}^k$. In particular, if $\Omega \subset \mathbb{R}^k$ is a Lebesgue measurable set, and define $\mathcal{B} = \{B \cap \Omega; B \in \mathcal{A}\}$, then $(\Omega, \mathcal{B}, m)$ is a measure space (where m is the restriction to $\mathcal{B}$ of the Lebesgue measure defined above).

3.2 Measurable Functions

In what follows we consider the measure space $(\mathbb{R}^k, \mathcal{A}, m)$ defined in the previous section. Note that similar considerations apply to any other measure spaces. Assume that $\mathbb{R} = \mathbb{R}^1$ is equipped with the usual topology.

Definition 3.5. *A function* $f : \mathbb{R}^k \to \mathbb{R}$ *is called* **measurable** *if for all* $\lambda \in \mathbb{R}$ *the set* $\{f > \lambda\} := \{x \in \mathbb{R}^k; f(x) > \lambda\}$ *is measurable (i.e., it belongs to* $\mathcal{A}$*).*

Remark 3.6. Equivalent definitions are obtained if the set $\{f > \lambda\}$ is replaced by $\{f \geq \lambda\}$, $\{f < \lambda\}$, or $\{f \leq \lambda\}$, $\lambda \in \mathbb{R}$. Indeed, if $\{f > \lambda\}$ is measurable for all $\lambda \in \mathbb{R}$ then so is

$$\{f \geq \lambda\} = \mathbb{R}^k \setminus \cap_{n=1}^{\infty} \{f > \lambda - 1/n\} \ \forall \lambda \in \mathbb{R},$$

hence so is

$$\{f < \lambda\} = \mathbb{R}^k \setminus \{f \geq \lambda\} \ \forall \lambda \in \mathbb{R},$$

and so on (the other implications are trivially satisfied).

Theorem 3.7. $f : \mathbb{R}^k \to \mathbb{R}$ *is measurable if and only if for every open set $D \subset \mathbb{R}$ the set $f^{-1}(D) := \{x \in \mathbb{R}^k; f(x) \in D\}$ is measurable.*

Proof. The set $D = (\lambda, \infty)$ is open for any $\lambda \in \mathbb{R}$. If $f^{-1}(D)$ is assumed to be measurable for any $\lambda \in \mathbb{R}$, then f is measurable, since $f^{-1}(D) = \{f > \lambda\}$. Conversely, let us assume that f is measurable. If $\emptyset \neq D \subset \mathbb{R}$ is an open set then it can be represented as a countable union of disjoint open intervals. Indeed, for $x \in D$ denote by $I(x)$ the maximal open interval containing x and included into D. If x, y are distinct points in D, then $I(x)$, $I(y)$ either coincide or are disjoint. Obviously, $D = \cup_{x \in D} I(x)$. Since each $I(x)$ contains a rational number, the number of distinct $I(x)$ must be countable so $D = \cup_{n=1}^{\infty} I_n$. Since f is measurable, we have $f^{-1}(I_n) \in \mathcal{A}$ for all $n \in \mathbb{N}$, which implies $f^{-1}(D) = \cup_{n=1}^{\infty} f^{-1}(I_n) \in \mathcal{A}$. $\qquad\square$

We say that a property (P) holds *almost everywhere* (abbreviated a.e.) in $\Omega \subset \mathbb{R}^k$ if it holds in $\Omega \setminus E$ with $m(E) = 0$; in other words, (P) holds for *almost all* (abbreviated a.a.) $x \in \Omega$.

Theorem 3.8. *Let $f, g : \mathbb{R}^k \to \mathbb{R}$. If f is measurable and $g = f$ a.e., then g is also measurable.*

Proof. Denote $E = \{g \neq f\}$. We have for any $\lambda \in \mathbb{R}$,

$$\{g > \lambda\} \cup E = \{f > \lambda\} \cup E \in \mathcal{A},$$

hence $G := \{g > \lambda\} \cup E \in \mathcal{A}$. Since $\{g > \lambda\}$ differs from G by a set of measure zero, it follows that $\{g > \lambda\} \in \mathcal{A}$. $\qquad\square$

Observe that the equality a.e. is an equivalence relation in the set of all measurable functions.

Theorem 3.9. *If $f : \mathbb{R}^k \to \mathbb{R}$ is measurable and $g : \mathbb{R} \to \mathbb{R}$ is continuous, then $g \circ f$ is measurable.*

Proof. As g is a continuous function, for any open set $D \subset \mathbb{R}$, $g^{-1}(D)$ is open, too. Hence, as f is measurable, we conclude that $(g \circ f)^{-1}(D) = f^{-1}(g^{-1}(D))$ is measurable for any open set $D \subset \mathbb{R}$. $\qquad\square$

Remark 3.10. It follows from the above result that, if f is measurable, then so are the functions λf ($\lambda \in \mathbb{R}$), $|f|^p$ ($p > 0$), $f^+ = \max\{f, 0\}$, $f^- = -\min\{f, 0\}$, etc.

Theorem 3.11. *If f, g are measurable, then so are $f + g$ and fg. If, in addition, $g \neq 0$ a.e., then f/g is measurable.*

Proof. For any $\lambda \in \mathbb{R}$ we have

$$\{f + g > \lambda\} = \cup_{q \in \mathbb{Q}}\{f > q > \lambda - g\} = \bigcup_{q \in \mathbb{Q}} \left(\{f > q\} \cap \{g > \lambda - q\} \right),$$

where $\mathbb{Q}$ is the set of rational numbers. It follows that $f + g$ is measurable. The function fg is also measurable since

$$fg = \frac{1}{4}\left[(f + g)^2 - (f - g)^2 \right].$$

In order to prove the last statement, it suffices to prove that $1/g$ is measurable. This follows from

$$\{1/g > \lambda\} = (\{g > 0\} \cap \{\lambda g < 1\}) \cup (\{g < 0\} \cap \{\lambda g > 1\}). \qquad \square$$

Theorem 3.12. *If $(f_n)_{n \in \mathbb{N}}$ is a sequence of measurable functions, then all of $\sup_{n \in \mathbb{N}} f_n$, $\inf_{n \in \mathbb{N}} f_n$, $\limsup_{n \to \infty} f_n$, and $\liminf_{n \to \infty} f_n$ are measurable. In particular, if $f_n \to f$ a.e. then f is measurable.*

Proof. For any $\lambda \in \mathbb{R}$ we have $\{\sup_{n \in \mathbb{N}} f_n > \lambda\} = \cup_{n \in \mathbb{N}}\{f_n > \lambda\}$ which implies that $\sup_{n \in \mathbb{N}} f_n$ is measurable. The function $\inf_{n \in \mathbb{N}} f_n$ is also measurable since it is equal to $-\sup_{n \in \mathbb{N}}(-f_n)$. The other statements follow from

$$\limsup_{n \to \infty} f_n = \inf_i \{\sup_{n \geq i} f_i\}, \quad \liminf_{n \to \infty} f_n = \sup_i \{\inf_{n \geq i} f_n\},$$

which coincide a.e. with $f = \lim f_n$ when $f_n \to f$ a.e. $\qquad \square$

Now, let us recall the definition of the *characteristic function* of a set E, denoted χ_E,

$$\chi_E(x) = \begin{cases} 1 & \text{if } x \in E, \\ 0 & \text{if } x \notin E. \end{cases}$$

Let $E \subset \mathbb{R}^k$. It is easily seen that χ_E is measurable if and only if E is measurable.

Definition 3.13. *A function $f : \mathbb{R}^k \to \mathbb{R}$ is called a* **simple function** *if it has the form*

$$f(x) = \sum_{i=1}^{p} y_i \chi_{M_i}(x), \tag{3.2.3}$$

where $p \in \mathbb{N}$, $y_i \in \mathbb{R}$, $i = 1, \ldots, p$, and the M_i's are disjoint, measurable subsets of $\mathbb{R}^k$, with $m(M_i) < \infty$, $i = 1, \ldots, p$.

Any simple function is measurable, as a finite linear combination of characteristic functions of measurable sets. Normally, in the above definition $y_1, \ldots, y_p$ are distinct numbers.

Theorem 3.14. *If $f : \mathbb{R}^k \to \mathbb{R}$ is a measurable function, then there exists a sequence of simple functions $(f_n)_{n \in \mathbb{N}}$ such that*

$$|f_n(x)| \le |f_{n+1}(x)|, \quad x \in \mathbb{R}^k, \ n = 1, 2, \ldots \quad (3.2.4)$$

and

$$\lim_{n \to \infty} f_n(x) = f(x), \quad x \in \mathbb{R}^k. \quad (3.2.5)$$

If, in addition, $f \ge 0$, $n = 1, 2, \ldots$, then one can find $f_n \ge 0$, $n = 1, 2, \ldots$

Proof. We assume first that f is a nonnegative measurable function. For a given $n \in \mathbb{N}$, define the following subsets of $\mathbb{R}^k$

$$M_j = \{\frac{j-1}{2^n} \le f < \frac{j}{2^n}\}, \quad j = 1, 2, \ldots, n2^n, \quad \text{and} \quad P_n = \{f \ge n\},$$

which are all measurable. Let

$$g_n(x) = \sum_{j=1}^{n2^n} \frac{j-1}{2^n} \chi_{M_j}(x) + n\chi_{P_n}(x), \quad x \in \mathbb{R}^k, \ n = 1, 2, \ldots$$

It is easily seen that $0 \le g_n \le g_{n+1}$ and $0 \le f(x) - g_n(x) \le 1/2^n$, whenever $f(x) \le n$, hence $g_n \to f$. Thus, the sequence (g_n) satisfies all the properties for a sequence (f_n) mentioned in the statement of the theorem, except for $m(P_n) < \infty$, $m(M_j) < \infty$ for all $n \in \mathbb{N}$, $j = 1, 2, \ldots, n2^n$ (see Definition 3.13). This inconvenience can be easily removed as follows. For any $n \in \mathbb{N}$, consider the closed cube C_n centered at the origin with side length n and define

$$
\begin{aligned}
f_n(x) &= g_n(x)\chi_{C_n}(x) \\
&= \sum_{j=1}^{n2^n} \frac{j-1}{2^n} \chi_{M_j \cap C_n}(x) + n\chi_{P_n \cap C_n}(x), \quad x \in \mathbb{R}^k.
\end{aligned}
$$

It is easily seen that (f_n) satisfies all the desired properties, including

$$0 \le f_n(x) \le f_{n+1}(x), \quad x \in \mathbb{R}^k, \ n = 1, 2, \ldots$$

For a general measurable function f one can use the decomposition $f = f^+ - f^-$, which implies $|f| = f^+ + f^-$. Since f^+ and f^- are both measurable and nonnegative, it follows from the proof above that there exist sequences (f_n^+) and (f_n^-) that satisfy the properties mentioned above and approximate f^+ and f^-, respectively. Then $(f_n = f_n^+ - f_n^-)$ is a sequence of simple functions satisfying (3.2.4) and (3.2.5).   □

Remark 3.15. Taking into account Theorems 3.12 and 3.14, one can say that a function $f : \mathbb{R}^k \to \mathbb{R}$ is measurable (in the sense of Definition 3.5) if and only if f is the limit of a sequence of simple functions (f_n), i.e., $f_n(x) \to f(x)$, as $n \to \infty$, for a.a. $x \in \mathbb{R}^k$. This equivalent condition can be used to define the notion of an X-valued measurable function, where X is a Banach space.

3.3   The Lebesgue Integral

If $f : \mathbb{R}^k \to \mathbb{R}$ is a simple function as in (3.2.3), the *Lebesgue integral* of f is defined by

$$\int_{\mathbb{R}^k} f(x)\, dx := \sum_{i=1}^{p} m(M_i) \cdot y_i. \qquad (3.3.6)$$

If Ω is a measurable subset of $\mathbb{R}^k$ then $g = f\chi_\Omega$ is also a simple function and we define

$$\int_\Omega f(x)\, dx := \int_{\mathbb{R}^k} f(x)\chi_\Omega(x)\, dx.$$

Denote by S the set of all simple functions $f : \mathbb{R}^k \to \mathbb{R}$. It is easily seen that S is a linear space over $\mathbb{R}$ with respect to the usual operations: addition of functions and scalar multiplication. We have the following statements:

- $\int_{\mathbb{R}^k}(\alpha f + \beta g)\, dx = \alpha \int_{\mathbb{R}^k} f\, dx + \beta \int_{\mathbb{R}^k} g\, dx \ \ \forall f, g \in S, \ \alpha, \beta \in \mathbb{R};$

- $f, g \in S, \ f \le g \implies \int_{\mathbb{R}^k} f\, dx \le \int_{\mathbb{R}^k} g\, dx;$

- If $\Omega_1, \Omega_2 \subset \mathbb{R}^k$ are disjoint measurable sets with $m(\Omega_i) < \infty$, $i = 1, 2$, then

$$\int_{\Omega_1 \cup \Omega_2} f\, dx = \int_{\Omega_1} f\, dx + \int_{\Omega_2} f\, dx;$$

- If $f \in S$, then so is $|f|$ and

$$\left| \int_{\mathbb{R}^k} f\, dx \right| \le \int_{\mathbb{R}^k} |f|\, dx.$$

The proofs are easy and are left to the reader.

In what follows we are concerned with the Lebesgue integration of non-negative measurable functions. Denote by S^+ the set of all nonnegative simple functions $f : \mathbb{R}^k \to \mathbb{R}$ (i.e., functions of the form (3.2.3), where each $y_i \geq 0$).

Definition 3.16. *A nonnegative measurable function $f : \mathbb{R}^k \to \mathbb{R}$ is called integrable in the sense of Lebesgue (or simply integrable) if*

$$\sup\left\{ \int_{\mathbb{R}^k} s \, dx;\ s \in S^+,\ s \leq f \right\} < +\infty,$$

and denote

$$\int_{\mathbb{R}^k} f \, dx := \sup\left\{ \int_{\mathbb{R}^k} s \, dx;\ s \in S^+,\ s \leq f \right\}.$$

If $\sup\{\int_{\mathbb{R}^k} s \, dx;\ s \in S^+,\ s \leq f\} = \infty$, *we write* $\int_{\mathbb{R}^k} f \, dx = \infty$.

Note that if f is a nonnegative simple function, i.e., a function of the form (3.2.3) with $y_i \geq 0$, $i = 1, \ldots, p$, then using this definition we reobtain $\int_{\mathbb{R}^k} f(x) \, dx = \sum_{i=1}^{p} m(M_i) \cdot y_i$.

If $f : \mathbb{R}^k \to \mathbb{R}$ is a nonnegative integrable function and $\Omega \subset \mathbb{R}^k$ is a measurable set, then $\int_{\Omega} f \, dx := \int_{\mathbb{R}^k} f \chi_{\Omega} \, dx$.

We have the following immediate statements for $f, g : \mathbb{R}^k \to \mathbb{R}$ nonnegative measurable functions and $\alpha \geq 0$:

- $f \leq g \Longrightarrow \int_{\mathbb{R}^k} f \, dx \leq \int_{\mathbb{R}^k} g \, dx$;

- If $\Omega_1 \subset \Omega_2 \subset \mathbb{R}^k$ are measurable sets, with $\Omega_1 \subset \Omega_2$, then $\int_{\Omega_1} f \, dx \leq \int_{\Omega_2} f \, dx$; We also have:

- If $f : \mathbb{R}^k \to \mathbb{R}$ is a nonnegative measurable function, then: $f = 0$ a.e. if and only if $\int_{\mathbb{R}^k} f \, dx = 0$.

Proof. Observe first that if $f = 0$ a.e., then for any $s \in S^+$, with $s \leq f$, we have $s = 0$ a.e., so $\int_{\mathbb{R}^k} s \, dx = 0$. Therefore $\int_{\mathbb{R}^k} f \, dx = 0$. Conversely, let us assume that $\int_{\mathbb{R}^k} f \, dx = 0$. Define $\Omega_n = \{x \in \mathbb{R}^k;\ f(x) \geq 1/n\}$, $n \in \mathbb{N}$. We have for all $n \in \mathbb{N}$

$$0 = \int_{\mathbb{R}^k} f \, dx \geq \int_{\mathbb{R}^k} \frac{1}{n} \chi_{\Omega_n} \, dx = \frac{1}{n} m(\Omega_n).$$

So $m(\Omega_n) = 0$ for all $n \in \mathbb{N} \Longrightarrow m(\{f > 0\}) = m(\cup_{n=1}^{\infty} \Omega_n) = 0 \Longrightarrow f = 0$ almost everywhere. $\square$

Let us now state the so-called *Monotone Convergence Theorem* or Beppo Levi's theorem.[2]

Theorem 3.17 (Monotone Convergence Theorem). *Let* $0 \leq f_1 \leq f_2 \leq \cdots \leq f_n \leq \cdots$ *be a sequence of measurable functions. Denote* $f(x) := \lim_{n \to \infty} f_n(x)$. *Then*

$$\lim_{n \to \infty} \int_{\mathbb{R}^k} f_n \, dx = \int_{\mathbb{R}^k} f \, dx \, .$$

Proof. Obviously, there exists

$$\lim_{n \to \infty} \int_{\mathbb{R}^k} f_n \, dx \leq \int_{\mathbb{R}^k} f \, dx \, .$$

In order to prove the converse inequality, let $s \in S^+$, $s \leq f$, and let $\varepsilon \in (0, 1)$. Define $M_n = \{x \in \mathbb{R}^k; \, f_n(x) \geq \varepsilon s(x)\}$, $n \in \mathbb{N}$. We have $\mathbb{R}^k = \cup_{n=1}^\infty M_n$. Indeed, if $x \in \mathbb{R}^k$ and $f(x) = 0$, then $s(x) = 0$, so $x \in M_1$. If $f(x) > 0$, then $f(x) > \varepsilon s(x)$, hence $x \in M_n$ for n large enough.

Next,

$$\int_{\mathbb{R}^k} f_n \, dx \geq \int_{M_n} f_n \, dx \geq \varepsilon \int_{M_n} s \, dx \, .$$

Since $M_n \subset M_{n+1}$ for all $n \in \mathbb{N}$, the last inequality implies

$$\lim_{n \to \infty} \int_{\mathbb{R}^k} f_n \, dx \geq \varepsilon \int_{\mathbb{R}^k} s \, dx \, ,$$

hence, as $\varepsilon \in (0, 1)$ was arbitrary,

$$\lim_{n \to \infty} \int_{\mathbb{R}^k} f_n \, dx \geq \int_{\mathbb{R}^k} s \, dx \quad \forall s \in S^+, \, s \leq f \, .$$

This implies

$$\lim_{n \to \infty} \int_{\mathbb{R}^k} f_n \, dx \geq \int_{\mathbb{R}^k} f \, dx \, ,$$

as claimed. $\square$

Remark 3.18. Combining Theorems 3.14 and 3.17, we infer that for any nonnegative integrable function $f : \mathbb{R}^k \to \mathbb{R}$, there exists an increasing sequence $(s_n)_\mathbb{N}$ in S^+ such that $s_n \to f$ pointwise (or a.e.) and $\int_{\mathbb{R}^k} s_n \, dx \to \int_{\mathbb{R}^k} f \, dx$. Using this observation, one can readily deduce that

---

[2]Beppo Levi, Italian mathematician, 1875–1961.

- if $f, g : \mathbb{R}^k \to \mathbb{R}$ are nonnegative integrable functions, then so is $f + g$ and

$$\int_{\mathbb{R}^k} (f + g) \, dx = \int_{\mathbb{R}^k} f \, dx + \int_{\mathbb{R}^k} g \, dx .$$

We also have

- $\int_{\mathbb{R}^k} \alpha f \, dx = \alpha \int_{\mathbb{R}^k} f \, dx \;\; \forall \alpha \geq 0$.

The next result is known as Fatou's lemma.[3]

Theorem 3.19. *Let $f_n : \mathbb{R}^k \to \mathbb{R}$ be a sequence of nonnegative measurable functions. Set $f = \liminf_{n \to \infty} f_n$. Then,*

$$\int_{\mathbb{R}^k} f \, dx \leq \liminf_{n \to \infty} \int_{\mathbb{R}^k} f_n \, dx . \tag{3.3.7}$$

Proof. Denote $g_n = \inf_{m \geq n} f_m$, $n \in \mathbb{N}$. Since (g_n) is an increasing sequence, we have

$$f = \sup_{n \in \mathbb{N}} g_n = \lim_{n \to \infty} g_n .$$

By the Monotone Convergence Theorem we have

$$\lim_{n \to \infty} \int_{\mathbb{R}^k} g_n \, dx = \int_{\mathbb{R}^k} f \, dx . \tag{3.3.8}$$

On the other hand, since $g_n \leq f_n$, $n \in \mathbb{N}$, we have

$$\int_{\mathbb{R}^k} g_n \, dx \leq \int_{\mathbb{R}^k} f_n \, dx, \quad n \in \mathbb{N} . \tag{3.3.9}$$

Combining (3.3.8) and (3.3.9) yields (3.3.7). $\qquad\qquad\qquad\square$

Now, we are going to define the Lebesgue integral for a general measurable function $f : \mathbb{R}^k \to \mathbb{R}$. One can use the decomposition $f = f^+ - f^-$. Obviously, f is measurable if and only if both f^+ and f^- are measurable.

Definition 3.20. *A measurable function $f : \mathbb{R}^k \to \mathbb{R}$ is called integrable if both f^+ and f^- are integrable and*

$$\int_{\mathbb{R}^k} f \, dx := \int_{\mathbb{R}^k} f^+ \, dx - \int_{\mathbb{R}^k} f^- \, dx .$$

---

[3]Pierre Joseph Louis Fatou, French mathematician, 1878–1929.

Denote by $\mathcal{L}(\mathbb{R}^k)$ the set of all (measurable and) integrable functions $f : \mathbb{R}^k \to \mathbb{R}$.

One can prove by elementary arguments the following statements:

- If $f : \mathbb{R}^k \to \mathbb{R}$ is measurable, then so is $|f|$ and

$$f \in \mathcal{L}(\mathbb{R}^k) \iff |f| \in \mathcal{L}(\mathbb{R}^k);$$

- If $f, g : \mathbb{R}^k \to \mathbb{R}$ are measurable, $g \in \mathcal{L}(\mathbb{R}^k)$ and $|f| \leq g$, then $f \in \mathcal{L}(\mathbb{R}^k)$;

- If $f \in \mathcal{L}(\mathbb{R}^k)$ and $\alpha \in \mathbb{R}$, then $\alpha f \in \mathcal{L}(\mathbb{R}^k)$ and

$$\int_{\mathbb{R}^k} \alpha f \, dx = \alpha \int_{\mathbb{R}^k} f \, dx;$$

  We also have

- If $f, g \in \mathcal{L}(\mathbb{R}^k)$, then $f + g \in \mathcal{L}(\mathbb{R}^k)$ and

$$\int_{\mathbb{R}^k} (f + g) \, dx = \int_{\mathbb{R}^k} f \, dx + \int_{\mathbb{R}^k} g \, dx.$$

Proof. Assume $f, g \in \mathcal{L}(\mathbb{R}^k)$. Then $f^+, f^-, g^+, g^-, f + g, (f + g)^+,$ $(f+g)^-$ are measurable, and $f^+, f^-, g^+, g^- \in \mathcal{L}(\mathbb{R}^k)$. From $(f+g)^+ \leq f^+ + g^+$ and $(f+g)^- \leq f^- + g^-$ we infer that $(f+g)^+, (f+g)^- \in \mathcal{L}(\mathbb{R}^k)$, which implies $f + g \in \mathcal{L}(\mathbb{R}^k)$. On the other hand,

$$(f + g)^+ - (f + g)^- = f + g = f^+ - f^- + g^+ - g^-,$$

so

$$(f + g)^+ + f^- + g^- = (f + g)^- + f^+ + g^+,$$

which involves only nonnegative integrable functions. Hence,

$$\int_{\mathbb{R}^k} (f + g)^+ \, dx + \int_{\mathbb{R}^k} f^- \, dx + \int_{\mathbb{R}^k} g^- \, dx$$
$$= \int_{\mathbb{R}^k} (f + g)^- \, dx + \int_{\mathbb{R}^k} f^+ \, dx + \int_{\mathbb{R}^k} g^+ \, dx,$$

which gives the desired equality. $\qquad \square$

- Let $f, g : \mathbb{R}^k \to \mathbb{R}$ be such that $f \in \mathcal{L}(\mathbb{R}^k)$ and $g = f$ a.e. Then, $g \in \mathcal{L}(\mathbb{R}^k)$ and $\int_{\mathbb{R}^k} g \, dx = \int_{\mathbb{R}^k} f \, dx$.

Proof. From $g = f$ a.e. we derive $g^+ = f^+ \geq 0$ a.e and $g^- = f^- \geq 0$ a.e., so

$$\int_{\mathbb{R}^k} g^+ \, dx = \int_{\mathbb{R}^k} f^+ \, dx, \quad \int_{\mathbb{R}^k} g^- \, dx = \int_{\mathbb{R}^k} f^- \, dx \,,$$

and the result follows.   □

- If $f, g \in \mathcal{L}(\mathbb{R}^k)$ and $f \leq g$ a.e., then $\int_{\mathbb{R}^k} f \, dx \leq \int_{\mathbb{R}^k} g \, dx$.
  The proof is easy.

- For every $f \in \mathcal{L}(\mathbb{R}^k)$ we have

$$\left| \int_{\mathbb{R}^k} f \, dx \right| \leq \int_{\mathbb{R}^k} |f| \, dx \,.$$

Proof. We know that $f \in \mathcal{L}(\mathbb{R}^k) \Rightarrow |f| \in \mathcal{L}(\mathbb{R}^k)$. We have

$$
\begin{aligned}
\int_{\mathbb{R}^k} f \, dx &= \int_{\mathbb{R}^k} f^+ \, dx - \int_{\mathbb{R}^k} f^- \, dx \\
&\leq \int_{\mathbb{R}^k} f^+ \, dx + \int_{\mathbb{R}^k} f^- \, dx \\
&= \int_{\mathbb{R}^k} |f| \, dx \,.
\end{aligned}
$$

Similarly,

$$-\int_{\mathbb{R}^k} f \, dx \leq \int_{\mathbb{R}^k} |f| \, dx \,,$$

so the result follows.   □

Theorem 3.21. *Let $f \in \mathcal{L}(\mathbb{R}^k)$. Then, for every $\varepsilon > 0$ there exists $\delta > 0$, such that for every measurable set $M \subset \mathbb{R}^k$ with $m(M) < \delta$, we have $\int_M |f| \, dx < \varepsilon$.*

Proof. For $n \in \mathbb{N}$ define

$$g_n(x) = \begin{cases} |f(x)| & \text{if } |f(x)| \leq n, \\ n & \text{if } |f(x)| > n. \end{cases}$$

Observe that, for every $n \in \mathbb{N}$, $0 \leq g_n \leq |f|$, so $g_n \in \mathcal{L}(\mathbb{R}^k)$. Moreover, (g_n) is an increasing sequence converging pointwise to $|f|$. By Beppo Levi,

$$\lim_{n \to \infty} \int_{\mathbb{R}^k} g_n \, dx = \int_{\mathbb{R}^k} |f| \, dx \,,$$

so, for a given $\varepsilon > 0$, there exists an $N \in \mathbb{N}$ such that

$$\int_{\mathbb{R}^k} (|f| - g_N) \, dx < \frac{\varepsilon}{2} \, . \tag{3.3.10}$$

Choosing $\delta = \varepsilon/(2N)$, we have

$$\forall M \in \mathcal{A} \text{ with } m(M) < \delta, \int_M g_N \, dx \leq \int_M N \, dx = N m(M) < \frac{\varepsilon}{2} \, . \tag{3.3.11}$$

Now, we derive from (3.3.10) and (3.3.11),

$$\int_M |f| \, dx = \int_M (|f| - g_N) \, dx + \int_M g_N \, dx < \varepsilon \, . \qquad \square$$

Recall that the equality a.e. is an equivalence relation in the linear space of measurable functions, in particular in $\mathcal{L}(\mathbb{R}^k)$. Denote by $L^1(\mathbb{R}^k)$ the quotient space $\mathcal{L}(\mathbb{R}^k)/\sim$, where $\sim$ stands for the equivalence relation we are talking about. In general, any equivalence class in $L^1(\mathbb{R}^k)/\sim$ is identified with a representative of the corresponding class, which is usually selected to be the most regular one. If $\Omega \subset \mathbb{R}^k$ is a measurable set, we can similarly define $L^1(\Omega) := \mathcal{L}(\Omega)/\sim$. Based on this identification, we can say that the above theory works for functions (in fact classes of functions) belonging to $L^1(\mathbb{R}^k)$ or to $L^1(\Omega)$. The next result is known as *Lebesgue's Dominated Convergence Theorem*.

Theorem 3.22 (Lebesgue's Dominated Convergence Theorem). *Let $\Omega \subset \mathbb{R}^k$ be a measurable set, possibly $\Omega = \mathbb{R}^k$. Let $(f_n)_{n \in \mathbb{N}}$ be a sequence in $L^1(\Omega)$ such that*

(a) $f_n(x) \to f(x)$ *a.e. on Ω;*

(b) $\exists g \in L^1(\Omega)$ *such that $|f_n(x)| \leq g(x)$ a.e. on Ω.*

   Then, $f \in L^1(\Omega)$ and $\lim_{n \to \infty} \int_\Omega |f_n(x) - f(x)| \, dx = 0$.

Proof. According to (a), f is measurable. Passing to the limit in (b) we get $|f| \leq g$ a.e., so $f \in L^1(\Omega)$. Set $h_n := |f_n - f|$. We have $h_n \to 0$ a.e. on Ω and $h_n \leq \tilde{g} := g + |f| \in L^1(\Omega)$. Applying Fatou's lemma to the sequence $(\tilde{g} - h_n)$, we get

$$\int_\Omega \tilde{g} \, dx \leq \liminf_{n \to \infty} \int_\Omega (\tilde{g} - h_n) \, dx = \int_\Omega \tilde{g} \, dx - \limsup_{n \to \infty} \int_\Omega h_n \, dx,$$

which implies

$$\limsup_{n\to\infty} \int_{\Omega} h_n \, dx \le 0 \, .$$

Thus

$$\lim_{n\to\infty} \int_{\Omega} h_n \, dx = 0 \, . \qquad\qquad \square$$

3.4 L^p Spaces

Throughout this section Ω denotes a measurable subset of $\mathbb{R}^k$ (possibly $\Omega = \mathbb{R}^k$). As usual, any class of measurable functions with respect to the equality a.e. will be identified with one of its representatives.

We have already defined the space $L^1(\Omega)$ as being the set of all functions $f : \Omega \to \mathbb{R}$ which are integrable over Ω, i.e., f is measurable and $\int_{\Omega} |f| \, dx < \infty$. This definition can be extended as follows:

$$L^p(\Omega) := \{ f : \Omega \to \mathbb{R}; \ f \text{ is measurable and } |f|^p \in L^1(\Omega) \} \, ,$$

for $1 \le p < \infty$. We also define

$L^\infty(\Omega) := \{ f : \Omega \to \mathbb{R}; \ f \text{ is measurable and there exists } C \ge 0$ such that $|f(x)| \le C$ a.e. on $\Omega \}$.

It is easily seen that, for every $1 \le p \le \infty$, $L^p(\Omega)$ is a linear space over $\mathbb{R}$.

Now, for $1 < p < \infty$ denote by q the *conjugate* of p, i.e.,

$$\frac{1}{p} + \frac{1}{q} = 1 \, .$$

Recall the so-called Young's inequality

$$ab \le \frac{a^p}{p} + \frac{b^q}{q} \, . \qquad\qquad (3.4.12)$$

This inequality follows from the fact that the log function is concave on $(0, \infty)$, so

$$\log\left(\frac{1}{p}a^p + \frac{1}{q}b^q\right) \ge \frac{1}{p}\log a^p + \frac{1}{q}\log b^q = \log(ab) \, .$$

Now, we set for $1 \le p < \infty$

$$\|f\|_{L^p(\Omega)} := \left(\int_{\Omega} |f(x)|^p \, dx\right)^{1/p} \quad \forall f \in L^p(\Omega) \, ,$$

and

$$\|f\|_{L^\infty(\Omega)} := \inf\{C;\ |f(x)| \leq C \text{ a.e. on } \Omega\} \quad \forall f \in L^\infty(\Omega).$$

We are going to prove that these are norms. To this purpose, we need the following auxiliary result which is known as *Hölder's inequality*.[4]

Lemma 3.23 (Hölder's Inequality). *Let $1 < p < \infty$. If $f \in L^p(\Omega)$ and $g \in L^q(\Omega)$, then $fg \in L^1(\Omega)$ and*

$$\int_\Omega |fg|\, dx \leq \|f\|_{L^p(\Omega)} \|g\|_{L^q(\Omega)}, \qquad (3.4.13)$$

where q is the conjugate of p.

Proof. If $f = 0$ a.e. on Ω, then (3.4.13) is trivially satisfied, so we can assume $\|f\|_{L^p(\Omega)} > 0$. By Young's inequality we have

$$|fg| \leq \frac{1}{p}|f|^p + \frac{1}{q}|g|^q \quad \text{a.e. on } \Omega.$$

This shows that $fg \in L^1(\Omega)$ and

$$\int_\Omega |fg|\, dx \leq \frac{1}{p}\|f\|_{L^p(\Omega)}^p + \frac{1}{q}\|g\|_{L^q(\Omega)}^q.$$

By replacing in this inequality f by αf with $\alpha > 0$, we obtain

$$\int_\Omega |fg|\, dx \leq \frac{\alpha^{p-1}}{p}\|f\|_{L^p(\Omega)}^p + \frac{1}{\alpha q}\|g\|_{L^q(\Omega)}^q,$$

whose right-hand side achieves its minimum for $\alpha = \|g\|_{L^q(\Omega)}^{q/p}/\|f\|_{L^p(\Omega)}$, thus (3.4.13) follows. $\qquad\square$

Theorem 3.24. $\|\cdot\|_{L^p(\Omega)}$ *is a norm in $L^p(\Omega)$ for all $1 \leq p \leq \infty$.*

Proof. The result is trivial for $p = 1$.
Now, if $f \in L^\infty(\Omega)$, then

$$|f(x)| \leq \|f\|_{L^\infty(\Omega)} \quad \text{a.e. on } \Omega. \qquad (3.4.14)$$

Indeed, we infer from the definition of $\|\cdot\|_{L^\infty(\Omega)}$ that, for each $n \in \mathbb{N}$, there exists a constant C_n such that

$$\|f\|_{L^\infty(\Omega)} \leq C_n < \|f\|_{L^\infty(\Omega)} + \frac{1}{n} \quad \text{and} \quad |f(x)| \leq C_n,$$

---

[4]Otto Ludwig Hölder, German mathematician, 1859–1937.

for $x \in \Omega \setminus A_n$ with $m(A_n) = 0$. Setting $A = \cup_{n=1}^{\infty} A_n$, we have $m(A) = 0$ and

$$|f(x)| \leq C_n, \quad x \in \Omega \setminus A.$$

As $C_n \to \|f\|_{L^\infty(\Omega)}$ we derive (3.4.14) by passing to the limit in the last inequality.

Using (3.4.14) one can easily prove that $\|\cdot\|_{L^\infty(\Omega)}$ is a norm in $L^\infty(\Omega)$. Now, let us consider the case $1 < p < \infty$. We have only to prove the triangle inequality (since the other axioms are trivially satisfied). For $f, g \in L^p(\Omega)$, we have

$$\|f + g\|_{L^p(\Omega)}^p = \int_\Omega |f + g|^{p-1} |f + g| \, dx$$

$$\leq \int_\Omega |f + g|^{p-1} |f| \, dx + \int_\Omega |f + g|^{p-1} |g| \, dx. \quad (3.4.15)$$

Noting that $|f + g|^{p-1} \in L^q(\Omega)$, we obtain by Hölder's inequality

$$\|f + g\|_{L^p(\Omega)}^p \leq \|f + g\|_{L^p(\Omega)}^{p-1} \left(\|f\|_{L^p(\Omega)} + \|g\|_{L^p(\Omega)} \right),$$

which implies

$$\|f + g\|_{L^p(\Omega)}^p \leq \|f\|_{L^p(\Omega)}^p + \|g\|_{L^p(\Omega)}^p. \qquad \square$$

Theorem 3.25. *For every $1 \leq p \leq \infty$, $L^p(\Omega)$ equipped with $\|\cdot\|_{L^p(\Omega)}$ is a Banach space.*

Proof. The fact that $\|\cdot\|_{L^p(\Omega)}$ is a norm was shown before (see Theorem 3.24). So we only need to prove that this norm is complete. We distinguish two cases.

Case 1: $1 \leq p < \infty$. Let $(f)_{n \in \mathbb{N}}$ be a Cauchy sequence in $L^p(\Omega)$. Then there exists a subsequence $(f_{n_m})_{m \in \mathbb{N}}$ which satisfies

$$\|f_{n_{m+1}} - f_{n_m}\|_{L^p(\Omega)} \leq \frac{1}{2^m}, \quad m = 1, 2, \dots \qquad (3.4.16)$$

Indeed, one may first choose $n_1 \in \mathbb{N}$ such that $\|f_m - f_n\|_{L^p(\Omega)} \leq 1/2 \ \forall m, n \geq n_1$; then choose $n_2 \in \mathbb{N}$, $n_2 \geq n_1$, such that $\|f_m - f_n\|_{L^p(\Omega)} \leq 1/2^2 \ \forall m, n \geq n_2$, and so on. We are going to show that there is a function $f \in L^p(\Omega)$ such that $\|f_{n_m} - f\|_{L^p(\Omega)} \to 0$, as $m \to \infty$. If we show this, the initial sequence (f_n) will be convergent in $L^p(\Omega)$, as a Cauchy sequence with a convergent subsequence. For simplicity, we redenote $f_m := f_{n_m}$, so (3.4.16) becomes

$$\|f_{m+1} - f_m\|_{L^p(\Omega)} \leq \frac{1}{2^m}, \quad m = 1, 2, \dots \qquad (3.4.17)$$

Set

$$g_n(x) = \sum_{i=1}^n |f_{i+1}(x) - f_i(x)|.$$

According to (3.4.17), we have

$$\|g_n\|_{L^p(\Omega)} \leq 1, \quad n = 1, 2, \dots$$

By the Monotone Convergence Theorem, $g_n(x)$ converges a.e. to a finite limit $g(x)$, and $g \in L^p(\Omega)$. Now, for $m \geq n \geq 2$ and for almost all $x \in \Omega$,

$$
\begin{aligned}
|f_m(x) - f_n(x)| &\leq |f_m(x) - f_{m-1}(x)| + \cdots + |f_{n+1}(x) - f_n(x)| \\
&= g_{m-1}(x) - g_{n-1}(x) \\
&\leq g(x) - g_{n-1}(x). \qquad (3.4.18)
\end{aligned}
$$

It follows that for almost all $x \in \Omega$, $(f_n(x))_{n \in \mathbb{N}}$ is Cauchy, so it converges to some $f(x)$. We also obtain for almost all $x \in \Omega$

$$|f(x) - f_n(x)| \leq g(x), \quad n = 2, 3, \dots$$

so, in particular, $f \in L^p(\Omega)$. As $|f_n - f|^p \to 0$ a.e. on Ω and $|f_n - f|^p \leq g^p \in L^1(\Omega)$, we are in a position to apply the Dominated Convergence Theorem to conclude that $\|f_n - f\|_{L^p(\Omega)} \to 0$.

Case 2: $p = \infty$. Let (f_n) be a Cauchy sequence in $L^\infty(\Omega)$. So, for any $j \in \mathbb{N}$, there exists $N_j \in \mathbb{N}$ such that

$$\|f_n - f_m\|_{L^\infty(\Omega)} \leq \frac{1}{j} \quad \forall n, m \geq N_j.$$

Hence, there exists a set M_j with $m(M_j) = 0$ such that

$$|f_n(x) - f_m(x)| \leq \frac{1}{j} \quad \forall x \in \Omega \setminus M_j, \; m, n \geq N_j. \qquad (3.4.19)$$

Obviously, the set $M = \cup_{j=1}^\infty M_j$ has measure zero. For each $x \in \Omega \setminus M$ the sequence $(f_n(x))$ is Cauchy and therefore convergent to some $f(x) \in \mathbb{R}$. Now, we deduce from (3.4.19)

$$|f_n(x) - f(x)| \leq \frac{1}{j} \quad \forall x \in \Omega \setminus M, \; n \geq N_j,$$

hence $f \in L^\infty(\Omega)$ and

$$\|f_n - f\|_{L^\infty(\Omega)} \le \frac{1}{j} \quad \forall n \ge N_j \,.$$

So (f_n) converges to f in $L^\infty(\Omega)$. $\square$

3.5   Exercises

1. A set $\Omega \subset \mathbb{R}^k$ is measurable $\iff$ for every $\varepsilon > 0$ there exists a closed set $F \subset \Omega$ such that $m(\Omega \setminus F) < \varepsilon$.

2. Let $\Omega \subset \mathbb{R}^k$ be a measurable set with $m(\Omega) < \infty$. Show that, for every $\varepsilon > 0$, there exists a compact set $K \subset \Omega$ such that $m(\Omega \setminus K) < \varepsilon$.

3. Let $A \subset \Omega \subset B \subset \mathbb{R}^k$, where A, B are measurable sets with $m(A) = m(B) < \infty$. Then Ω is measurable.

4. Let $h \in \mathbb{R}^k \setminus \{0\}$ and $\alpha \in \mathbb{R}$. Show that for every measurable set $\Omega \subset \mathbb{R}^k$ we have

   (a) $\Omega_h := \{x + h; \, x \in \Omega\}$ is measurable and $m(\Omega_h) = m(\Omega)$ (translation invariance);

   (b) $\alpha\Omega := \{\alpha x; \, x \in \Omega\}$ is measurable and $m(\alpha\Omega) = |\alpha|^k m(\Omega)$.

5. Let $h \in \mathbb{R}^k \setminus \{0\}$ and $\alpha \in \mathbb{R} \setminus \{0\}$. If $f \in L^1(\mathbb{R}^k)$, then so are the functions $x \mapsto f(x - h)$, $x \mapsto f(\alpha x)$ and

   $$\int_{\mathbb{R}^k} f(x - h)\, dx = \int_{\mathbb{R}^k} f(x)\, dx, \quad \int_{\mathbb{R}^k} f(\alpha x)\, dx$$
   $$= \frac{1}{|\alpha|^k} \int_{\mathbb{R}^k} f(x)\, dx \,.$$

6. Let $-\infty < a < b < +\infty$ and let $f : [a, b] \to \mathbb{R}$ be a bounded function. If f is Riemann integrable then $f \in L^1(a, b) := L^1((a, b); \mathbb{R})$, and the two integrals coincide:

   $$(L) \int_a^b f(x)\, dx = (R) \int_a^b f(x)\, dx \,.$$

Use the (Dirichlet) function $D : [0,1] \to \mathbb{R}$,

$$D(x) = \begin{cases} 1 & \text{if } x \in \mathbb{Q} \cap [0,1], \\ 0 & \text{if } x \in [0,1] \setminus \mathbb{Q}, \end{cases}$$

to show that the converse implication is not true in general.

7. Let $f_n : [0,1] \to \mathbb{R}$ be defined by

$$f_n(x) = \frac{n x^{n-1}}{1+x}, \quad x \in [0,1], \ n \in \mathbb{N}.$$

Show that

$$\lim_{n \to \infty} \int_0^1 f_n(x)\, dx = \frac{1}{2}.$$

8. Show that $f : [1, \infty) \to \mathbb{R}$ defined by

$$f(x) = x^{-2} \ln x, \quad x \in [1, \infty),$$

is Lebesgue integrable and

$$\int_1^\infty f(x)\, dx = 1.$$

9. Show that

$$\lim_{n \to \infty} \int_0^n \left(1 + \frac{x}{n}\right)^n e^{-2x}\, dx = 1.$$

10. Let $f : [0, \infty) \to \mathbb{R}$ be a continuous function such that

$$\lim_{x \to \infty} f(x) = a,$$

where $a \in \mathbb{R}$. Show that, for every $b \in (0, \infty)$,

$$\lim_{n \to \infty} \int_a^b f(nx)\, dx = ab.$$

11. Let $f : [0,1] \to \mathbb{R}$ be defined by

$$f(x) = \begin{cases} 0 & \text{if } x = 0, \\ \sqrt{n} & \text{if } x \in \left(\frac{1}{n+1}, \frac{1}{n}\right], \ n \in \mathbb{N}. \end{cases}$$

Show that

(a) f is not Riemann integrable on $[0,1]$;

(b) $f \in L^p(0,1)$ for $1 \leq p < 2$, and $f \notin L^p(0,1)$ for $2 \leq p \leq \infty$.

12. Show that the following functions are not Lebesgue integrable:

(a) $f(x) = \frac{1}{x}$, $x \in (0,1)$;

(b) $g(x) = \sin x + \cos x$, $x \in (0,\infty)$.

13. Let $f \in C[0,1] := C([0,1]; \mathbb{R})$, such that $f(0) = 0$, and f is differentiable at $x = 0$. Then prove that $g : (0,1) \to \mathbb{R}$, defined by

$$g(x) = x^{-3/2} f(x), \quad x \in (0,1),$$

belongs to $L^1(0,1)$.

14. If $f \in L^1(0,1)$, show that $\int_0^1 x^n f(x)\, dx \to 0$ as $n \to \infty$.

15. Let $\Omega \subset \mathbb{R}^k$ be a measurable set with $m(\Omega) < \infty$ and let $1 \leq p < q \leq \infty$. Prove that $L^q(\Omega) \subset L^p(\Omega)$ and

$$\|f\|_{L^p(\Omega)} \leq m(\Omega)^{(q-p)/pq} \|f\|_{L^q(\Omega)} \quad \forall f \in L^q(\Omega).$$

16. Let $\Omega \subset \mathbb{R}^k$ be a measurable set with $m(\Omega) < \infty$ and let $f \in L^\infty(\Omega)$. Prove that

$$\lim_{p \to \infty} \|f\|_{L^p(\Omega)} = \|f\|_{L^\infty(\Omega)}.$$

Chapter 4

Continuous Linear Operators and Functionals

In this chapter we discuss linear operators between linear spaces, but our presentation is restricted at this stage to the space of continuous (bounded) linear operators between normed spaces. When the target space is either $\mathbb{R}$ or $\mathbb{C}$, they are called (continuous linear) *functionals* and are used to define dual spaces and weak topologies.

Unless otherwise specified, this chapter only considers linear spaces over the field $\mathbb{K}$, with $\mathbb{K}$ being $\mathbb{R}$ or $\mathbb{C}$. When two or more linear spaces are involved then all of them will be over the same field.

4.1 Definitions, Examples, Operator Norm

We begin this section with some basic definitions.

Definition 4.1. *Let X, Y be linear spaces and let $A : D(A) \subset X \to Y$. A is called a* **linear operator** *if $D(A)$ is a linear subspace of X and*

$$A(\alpha x + \beta y) = \alpha A x + \beta A y, \quad \forall \alpha, \beta \in \mathbb{K}, \ \forall x, y \in D(A).$$

We denote the range of A by $R(A)$, i.e., $R(A) = \{Ax; \ x \in D(A)\}$. The range $R(A)$ is a linear subspace of Y.

We say that A is **injective** or **one-to-one** if $N(A)$, the nullspace of A, defined by $N(A) = \{x \in D(A); \ Ax = 0\}$, is precisely $\{0\}$. The operator A is called **surjective** or **onto** if $R(A) = Y$.

© Springer Nature Switzerland AG 2019

G. Moroșanu, *Functional Analysis for the Applied Sciences*,
Universitext, https://doi.org/10.1007/978-3-030-27153-4_4

Example 1.

Let $X = \mathbb{R}^n$, $Y = \mathbb{R}^m$ with $n, m \in \mathbb{N}$. Let M be an $m \times n$ matrix with real entries, then $A : D(A) = X \to Y$ defined by

$$Au = Mu \qquad \forall u = (u_1, \ldots, u_n)^T \in X$$

is a linear operator, and in fact all linear maps between these spaces can be represented in this way. Here we consider that the elements of both X and Y are column vectors. If $m = 1$ then A is a linear form on X, as defined in Chap. 1.

Example 2.

For $X = Y = C[a, b] := C([a, b]; \mathbb{R})$ with $-\infty < a < b < \infty$, the derivative operator $Af = f'$ is defined on $D(A) = C^1[a, b]$ (which is the set of all continuously differentiable functions $f : [a, b] \to \mathbb{R}$), and its range is $R(A) = C[a, b] = Y$, so A is surjective. Note that A is not injective because its nullspace $N(A) := \{f \in D(A); Af = 0\} \neq \{0\}$ (more precisely, $N(A)$ consists of all constant functions).

Example 3.

For $X = Y = C[a, b]$, $-\infty < a < b < \infty$, the antiderivative operator $(Af)(t) = \int_a^t f(s)\,ds$ is defined on $D(A) = C[a, b] = X$. It is injective because $Af = 0$ implies $f = 0$. However A is not surjective because $(Af)(a) = 0$ for all $f \in D(A) = C[a, b]$, and thus $R(A)$ is a proper subset of $Y = C[a, b]$.

Proposition 4.2. *Let $(X, \|\cdot\|_X)$, $(Y, \|\cdot\|_Y)$ be normed (linear) spaces and let $A : X \to Y$ be a linear operator. Then the following are equivalent*

1. *A is continuous on X;*

2. *A is continuous at $x = 0$;*

3. *A maps bounded subsets of X to bounded subsets of Y;*

4. *There exists $c > 0$ such that $\|Au\|_Y \leq c\|u\|_X$ for all $u \in X$.*

Proof. An exercise. $\square$

Remark 4.3. If X, Y are finite dimensional spaces then both of them can be equipped with norms, and every linear operator between the two spaces is continuous (prove it!). In fact, any such operator can be represented by a matrix which depends on the bases of the two spaces. So continuity of linear operators is interesting only in the case of infinite dimensional linear spaces.

Remark 4.4. A linear operator $A : D(A) \subset X \to Y$ is said to be **bounded** if

$$\sup \{\|Ax\|_Y; \; x \in D(A), \|x\|_X \le 1\} < \infty. \tag{4.1.1}$$

Otherwise, A is called **unbounded**.

Obviously, any continuous linear operator from $(X, \|\cdot\|_X)$ to $(Y, \|\cdot\|_Y)$ is bounded. Conversely, if $A : D(A) \subset X \to Y$ is a bounded linear operator, then denoting by $\hat{c}$ the supremum in (4.1.1) we have

$$\|Ax\|_Y \le \hat{c} \|x\|_X \quad \forall x \in D(A), \tag{4.1.2}$$

so A is continuous from $(D(A), \|\cdot\|_X)$ to $(Y, \|\cdot\|_Y)$ (see Proposition 4.2). That is why continuous linear operators are also called bounded.

Note that if A is a continuous (bounded) linear operator from $(D(A), \|\cdot\|_X)$ to $(Y, \|\cdot\|_Y)$, then A can be extended by continuity to a continuous linear operator $A_1 : D(A_1) = X_1 \to Y_1$, where X_1, Y_1 denote the completions of $D(A)$ and Y with respect to $\|\cdot\|_X$ and $\|\cdot\|_Y$, respectively.

For $(X, \|\cdot\|_X)$, $(Y, \|\cdot\|_Y)$ normed spaces, denote

$$L(X, Y) = \{A : X \to Y; \; A \text{ is linear and continuous}\}.$$

Obviously, $L(X, Y)$ is a linear space. It is a normed space with the so-called **operator norm**

$$\|A\| = \sup \{\|Au\|_Y; \; u \in X, \; \|u\|_X \le 1\}.$$

Clearly, we have

$$\|Au\|_Y \le \|A\| \cdot \|u\|_X \quad \forall u \in X.$$

If $(Z, \|\cdot\|_Z)$ is another normed space, and $A \in L(X, Y)$, $B \in L(Z, X)$, then $AB \in L(Z, Y)$ and

$$\|AB\| \le \|A\| \cdot \|B\|,$$

where AB denotes the composition $A \circ B$.

In the case $X = Y$ we simply write $L(X) = L(X, X)$.

Examples. If $X = C[0, 1]$ equipped with the usual sup-norm, the antiderivative operator $A : X \to X$, $(Af)(t) = \int_0^t f(s)\, ds$, $t \in [0, 1]$, $f \in X$, is linear and continuous (hence bounded) with $\|A\| = 1$. On the other hand, for the same space X, the derivative operator $B : D(B) = C^1[0, 1] \subset X \to X$, $Bf = f'$, is linear but unbounded because for $f_n(t) = t^n$, $t \in [0, 1]$, $n \in \mathbb{N}$, we have $\|f_n\| = 1$, while $\|Bf_n\| = n \to \infty$.

Remark 4.5. If $X \neq \{0\}$, then

$$\|A\| = \sup\left\{\|Au\|_Y;\ u \in X,\ \|u\|_X = 1\right\}. \tag{4.1.3}$$

Proof. If we denote the right-hand side by a, then clearly

$$a \leq \|A\|. \tag{4.1.4}$$

Now, from the inequality

$$\|A(\|u\|_X^{-1} u)\|_Y \leq a \quad \forall u \in X \setminus \{0\}$$

we derive

$$\|Au\|_Y \leq a\|u\|_X \quad \forall u \in X. \tag{4.1.5}$$

By taking the supremum in (4.1.5) over all $u \in X, \|u\|_X \leq 1$, we find $\|A\| \leq a$, which combined with (4.1.4) proves (4.1.3). $\square$

Theorem 4.6. *If $(X, \|\cdot\|_X)$ is a normed space and $(Y, \|\cdot\|_Y)$ is a Banach space, then $L(X, Y)$ is a Banach space with respect to the operator norm.*

Proof. We know that $L(X, Y)$ is a normed space, so we have to show that it is complete. For the sake of simplicity we redenote by $\|\cdot\|$ both the norms $\|\cdot\|_X$ and $\|\cdot\|_Y$. Consider a Cauchy sequence (A_n) in $L(X, Y)$, i.e.,

$$\forall \varepsilon > 0 \ \exists N_\varepsilon \ \text{such that} \ \|A_n - A_m\| < \varepsilon \quad \forall n, m > N_\varepsilon.$$

For the same ε, we have

$$\|A_n v - A_m v\| \leq \varepsilon \|v\| \quad \forall v \in X,\ n, m > N_\varepsilon.$$

Now, $(A_n v)$ converges in Y since Y is Banach, so we have an operator $A : X \to Y$, $Av = \lim_{n \to \infty} A_n v$, and because each A_n is linear, A is as well. Since for all $v \in X$

$$
\begin{aligned}
\|Av\| &\leq \|Av - A_{N_\epsilon+1}v\| + \|A_{N_\epsilon+1}v\| \\
&\leq \varepsilon \|v\| + \|A_{N_\epsilon+1}\| \cdot \|v\| \\
&= (\varepsilon + \|A_{N_\epsilon+1}\|)\|v\|,
\end{aligned}
$$

we see that A is continuous, so $A \in L(X, Y)$.
Since

$$
\|A_n v - Av\| \leq \varepsilon
$$

for $v \in X$ such that $\|v\| \leq 1$ and $n > N_\varepsilon$, we get

$$
\|A_n - A\| \leq \varepsilon \quad \forall n > N_\varepsilon,
$$

which implies that $A_n \to A$ in $L(X, Y)$. $\square$

4.2 Main Principles of Functional Analysis

In this section we present some important principles of Functional Analysis: the Uniform Boundedness Principle, the Open Mapping Theorem, and the Closed Graph Theorem. We begin with the Uniform Boundedness Principle, which was proven by Banach and Steinhaus.[1]

Theorem 4.7 (Banach–Steinhaus, Uniform Boundedness Principle). *Let $(X, \|\cdot\|_X)$ and $(Y, \|\cdot\|_Y)$ be Banach spaces and let $\{T_i\}_{i \in I} \subset L(X, Y)$ be a collection of operators satisfying*

$$
\sup_{i \in I} \|T_i x\|_Y < \infty \quad \forall x \in X. \tag{4.2.6}
$$

Then,

$$
\sup_{i \in I} \|T_i\| < \infty. \tag{4.2.7}
$$

Proof. Denote

$$
X_n = \{x \in X; \sup_{i \in I} \|T_i x\|_Y \leq n\}, \ n \in \mathbb{N}.
$$

---

[1] Hugo Steinhaus, Polish mathematician, 1887–1972.

Obviously, X_n is a closed set for every $n \in \mathbb{N}$, and by (4.2.6) we have

$$X = \bigcup_{n=1}^{\infty} X_n.$$

It follows by Baire's Theorem (Theorem 2.10) that there exists an $n_0 \in \mathbb{N}$ such that $\text{Int} \, X_{n_0} \neq \emptyset$, i.e., there is a ball $B(x_0, r_0) \subset X_{n_0}$, $r_0 > 0$. Hence,

$$\|T_i(x_0 + r_0 w)\|_Y \leq n_0 \quad \forall i \in I, \, \forall w \in B(0,1),$$

which implies

$$r_0 \|T_i\| \leq n_0 + \|T_i x_0\|_Y \quad \forall i \in I.$$

This shows that (4.2.7) holds true (see also (4.2.6)). $\qquad\square$

Theorem 4.8 (Open Mapping Theorem). *Let* $(X, \|\cdot\|_X)$, $(Y, \|\cdot\|_Y)$ *be Banach spaces. If* $A : D(A) \subset X \to Y$ *is a linear, continuous, and surjective operator, then* A *maps open sets in* X *to open sets in* Y.

Proof. It suffices to prove that there exists a constant $r > 0$ such that

$$B_Y(0, r) \subset A(B_X(0, 1)), \tag{4.2.8}$$

where $B_X(0,1)$, $B_Y(0,r)$ denote the open balls in X and Y centered at 0 with radii 1 and r, respectively. In order to prove (4.2.8) we shall first show the existence of a constant $r_1 > 0$ such that

$$B_Y(0, r_1) \subset \text{Cl}\left(A(B_X(0,1))\right). \tag{4.2.9}$$

Denote $Y_n = n \, \text{Cl}\left(A(B_X(0,1))\right)$, $n \in \mathbb{N}$. Since A is surjective, we have $Y = \cup_{n \in \mathbb{N}} Y_n$. By Baire's Theorem (Theorem 2.10) $\text{Int} \, Y_{n_0} \neq \emptyset$ for some $n_0 \in \mathbb{N}$, hence $\text{Int} \, \text{Cl}\left(A(B_X(0,1))\right) \neq \emptyset$. So, for some $y_0 \in Y$ and some $r_1 > 0$, we have

$$B_Y(y_0, 2r_1) \subset \text{Cl}\left(A(B_X(0,1))\right). \tag{4.2.10}$$

Adding the fact that $-y_0 \in \text{Cl}\left(A(B_X(0,1))\right)$ to (4.2.10), we obtain

$$\begin{aligned} B_Y(0, 2r_1) &\subset \text{Cl}\left(A(B_X(0,1))\right) + \text{Cl}\left(A(B_X(0,1))\right) \\ &= 2\,\text{Cl}\left(A(B_X(0,1))\right) \end{aligned}$$

(since $\text{Cl}\left(A(B_X(0,1))\right)$ is a convex set), hence (4.2.9) holds true. Now we are going to prove (4.2.8) by using (4.2.9) with $r_1 = 2r$, i.e.,

$$B_Y(0, 2r) \subset \text{Cl}\left(A(B_X(0,1))\right). \tag{4.2.11}$$

Choose an arbitrary $y \in B_Y(0, r)$. By (4.2.11) we have

$$\forall \varepsilon > 0 \ \exists v \in B_X(0, 1/2) \ \text{such that} \ \|y - Av\|_Y < \varepsilon. \tag{4.2.12}$$

In particular, for $\varepsilon = r/2$ there exists a $v_1 \in B_X(0, 1/2)$ with

$$\|y - Av_1\|_Y < \frac{r}{2^1}.$$

Now choosing $y - Av_1$ instead of y and $\varepsilon = 1/2^2$ in (4.2.12), we can find some $v_2 \in B_X(0, 1/2^2)$ with

$$\|(y - Av_1) - Av_2\|_Y < \frac{r}{2^2}.$$

Continuing the process we find $v_n \in B_X(0, 1/2^n)$ such that

$$\|y - A(v_1 + v_2 + \cdots + v_n)\|_Y < \frac{r}{2^n}. \tag{4.2.13}$$

Obviously, $x_n = v_1 + v_2 + \cdots + v_n$ defines a Cauchy sequence in X, hence x_n converges to some $x \in X$ with $\|x\|_X < 1$ and $y = Ax$ since $A \in L(X, Y)$ (see (4.2.13)). As y was an arbitrary vector in $B_Y(0, r)$ the proof of (4.2.8) is complete. $\qquad\square$

Remark 4.9. If $(X, \|\cdot\|_X)$, $(Y, \|\cdot\|_Y)$ are Banach spaces and $A \in L(X, Y)$ is bijective, then $A^{-1} \in L(Y, X)$. This follows from (4.2.8).

Theorem 4.10 (Closed Graph Theorem). *Let $(X, \|\cdot\|_X)$, $(Y, \|\cdot\|_Y)$ be Banach spaces. If $A : X \to Y$ is a linear operator and its graph $G(A) := \{(x, Ax); x \in X\}$ is closed in $X \times Y$ (in other words, A is a* **closed operator***), then $A \in L(X, Y)$.*

Proof. Define on X the norm

$$\|x\|_A = \|x\|_X + \|Ax\|_Y, \quad x \in X,$$

which is called the graph norm. Since $G(A)$ is a closed set in $(X, \|\cdot\|_X) \times (Y, \|\cdot\|_Y)$, it follows that $(X, \|\cdot\|_A)$ is a Banach space. Obviously,

$$\|x\|_X \leq \|x\|_A \quad \forall x \in X,$$

so the identity operator $I : (X, \| \cdot \|_A) \to (X, \| \cdot \|_X)$ is continuous. So, by Remark 4.9, its inverse $I^{-1} = I \in L((X, \| \cdot \|_X), (X, \| \cdot \|_A))$, i.e., there exists a constant $C > 0$ such that

$$\|x\|_A \leq C\|x\|_X \quad \forall x \in X.$$

In particular,

$$\|Ax\|_Y \leq C\|x\|_X \quad \forall x \in X,$$

which means A is continuous from $(X, \| \cdot \|_X)$ to $(Y, \| \cdot \|_Y)$. $\square$

4.3   Compact Linear Operators

If X, Y are normed spaces and $A : X \to Y$ is a linear operator then A is called **compact** or **completely continuous** if A takes bounded sets of X into relatively compact subsets of Y.

Example.
Let $X = Y = C[a, b]$, $-\infty < a < b < +\infty$, equipped with the usual sup-norm, and let $A : X \to X$ be defined by

$$(Af)(t) = \int_a^b k(t, s) f(s)\, ds \quad \forall f \in X, \ \forall t \in [a, b],$$

where $k \in C([a, b] \times [a, b])$.
Obviously A is a linear operator. Moreover, it follows from Arzelà–Ascoli's Criterion that A is a compact operator. The key argument here is that the equicontinuity condition is a consequence of the uniform continuity of k.

A compact linear operator is clearly continuous (see Proposition 4.2). Denote by

$$K(X, Y) = \{A \in L(X, Y); \ A \text{ is compact} \}.$$

It is clear that $K(X, Y)$ is a linear subspace of $L(X, Y)$. Moreover, we have the following theorem.

Theorem 4.11. *If X is a normed space and Y is a Banach space, then $K(X, Y)$ is a closed linear subspace of $L(X, Y)$, i.e., $K(X, Y)$ is a Banach space with respect to the operator norm (see Theorem 4.6).*

Proof. We shall denote by $\|\cdot\|$ all the three norms of X, Y, and $L(X,Y)$. Let (A_n) be a sequence in $L(X,Y)$ which converges to some $A \in L(X,Y)$, namely $\|A_n - A\| \to 0$. So, for $\varepsilon > 0$ there exists $m \in \mathbb{N}$ sufficiently large such that

$$\|A_m - A\| < \frac{\varepsilon}{3r} \, . \tag{4.3.14}$$

Let (x_n) be a sequence in the ball $B(0, r) \subset X$, where $r > 0$ is arbitrary but fixed. Since A_m is compact there exists a subsequence of (x_n), say $(x_{n_k})_{k \geq 1}$, such that $(Ax_{n_k})_{k \geq 1}$ is convergent, hence Cauchy. Thus, for any $\varepsilon > 0$ (which can be the same as above), there exists $N \in \mathbb{N}$ such that

$$\|A_m x_{n_k} - A_m x_{n_j}\| < \frac{\varepsilon}{3} \quad \forall k, j > N \, . \tag{4.3.15}$$

Using (4.3.14) and (4.3.15) we deduce

$$
\begin{aligned}
&\|Ax_{n_k} - Ax_{n_j}\| \\
\leq\ & \|Ax_{n_k} - A_m x_{n_k}\| + \|A_m x_{n_k} - A_m x_{n_j}\| + \|A_m x_{n_j} - Ax_{n_j}\| \\
\leq\ & \|A - A_m\| \cdot \|x_{n_k}\| + \|A_m x_{n_k} - A_m x_{n_j}\| + \|A_m - A\| \cdot \|x_{n_j}\| \\
<\ & r \cdot \frac{\varepsilon}{3r} + \frac{\varepsilon}{3} + r \cdot \frac{\varepsilon}{3r} = \varepsilon \, ,
\end{aligned}
$$

in other words, (Ax_{n_k}) is Cauchy, hence convergent, and therefore $A \in K(X,Y)$. $\qquad\square$

Remark 4.12. It is worth pointing out that if $A \in K(X,Y)$, where X is a normed space and Y is a Hilbert space (see Chap. 6), then there exists a sequence $(A_n)_{n \geq 1}$ in $L(X,Y)$, such that the range of A_n is finite dimensional (hence A_n is compact) for all $n \geq 1$ and $\|A_n - A\| \to 0$. For the proof of this nice result see Brezis[2] [6, Remark 1, pp. 157–158].

4.4 Linear Functionals, Dual Spaces, Weak Topologies

We begin this section by defining the important concept of a dual space.

Definition 4.13. *Let* $(X, \|\cdot\|)$ *be a normed space. Define the* **dual** *of* X, *denoted* X^*, *by*

$$X^* = \{f : X \to \mathbb{K}; \ f \text{ is linear and continuous}\},$$

---

[2]Haim Brezis, French mathematician, born 1944.

so X^* is in fact $L(X, \mathbb{K})$. The elements of X^* are called functionals.

Since $(\mathbb{K}, |\cdot|)$ is a Banach space, X^* is also a Banach space with respect to

$$\|f\| = \sup\{|f(v)|; \, v \in X, \, \|v\| \leq 1\}.$$

By definition

$$|f(v)| \leq \|f\| \cdot \|v\| \quad \forall v \in X, \, \forall f \in X^*.$$

Example 1.
Let X be the linear space of all sequences of real numbers $(x_n)_{n \geq 1}$ satisfying

$$\sum_{n=1}^{\infty} |x_n| < \infty.$$

X is usually denoted by l^1 and is a Banach space (over $\mathbb{R}$) with respect to the norm

$$\|(x_n)\| = \sum_{n=1}^{\infty} |x_n|.$$

See Exercise 2.19.
It is easily seen that any functional $f \in X^*$ has the form

$$f((x_n)) = \sum_{n=1}^{\infty} a_n x_n,$$

where (a_n) is a bounded sequence in $\mathbb{R}$. X^* is usually denoted by l^∞ and is a Banach space with the norm

$$\|(a_n)\|_\infty = \sup_{n \geq 1} |a_n|.$$

Example 2.
Let $X = C[a, b]$, $-\infty < a < b < +\infty$, with the sup-norm, denoted $\|\cdot\|$. For a fixed $v \in X$ define $f : X \to \mathbb{R}$ by

$$f(u) = \int_a^b u(t)v(t)\,dt \quad \forall u \in X.$$

We see that f is linear and also continuous because

$$|f(u)| \leq (b - a)\|v\| \cdot \|u\| \quad \forall u \in X,$$

and therefore $f \in X^*$.

Now, consider the same space $X = C[a, b]$ equipped with another norm, namely the L^2-norm, and the same functional f, which can be expressed as the scalar product

$$f(u) = (u, v)_{L^2(a,b)} \quad \forall u \in X .$$

Again, f is linear and by the Bunyakovsky–Cauchy–Schwarz inequality

$$|f(u)| \le \|v\|_{L^2(a,b)} \cdot \|u\|_{L^2(a,b)} \quad \forall u \in X ,$$

so $f \in (X, \| \cdot \|_{L^2(a,b)})^*$.

Question: Given $f \in (X, \| \cdot \|_{L^2(a,b)})^*$, does there exist $v \in X = C[a, b]$ such that $f(u) = (u, v)_{L^2(a,b)}$ for all $u \in X$? We shall show later (Theorem 6.10) that there exists such a v in the $L^2(a, b)$, but not necessarily in $X = C[a, b]$.

In what follows we present the Hahn[3]–Banach Theorem on the extension of linear (not necessarily continuous) $\mathbb{R}$-valued functionals.

Theorem 4.14 (Hahn–Banach). *Let X be a real linear space, and let $p : X \to \mathbb{R}$ be a map which satisfies*

$$p(x + y) \le p(x) + p(y) \quad \forall x, y \in X ,$$

$$p(\alpha x) = \alpha p(x) \quad \forall \alpha > 0, \ x \in X .$$

If Y is a linear subspace of X and $f : Y \to \mathbb{R}$ is a linear functional satisfying

$$f(x) \le p(x) \quad \forall x \in Y ,$$

then there exists a linear functional $g : X \to \mathbb{R}$ such that

$$g(x) = f(x) \quad \forall x \in Y ,$$

$$g(x) \le p(x) \quad \forall x \in X .$$

Proof. The case $Y = X$ is trivial, so we assume that Y is a proper subspace of X. Consider the collection $\mathcal{E}$ of all linear extensions of f in the above sense, i.e., $h \in \mathcal{E}$ if and only if $D(h)$ is a linear subspace of X, $Y \subset D(h)$, h is linear, h extends f, and $h(x) \le p(x) \ \forall x \in D(h)$. Clearly $f \in \mathcal{E}$ so $\mathcal{E}$ is nonempty. Define on $\mathcal{E}$ the order relation

$$h_1 \preceq h_2 \iff D(h_1) \subset D(h_2) \ \text{ and } \ h_2(x) = h_1(x) \ \forall x \in D(h_1) .$$

---

[3]Hans Hahn, Austrian mathematician, 1879–1934.

We wish to apply Zorn's Lemma, so let $\mathcal{G} = \{h_i\}_{i \in I}$ be a totally ordered subset of $\mathcal{E}$ and consider the functional h defined by

$$D(h) = \cup_{i \in I} D(h_i), \ h(x) = h_i(x) \quad \text{if } x \in D(h_i) \text{ for some } i \in I.$$

Obviously, h is well defined and belongs to $\mathcal{E}$ and is an upper bound for $\mathcal{G}$. Hence $\mathcal{E}$ is inductive, so by Zorn's Lemma $\mathcal{E}$ has a maximal element $g \in \mathcal{E}$.

To complete the proof let us show that $D(g) = X$. Assume by contradiction that this is not the case, so $\exists x_0 \in X \setminus D(g)$. Consider $Z = \text{Span} \left(\{x_0\} \cup D(g) \right)$, and define on Z a linear functional $\tilde{g}$ of the form

$$\tilde{g}(tx_0 + x) = \alpha t + g(x), \quad t \in \mathbb{R}, \ x \in D(g),$$

where α is a real parameter. We shall prove that there exists an α such that $\tilde{g} \in \mathcal{E}$, i.e.,

$$\alpha t + g(x) \leq p(tx_0 + x) \quad \forall x \in D(g), \ t \in \mathbb{R}. \tag{4.4.16}$$

In particular,

$$\begin{aligned} g(x) + \alpha &\leq p(x + x_0) \quad \forall x \in D(g), \\ g(y) - \alpha &\leq p(y - x_0) \quad \forall y \in D(g), \end{aligned}$$

hence α should satisfy

$$g(y) - p(y - x_0) \leq \alpha \leq p(x + x_0) - g(x) \quad \forall x, y \in D(g),$$

which is equivalent to

$$\sup_{y \in D(g)} [g(y) - p(y - x_0)] \leq \alpha \leq \inf_{x \in D(g)} [p(x + x_0) - g(x)].$$

Such an α exists indeed since

$$\begin{aligned} g(y) - p(y - x_0) &\leq p(x + x_0) - g(x) \\ &\Leftrightarrow g(x + y) \leq p(x + x_0) + p(y - x_0) = p(x + y), \end{aligned}$$

which is clearly valid for all $x, y \in D(g)$. It is easy to check that $\tilde{g}$ with this alpha satisfies (4.4.16), so $\tilde{g} \in \mathcal{E}$. But $\tilde{g}$ is a proper extension of g (since $D(g)$ is a proper subset of $D(\tilde{g}) = Z$) and this contradicts the maximality of g. $\qquad\qquad\square$

Corollary 4.15. *Let $(X, \|\cdot\|)$ be a normed space and let Y be a linear subspace of X. If $f \in Y^* := (Y, \|\cdot\|)^*$, then there exists an extension g of f such that $g \in X^* := (X, \|\cdot\|)^*$ and*

$$\|g\|_{X^*} = \|f\|_{Y^*}.$$

Proof. If $\mathbb{K} = \mathbb{R}$ then we can apply the Hahn–Banach Theorem with $p(x) = \|f\|_{Y^*}\|x\|$ to derive the existence of a linear extension $g : X \to \mathbb{R}$ satisfying

$$g(x) \leq \|f\|_{Y^*}\|x\| \quad \forall x \in X.$$

Since $-g(x) = g(-x)$ satisfies a similar inequality, we have $g \in X^*$ and

$$\|g\|_{X^*} \leq \|f\|_{Y^*}.$$

Obviously, the converse inequality is also satisfied, so $\|g\|_{X^*} = \|f\|_{Y^*}$. If $\mathbb{K} = \mathbb{C}$ define

$$q(x) := \operatorname{Re} f(x) \quad \forall x \in Y.$$

Then,

$$f(x) = q(x) - iq(ix) \quad \forall x \in Y,$$

and

$$|q(x)| \leq \|f\|_{Y^*}\|x\| \quad \forall x \in Y. \tag{4.4.17}$$

Now, if we regard X, Y as real linear spaces and take into account (4.4.17), we deduce from the first part of the proof the existence of a continuous linear functional $h : X \to \mathbb{R}$ which extends q and satisfies

$$|h(x)| \leq \|f\|_{Y^*}\|x\| \quad \forall x \in X. \tag{4.4.18}$$

Set

$$g(x) = h(x) - ih(ix), \quad x \in X.$$

Functional $g : X \to \mathbb{C}$ is an extension of f and is linear on the complex space X. Let us prove that

$$|g(x)| \leq \|f\|_{Y^*}\|x\| \quad \forall x \in X.$$

Indeed, for each $x \in X$, $g(x)$ can be written as $g(x) = re^{i\theta}$, $r \geq 0$, so

$$
\begin{aligned}
|g(x)| = r &= \operatorname{Re}\left(e^{-i\theta}g(x)\right) \\
&= \operatorname{Re} g\left(e^{-i\theta}x\right) \\
&= h(e^{-i\theta}x) \\
\text{(by (4.4.18))} \quad &\leq \|f\|_{Y^*}\|x\| \quad \forall x \in X.
\end{aligned}
$$

Therefore, $g \in X^*$, and $\|g\|_{X^*} \leq \|f\|_{Y^*}$. As the converse inequality is trivially satisfied, we have $\|g\|_{X^*} = \|f\|_{Y^*}$. $\square$

Remark 4.16. In fact, even Theorem 4.14 above can be extended to the complex case $\mathbb{K} = \mathbb{C}$ by a similar procedure.

Corollary 4.17. *Let* $(X, \|\cdot\|)$ *be a normed space. Then for every* $x_0 \in X \setminus \{0\}$ *there exists a functional* $g \in X^*$ *such that*

$$\|g\|_{X^*} = 1 \quad and \quad g(x_0) = \|x_0\|.$$

Proof. Apply Corollary 4.15 with $Y = \mathrm{Span}\{x_0\}$ and $f : Y \to \mathbb{K}$ defined by

$$f(x) = t\|x_0\| \quad \text{for } x = tx_0, \ t \in \mathbb{K}.$$

$\square$

Corollary 4.18. *Let* $(X, \|\cdot\|)$ *be a normed space. Then for every* $x \in X$ *we have*

$$\|x\| = \sup\{|f(x)|; \ f \in X^*, \ \|f\|_{X^*} \leq 1\}, \tag{4.4.19}$$

where the sup is attained.

Proof. For $x = 0$ (4.4.19) is obvious. Let $x \in X \setminus \{0\}$ and denote by a the right-hand side of (4.4.19). Clearly, $a \leq \|x\|$. In fact, $a = \|x\|$ by virtue of Corollary 4.17. $\square$

Remark 4.19. Let $(X, \|\cdot\|)$ be a normed space. Define

$$J(x) = \{x^* \in X^*; \ \|x^*\|_{X^*} = \|x\|, \ x^*(x) = \|x\|^2\}.$$

From Corollary 4.17 we see that $J(x)$ is nonempty for all $x \in X$. In general, $J(x)$ is not a singleton, but there are cases when this happens for all $x \in X$ (e.g., if X is a Hilbert space, as will be shown later). The set-valued map $x \mapsto J(x)$ is called the **duality map** from X to X^*.

Recall that, given a normed space $(X, \|\cdot\|)$, the **strong (norm) topology** of X is the metric topology generated by $d(x, y) = \|x - y\|$ for $x, y \in X$. In fact, we can consider that X is a Banach space (in other words, $\|\cdot\|$ is complete, or d is complete), otherwise we can use the completion procedure (see Theorem 2.8) to reach this framework.

Definition 4.20. *The* **weak topology** *of X is the one generated by neighborhoods of the origin of the form*

$$V_{x_1^*, x_2^*, \ldots, x_m^*; \varepsilon} = \{x \in X;\ |x_j^*(x)| < \varepsilon,\ j = 1, 2, \ldots, m\},$$

for all finite systems of functionals $\{x_1^*, x_2^*, \ldots, x_m^*\}$ and for all $\varepsilon > 0$. We write $x_n \overset{w}{\to} x$ or $x_n \rightharpoonup x$ to mean convergence in the weak topology, i.e., $x^*(x_n) \to x^*(x)$ for all $x^* \in X^*$.

Remark 4.21. If $x_n \to x$, i.e., $\|x_n - x\| \to 0$, then $x_n \overset{w}{\to} x$. Indeed, for all $x^* \in X^*$,

$$|x^*(x_n) - x^*(x)| = |x^*(x_n - x)|$$
$$\leq \|x^*\| \cdot \|x_n - x\|,$$

which tends to 0. The converse is not true in general, and we shall see some examples later.

However, if X is finite dimensional then strong and weak convergence are equivalent. Indeed, by choosing particular functionals, one can see that weak convergence reduces to convergence on coordinates.

Definition 4.22. *In X^*, besides the strong topology and the weak topology, defined by means of functionals from $X^{**} := (X^*)^*$ (the* **bidual** *of X), we have the so-called* **weak-star topology** *w^*, starting from another neighborhood basis consisting of*

$$V_{x_1, x_2, \ldots, x_m; \varepsilon} = \{x^* \in X^*;\ |x^*(x_j)| < \varepsilon,\ j = 1, 2, \ldots, m\},$$

for all finite systems $\{x_1, x_2, \ldots, x_m\} \subset X$, and for all $\varepsilon > 0$. So convergence $x_n^* \overset{w^*}{\to} x^*$ means $x_n^*(x) \to x^*(x)$ for all $x \in X$, i.e., pointwise convergence for a sequence of functionals. In general this is different than w-convergence.

In general X is embedded into X^{**}, which is to say that there is an injection $x \overset{i}{\mapsto} f_x$ defined by $f_x(x^*) = x^*(x)$ for all $x^* \in X^*$. Clearly, $i \in L(X, X^{**})$ since

$$|f_x(x^*)| \leq \|x^*\| \cdot \|x\|.$$

Moreover, using Corollary 4.17, we see that i is an isometry.

If $i : X \to X^{**}$ is onto (surjective), then X is said to be **reflexive**. In particular Hilbert spaces are reflexive, as will be shown later.

Remark 4.23. It is easily seen that if X is reflexive then $w = w^*$ on X^*.

4.5   Exercises

1. Let X, Y be linear spaces. Find a necessary and sufficient condition for a subset $G \subset X \times Y$ to be the graph of a linear operator from X into Y.

2. Let X, Y be normed spaces over $\mathbb{R}$. If $: X \to Y$ is a continuous operator satisfying the condition

$$A(x_1 + x_2) = Ax_1 + Ax_2 \quad \forall x_1, x_2 \in X,$$

then A is linear (hence $A \in L(X, Y)$).

3. Let $-\infty < a < b < +\infty$. Find the operator norm of $A \in L(X)$ given by

$$(Af)(t) = tf(t), \quad t \in [a, b], \ f \in X,$$

when

(i) $X = C[a, b]$ with the sup-norm;

(ii) $X = L^p(a, b)$, with the usual norm, for some $1 \le p < \infty$.

4. Let $X = C[a, b]$, where $-\infty < a < b < +\infty$. Assume that X is equipped with the usual sup- norm and consider the operator A defined by

$$(Af)(t) = \int_a^t g(s)f(s)\, ds, \quad f \in X, \ t \in [a, b],$$

where g is a given function in $L^1(a, b)$ with $g(s) \ge 0$ for almost all $s \in (a, b)$. Show that A is a compact linear operator from X into itself (i.e., $A \in K(X) \subset L(X)$) and calculate $\|A\|$.

5. Let $(X, \| \cdot \|_X)$, $(Y, \| \cdot \|_Y)$ be normed spaces. Show that a linear operator $A : X \to Y$ is continuous if and only if the following implication holds

$(*)$ $x_n \in X$, $\|x_n\|_X \to 0 \implies (\|Ax_n\|_Y)$ is a bounded sequence.

6. Let $(X, \| \cdot \|_X)$ be a normed space and let $(Y, \| \cdot \|_Y)$ be a Banach space. Show that, for any sequence $(A_n)_{n \in \mathbb{N}}$ in $L(X, Y)$ satisfying $\|A_n\| \le a_n$ $\forall n \in \mathbb{N}$ with $\sum_{n=1}^\infty a_n < \infty$, the series $\sum_{n=1}^\infty A_n$ is convergent in $L(X, Y)$.

7. Let $(X, \|\cdot\|)$ be a Banach space. Show that

   (i) for all $A \in L(X)$ the series

   $$I + \frac{1}{1!}A + \frac{1}{2!}A^2 + \cdots + \frac{1}{n!}A^n + \cdots$$

   is convergent in $L(X)$ with its usual operator norm (the sum of the series being denoted e^A). Here I denotes the identity operator on X.

   (ii) for all $A \in L(X)$ with $\|A\| < 1$, $I - A$ is invertible and $(I - A)^{-1} \in L(X)$.

8. Let $(X, \|\cdot\|)$ be a Banach space. For every pair of operators $A, B \in L(X)$ that commute (i.e., $AB = BA$) one has $e^A e^B = e^{A+B}$ (for the notation see the previous exercise).

9. Let $(X, \|\cdot\|_X)$, $(Y, \|\cdot\|_Y)$ be Banach spaces. Let $(T_n)_{n \in \mathbb{N}}$ be a sequence in $L(X, Y)$ which is pointwise convergent, i.e.,

   $$\forall x \in X \; \exists y_x \in Y \quad \text{such that} \quad \|T_n x - y_x\|_Y \to 0.$$

   Define $T : X \to Y$ by $Tx = y_x$, $x \in X$. Show that

   (a) $(\|T_n\|)_{n \in \mathbb{N}}$ is bounded in $(\mathbb{K}, |\cdot|)$;

   (b) $T \in L(X, Y)$;

   (c) $\|T\| \le \liminf \|T_n\|$.

10. Let $(X, \|\cdot\|)$ be a Banach space and let S be a nonempty subset of X such that for all $f \in X^*$ the set

    $$f(S) = \{f(x); \; x \in S\} \quad \text{is bounded in } (\mathbb{K}, |\cdot|)\}.$$

    Prove that S is bounded in $(X, \|\cdot\|)$.

11. Let X be a Banach space and let $A : X \to X^*$ be a linear operator satisfying

    $$(Ax)(y) = (Ay)(x) \quad \forall x, y \in X.$$

    Show that A is a continuous operator, i.e. $A \in L(X, X^*)$.

12. Let $(X, \|\cdot\|_X)$, $(Y, \|\cdot\|_Y)$ be Banach spaces. If $A : D(A) \subset X \to Y$ is a linear closed operator with $D(A)$ closed in $(X, \|\cdot\|_X)$, then prove there exists a constant $C > 0$ such that

    $$\|Ax\|_Y \le C\|x\|_X \quad \forall x \in D(A).$$

13. Let X be a Banach space and let $A : X \to X^*$ be a linear operator satisfying

$$(Ax)(x) \geq 0 \quad \forall x \in X.$$

Show that $A \in L(X, X^*)$.

14. Let X be a linear space, equipped with two norms, $\|\cdot\|_1$ and $\|\cdot\|_2$, such that X is Banach for both norms. Assume there exists a constant $C > 0$ such that

$$\|x\|_2 \leq C\|x\|_1 \quad \forall x \in X.$$

Show that $\|\cdot\|_1$ and $\|\cdot\|_2$ are equivalent.

15. Let X be an n-dimensional linear space, with $n \in \mathbb{N}$. Let $B = \{u_1, u_2, \ldots, u_n\}$ be a basis in X.

For any linear functional $f : X \to \mathbb{K}$ we have

$$f(u) = \sum_{i=1}^{n} \alpha_i f_i \quad \forall u = \sum_{i=1}^{n} \alpha_i u_i \in X,$$

where $f_i := f(u_i)$, $i = 1, 2, \ldots, n$. Obviously, any such f is continuous with respect to any norm of X, i.e., $f \in X^*$.

Compute the norm of f, $\|f\|_{X^*}$, explicitly, in terms of the f_i's, when the norm of X is defined by

(i) $\|u\|_\infty = \max_{1 \leq i \leq n} |\alpha_i| \quad \forall u = \sum_{i=1}^{n} \alpha_i u_i \in X$;

(ii) $\|u\|_1 = \sum_{i=1}^{n} |\alpha_i| \quad \forall u = \sum_{i=1}^{n} \alpha_i u_i \in X$;

(iii) $\|u\|_p = \left(\sum_{i=1}^{n} |\alpha_i|^p \right)^{1/p} \quad \forall u = \sum_{i=1}^{n} \alpha_i u_i \in X$, where $p \in (1, \infty)$.

16. Let $X = \{u \in C[0,1]; u(0) = 0\}$ with the usual sup-norm. Let $f : X \to \mathbb{R}$ be defined by

$$f(u) = \int_0^1 u(s)\, ds \quad \forall u \in X.$$

Show that $f \in X^*$ and compute $\|f\|_{X^*}$. Can one find some $u \in X$ such that $\|u\|_{\sup} = 1$ and $f(u) = \|f\|_{X^*}$?

Chapter 5

Distributions, Sobolev Spaces

In this chapter we first present test functions, which are then used to introduce scalar distributions. The space $\mathcal{D}'(\Omega)$ of distributions is analyzed in detail and some related applications are discussed: the interpretation of the density of a mass concentrated at a point by means of the Dirac distribution, solving the Poisson equation in $\mathcal{D}'(\Omega)$, solving ordinary differential equations in $\mathcal{D}'(\mathbb{R})$, solving the equation of the vibrating string with non-smooth initial displacement function, and the boundary controllability for a problem associated with the same wave equation. We also introduce and discuss Sobolev spaces. In order to introduce vector distributions we shall present in a separate section the Bochner integral for vector functions. Vector distributions and $W^{k,p}(a, b; X)$ spaces are then presented. These will later be used in solving problems associated with parabolic and hyperbolic PDE's.

5.1   Test Functions

Let $\Omega \subset \mathbb{R}^k$ be a nonempty open set in $\mathbb{R}^k$ (which is equipped with the usual topology).

For $u : \Omega \to \mathbb{R}$ define the **support** of u by

$$\operatorname{supp} u = \overline{\{x \in \Omega;\ u(x) \neq 0\}}.$$

For a given $m \in \mathbb{N}$, let $C^m(\Omega)$ denote the set of all functions $u : \Omega \to \mathbb{R}$ such that u, and all its n-th order partial derivatives, $1 \leq n \leq m$, exist

G. Moroşanu, *Functional Analysis for the Applied Sciences*,
Universitext, https://doi.org/10.1007/978-3-030-27153-4_5

and are continuous. Further let $C_0^m(\Omega) = \{u \in C^m(\Omega);\ \text{supp}\, u$ is a compact (bounded) set $\subset \Omega\}$. For $m = \infty$ extend the definitions above in the obvious way. The elements (functions) in $C_0^\infty(\Omega)$ are called **test functions** since they serve as arguments of distributions that will be defined later.

A typical example of a test function is $\phi : \Omega = \mathbb{R}^k \to \mathbb{R}$ defined by

$$\phi(x) = \begin{cases} \exp\left(\frac{1}{\|x\|_2^2 - 1}\right), & \|x\|_2 < 1, \\ 0, & \|x\|_2 \geq 1, \end{cases}$$

where $\|\cdot\|_2$ is the Euclidean norm. The function $\phi \in C^\infty(\mathbb{R}^k)$ (prove it!) and so $\phi \in C_0^\infty(\mathbb{R}^k)$ with $\text{supp}\, \phi = \overline{B(0,1)}$. For later use we also define

$$\omega(x) = C\phi(x) \quad \text{with} \ \ C > 0 \ \ \text{such that} \quad \int_{\mathbb{R}^k} \omega(x)\, dx = 1. \quad (5.1.1)$$

Obviously, $C_0^\infty(\Omega)$ is a real linear space with respect to the usual operations (addition of functions and scalar multiplication).

In what follows, we introduce the usual topology on $C_0^\infty(\Omega)$. To this purpose, we must first discuss a few important concepts.

Seminorms, Locally Convex Spaces, Inductive Limit

Let X be a linear space over $\mathbb{K}$ (as usual, $\mathbb{K}$ is either $\mathbb{R}$ or $\mathbb{C}$). A function $p : X \to \mathbb{R}$ is called a *seminorm* if the following conditions are satisfied:

(j) $p(x + y) \leq p(x) + p(y),\ \ x, y \in X,$

(jj) $p(\alpha x) = |\alpha| p(x),\ \ \alpha \in \mathbb{K},\ x \in X.$

If p is a seminorm, then $p(0) = 0$, while the case when $p(x) = 0$ for some $x \neq 0$ is not excluded. We also have

$$|p(x_1) - p(x_2)| \leq p(x_1 - x_2)\ \ \forall x_1, x_2 \in X, \quad\quad (5.1.2)$$

which in particular shows that $p(x) \geq 0$ for all $x \in X$. Indeed, by (j), $p(x_1) - p(x_2) \leq p(x_1 - x_2)$ and so $p(x_1 - x_2) = p(x_2 - x_1) \geq p(x_2) - p(x_1)$. Obviously, (5.1.2) follows from these two inequalities.

We will use seminorms to equip X with a topology. If p is a seminorm and M is the set $\{x \in X;\ p(x) < \varepsilon\}$, where ε is a positive constant, then, obviously, $0 \in M$ and M is *convex, balanced* (i.e., $x \in M$ and

$|\alpha| \leq 1$ implies $\alpha x \in M$), and *absorbing* (i.e., for any $x \in X$ there exists an $\alpha > 0$ such that $\alpha^{-1}x \in M$).

Let $\mathcal{F} = \{p_i : X \to \mathbb{R}; i \in I\}$ be a family of seminorms satisfying the axiom of separation: *for any $y \in X$, $y \neq 0$, there exists $j \in I$ such that $p_j(y) \neq 0$.* Consider the collection $\mathcal{V}(0)$ of all sets which are finite intersections of sets $\{x \in X; p_i(x) < \varepsilon_i\}$, $i \in I$, $\varepsilon > 0$. Such an intersection looks like

$$V = \{x \in X; p_i(x) < \varepsilon_i, \ i = 1, \ldots, n\},$$

where $\{p_1, \ldots, p_n\} \subset \mathcal{F}$ and $\{\varepsilon_1, \ldots, \varepsilon_n\} \subset (0, \infty)$. Obviously, V is a convex, balanced, and absorbing set. Each $V \in \mathcal{V}(0)$ is considered to be a neighborhood of $0 \in X$ and $y + V := \{y + v; v \in V\}$ a neighborhood of any $y \in X$.

Now, a set $D \subset X$ which is a neighborhood of any $y \in D$ is called open. Indeed, the collection τ of all such sets, plus $\emptyset \subset X$, satisfies the axioms of a topology, so (X, τ) is a topological space.

Using the separating property of $\mathcal{F}$ we can infer that singletons are closed sets. Indeed, let $y \in X$ be a given point. For each $x \in X$, $x \neq y$, let D_x be an open set containing x but not y. Then $D = \cup_{x \neq y} D_x$ is open and its complement is $\{y\}$, so the singleton $\{y\}$ is closed, as claimed. Note that if $\mathcal{F}$ does not satisfy the axiom of separation then the closedness of singletons is not guaranteed.

It is easily seen that the mappings $X \times X \ni (x, y) \mapsto x + y \in X$ and $\mathbb{K} \times X \ni (\alpha, x) \mapsto \alpha x \in X$ are both continuous, so X is a topological linear space.

Definition. A topological linear space X is called *locally convex* if every open set containing 0 includes a convex, balanced, and absorbing open subset.

To summarize, we can say that any linear space X equipped (as above) with the topology generated by a family of seminorms $\{p_i; i \in I\}$ satisfying the axiom of separation is a locally convex space in which any seminorm p_i is continuous (cf. (5.1.2)).

Conversely, any locally convex space X is a topological linear space whose topology is generated by a collection of seminorms. In order to show this, we define for a convex, balanced, and absorbing set $M \subset X$ the so-called Minkowski functional:

$$p_M(x) = \inf\{\alpha; \ \alpha > 0, \ \alpha^{-1}x \in M\}, \quad x \in X.$$

Observe that $M = \{x \in X; p_M(x) \leq 1\}$. p_M is a seminorm on X. Indeed, by the convexity of M and the obvious relations

$$\frac{x}{p_M(x) + \varepsilon} \in M, \quad \frac{y}{p_M(y) + \varepsilon} \in M, \quad \varepsilon > 0,$$

we deduce

$$\frac{p_M(x) + \varepsilon}{p_M(x) + p_M(y) + 2\varepsilon} \cdot \frac{x}{p_M(x) + \varepsilon}$$
$$+ \frac{p_M(y) + \varepsilon}{p_M(x) + p_M(y) + 2\varepsilon} \cdot \frac{y}{p_M(y) + \varepsilon} \in M.$$

Hence

$$p_M(x + y) \leq p_M(x) + p_M(y) + 2\varepsilon \; \forall \varepsilon > 0$$
$$\Rightarrow p_M(x + y) \leq p_M(x) + p_M(y).$$

We also have $p_M(\alpha x) = |\alpha| p_M(x)$, since M is balanced.

So, the topology of a given locally convex space X is the one generated by the collection of seminorms obtained as the Minkowski functionals associated with convex, balanced, and absorbing open subsets of X.

Definition. Let X be a linear space over $\mathbb{K}$ and let $\{X_\alpha; \alpha \in J\}$ be a collection of linear subspaces of X such that $X = \cup_{\alpha \in J} X_\alpha$. Suppose that each X_α is a locally convex space such that, if $X_{\alpha_1} \subset X_{\alpha_2}$, then the topology of X_{α_1} coincides with the relative topology of X_{α_1} as a subset of X_{α_2}. Every convex, balanced, and absorbing set $D \subset X$ is considered open $\Longleftrightarrow D \cap X_\alpha$ is an open set of X_α containing $0 \in X_\alpha$ for all $\alpha \in J$. If X is a locally convex space with respect to the topology defined in this way, then X is called the *inductive limit* of the X_α's.

Now let us return to $C_0^\infty(\Omega)$. For any compact $K \subset \Omega$ define the set

$$\mathcal{D}_K(\Omega) = \{\phi \in C_0^\infty(\Omega); \operatorname{supp} \phi \subset K\},$$

which is a linear subspace of $C_0^\infty(\Omega)$.

For $m \in \mathbb{N}_0 = \mathbb{N} \cup \{0\}$ and $K \subset \Omega$ compact,

$$p_{K,m}(\phi) = \sup_{x \in K, |\alpha| \leq m} |D^\alpha \phi(x)|$$

is a seminorm on $\mathcal{D}_K(\Omega)$, where $\alpha = (\alpha_1, \alpha_2, \ldots, \alpha_k) \in \mathbb{N}_0^k$, $|\alpha| = \alpha_1 + \alpha_2 + \cdots + \alpha_k$, and the α-derivative of ϕ is defined as

$$D^\alpha \phi = \frac{\partial^{|\alpha|} \phi}{\partial x_1^{\alpha_1} \partial x_2^{\alpha_2} \cdots \partial x_k^{\alpha_k}}.$$

Note that the order of differentiation is not important since ϕ is a smooth function. If $\alpha = (0, 0, \ldots, 0)$, then $D^\alpha \phi = \phi$ by convention. Then $\mathcal{D}_K(\Omega)$ is a locally convex space and, if $K_1 \subset K_2$ the topology of $\mathcal{D}_{K_1}(\Omega)$ coincides with the relative topology of $\mathcal{D}_{K_1}(\Omega)$ as a subset of $\mathcal{D}_{K_2}(\Omega)$. Then $C_0^\infty(\Omega)$ can be regarded as the inductive limit of the $\mathcal{D}_K(\Omega)$'s, where K ranges over all compact subsets of Ω. The space $C_0^\infty(\Omega)$, topologized in this way, is denoted by $\mathcal{D}(\Omega)$.

One of the seminorms defining the topology of $\mathcal{D}(\Omega)$ is

$$p(\phi) = \sup_{x \in \Omega} |\phi(x)|, \quad \phi \in C_0^\infty(\Omega).$$

If $D = \{\phi \in C_0^\infty(\Omega); p(\phi) < 1\}$ and K is a compact subset of Ω, then $D \cap \mathcal{D}_K(\Omega) = \{\phi \in \mathcal{D}_K(\Omega); p_K(\phi) := \sup_{x \in K} |\phi(x)| < 1\}$.

Theorem 5.1. *Convergence of a sequence $\phi_n \to 0$ in $\mathcal{D}(\Omega)$ means that the following conditions are satisfied:*

(a) *there exists a compact set $K \subset \Omega$ such that $\operatorname{supp} \phi_n \subset K$ for all n;*

(b) *$D^\alpha \phi_n \to 0$ uniformly on K as $n \to \infty$ for all $\alpha \in \mathbb{N}_0^k$.*

Proof. If (a) is satisfied, then (b) is satisfied, too. So all we need to do is to prove (a). Assume by contradiction that (a) is not satisfied. So there exists a sequence $(x_j)_{j \geq 1}$ in Ω with no cluster point in Ω and a subsequence $(\phi_{n_j})_{j \geq 1}$ such that $\phi_{n_j}(x_j) \neq 0$ for all $j \geq 1$. Define a seminorm $p : C_0^\infty(\Omega) \to \mathbb{R}$ by

$$p(\phi) = 2 \sum_{j=1}^\infty \sup_{x \in K_j \setminus K_{j-1}} \frac{|\phi(x)|}{|\phi_{n_j}(x_j)|}, \quad \phi \in C_0^\infty(\Omega),$$

where the sequence of compacts $K_1 \subset K_2 \subset \cdots \subset \Omega$ satisfies $\cup_{j \geq 1} K_j = \Omega$ and $x_j \in K_j \setminus K_{j-1}$, $j = 1, 2, \ldots$, $K_0 = \emptyset$. Clearly, the set $V = \{\phi \in C_0^\infty(\Omega); p(\phi) < 1\}$ is a neighborhood of $0 \in C_0^\infty(\Omega)$ and none of the ϕ_{n_j} belongs to V, which gives a contradiction. $\qquad\square$

Obviously, the convergence $\phi_n \to \phi$ in $\mathcal{D}(\Omega)$ means that (ϕ_n) satisfies condition (a) with some compact $K \subset \Omega$, and $D^\alpha \phi_n \to D^\alpha \phi$ uniformly on K as $n \to \infty$ for all $\alpha \in \mathbb{N}_0^k$.

Example 1. For $\Omega = \mathbb{R}^k$ let $\phi_n(x) = \frac{1}{n} \omega(x)$, where ω is the test function defined by (5.1.1). Then $K = \overline{B(0, 1)}$ and all derivatives of ϕ_n converge uniformly to 0, so $\phi_n \to 0$ in $\mathcal{D}(\Omega)$.

Example 2. For $\Omega = \mathbb{R}^k$ let $\psi_n(x) = \frac{1}{n}\omega(\frac{1}{n}x)$ for $x \in \mathbb{R}^k$. $D^\alpha\psi_n \to 0$ uniformly as $n \to \infty$ for all $\alpha \in \mathbb{N}_0^k$, but there is no K satisfying (a). In fact $\operatorname{supp}\psi_n = \overline{B(0,n)}$, therefore ψ_n does not converge in $\mathcal{D}(\Omega)$.

5.2 Friedrichs' Mollification

Friedrichs' mollification will allow us to associate with "bad functions" very good approximate functions.[1]

Consider again the test function $\omega : \mathbb{R}^k \to \mathbb{R}$ defined in the previous section, i.e.,

$$\omega(x) = \begin{cases} C\exp\left(\frac{1}{\|x\|_2^2-1}\right), & \|x\|_2 < 1, \\ 0, & \|x\|_2 \geq 1, \end{cases}$$

with $C > 0$ such that $\int_{\mathbb{R}^k}\omega(x)\,dx = \int_{B(0,1)}\omega(x)\,dx = 1$.

Definition 5.2. *For $\varepsilon > 0$ define $\omega_\varepsilon(x) = \frac{1}{\varepsilon^k}\omega(\frac{1}{\varepsilon}x)$ for all $x \in \mathbb{R}^k$. This is called the* **mollifier***.*

The mollifier ω_ε has the following properties:

1. $\omega_\varepsilon \in C^\infty(\mathbb{R}^k)$;

2. $\operatorname{supp}\omega_\varepsilon = \overline{B(0,\varepsilon)}$;

3. $\int_{\mathbb{R}^k}\omega_\varepsilon(x)\,dx = \int_{B(0,\varepsilon)}\omega_\varepsilon(x)\,dx = 1$.

Definition 5.3. *Let $f \in L^1_{\mathrm{loc}}(\mathbb{R}^k)$, i.e., f is a real measurable function and $f \in L^1(K)$ for any compact $K \subset \mathbb{R}^k$. For $\varepsilon > 0$ define $f_\varepsilon(x)$ the* **Friedrichs' mollification** *of f as*

$$f_\varepsilon(x) = (\omega_\varepsilon * f)(x),$$

where $*$ denotes the convolution product

$$= \int_{\mathbb{R}^k}\omega_\varepsilon(x-y)f(y)\,dy\,,$$

---

[1]Kurt Otto Friedrichs, German-American mathematician, 1901–1982.

and by changing variables,

$$= \int_{\mathbb{R}^k} \omega_\varepsilon(y) f(x-y) \, dy$$

$$= \int_{B(0,\varepsilon)} \omega_\varepsilon(y) f(x-y) \, dy$$

for almost all $x \in \mathbb{R}^k$.

If $f \in L^1_{\mathrm{loc}}(\Omega)$, then f can be extended as $f = 0$ for $x \in \mathbb{R}^k \setminus \Omega$, and we can define f_ε as before.

For $\varepsilon > 0$ and $f \in L^1_{\mathrm{loc}}(\mathbb{R}^k)$, we have

1. $f_\varepsilon \in C^\infty(\mathbb{R}^k)$;

2. supp $f_\varepsilon \subset$ supp $f + \overline{B(0,\varepsilon)}$, i.e., not much larger than supp f ;

3. If f has compact support, so does f_ε.

Proposition 5.4. *If $f \in C_0(\Omega)$, then $f_\varepsilon(x) \to f(x)$ uniformly as $\varepsilon \to 0^+$ in Ω, where $C_0(\Omega) = \{u \in C(\Omega);\ u$ has compact (bounded) support $\subset \Omega\}$.*

Proof. Set $K = $ supp f and $K' = K + \overline{B(0,\varepsilon_0)}$, where $\varepsilon_0 > 0$. Then supp $f_\varepsilon \subset K' \subset \Omega$ for $0 < \varepsilon \le \varepsilon_0$, if ε_0 is small enough. For $0 < \varepsilon \le \varepsilon_0$ and $x \in K'$,

$$|f_\varepsilon(x) - f(x)| = \left| \int_{\mathbb{R}^k} f(x-y)\omega_\varepsilon(y) \, dy - \int_{\mathbb{R}^k} f(x)\omega_\varepsilon(y) \, dy \right|$$

$$\le \int_{B(0,\varepsilon)} |f(x-y) - f(x)|\, \omega_\varepsilon(y) \, dy \,.$$

f is continuous on K', hence uniformly continuous on K', so for any $\eta > 0$, $|f(x-y) - f(x)| < \eta$ for all $y \in B(0,\varepsilon)$ with $\varepsilon > 0$ small. Thus $\sup_{x \in \Omega} |f_\varepsilon(x) - f(x)| \le \eta$ for all $\varepsilon > 0$ sufficiently small, hence $f_\varepsilon \to f$ uniformly in Ω as $\varepsilon \to 0^+$. $\square$

Theorem 5.5. *If $f \in L^p(\Omega)$ for some $1 \le p < \infty$, then (the restriction to Ω of) f_ε is in $L^p(\Omega)$ for all $\varepsilon > 0$ and*

1. $\|f_\varepsilon\|_{L^p(\Omega)} \le \|f\|_{L^p(\Omega)}$ *for all $\varepsilon > 0$*,

2. $\|f_\varepsilon - f\|_{L^p(\Omega)} \to 0$ *as $\varepsilon \to 0^+$*.

Proof. It suffices to consider $\Omega = \mathbb{R}^k$, because we can extend f to $\mathbb{R}^k$ as before, and the two conclusions of the theorem for the extension of f will imply the same conclusions for $f \in L^p(\Omega)$.

Consider first the case $p = 1$, i.e., $f \in L^1(\mathbb{R}^k)$. Note that

$$(x, y) \mapsto |f(y)| \, \omega_\varepsilon(x - y) \tag{5.2.3}$$

is measurable on $\mathbb{R}^k \times \mathbb{R}^k$ and

$$\int_{\mathbb{R}^k} |f(y)| \, \omega_\varepsilon(x - y) \, dx = |f(y)| \underbrace{\int_{\mathbb{R}^k} \omega_\varepsilon(x - y) \, dx}_{=1}$$

$$= |f(y)|$$

for almost all $y \in \mathbb{R}^k$. Next,

$$\int_{\mathbb{R}^k} \int_{\mathbb{R}^k} |f(y)| \, \omega_\varepsilon(x - y) \, dx \, dy = \int_{\mathbb{R}^k} |f(y)| \, dy = \|f\|_{L^1(\mathbb{R}^k)} < \infty. \tag{5.2.4}$$

Thus, by Fubini-Tonelli's Theorem (see, e.g., [51, p. 18]), function (5.2.3) is a member of $L^1(\mathbb{R}^k \times \mathbb{R}^k)$ and

$$\int_{\mathbb{R}^k} |f_\varepsilon(x)| \, dx = \int_{\mathbb{R}^k} \left| \int_{\mathbb{R}^k} \omega_\varepsilon(x - y) f(y) \, dy \right| dx$$

$$\leq \int_{\mathbb{R}^k} |f(y)| \underbrace{\int_{\mathbb{R}^k} \omega_\varepsilon(x - y) \, dx}_{=1} \, dy$$

$$= \|f\|_{L^1(\mathbb{R}^k)},$$

so that

$$\|f_\varepsilon\|_{L^1(\mathbb{R}^k)} \leq \|f\|_{L^1(\mathbb{R}^k)},$$

as claimed.

We now consider the case $1 < p < \infty$ for the same function (5.2.3). Then $f_\varepsilon \in L^p(\mathbb{R}^k)$ and, denoting by p' the conjugate of p (i.e., $(1/p) + (1/p') = 1$), we have

$$|f_\varepsilon(x)| \leq \int_{\mathbb{R}^k} |f(y)| \, \omega_\varepsilon(x - y) \, dy$$

$$= \int_{\mathbb{R}^k} \omega_\varepsilon(x - y)^{1/p'} \omega_\varepsilon(x - y)^{1/p} |f(y)| \, dy$$

which by Hölder's inequality

$$\leq \left(\underbrace{\int_{\mathbb{R}^k} \omega_\varepsilon(x-y)\,dy}_{=1} \right)^{1/p'} \left(\int_{\mathbb{R}^k} \omega_\varepsilon(x-y)|f(y)|^p dy \right)^{1/p}$$

so that

$$|f_\varepsilon(x)|^p \leq \int_{\mathbb{R}^k} \omega_\varepsilon(x-y)|f(y)|^p dy$$

and integrating

$$\int_{\mathbb{R}^k} |f_\varepsilon(x)|^p dx \leq \int_{\mathbb{R}^k} \left(\int_{\mathbb{R}^k} \omega_\varepsilon(x-y)|f(y)|^p dy \right) dx$$
$$= \int_{\mathbb{R}^k} \left(|f(y)|^p \underbrace{\int_{\mathbb{R}^k} \omega_\varepsilon(x-y)\,dx}_{=1} \right) dy$$
$$= \int_{\mathbb{R}^k} |f(y)|^p dy$$
$$= \|f\|^p_{L^p(\mathbb{R}^k)}$$

so that

$$\|f_\varepsilon\|_{L^p(\mathbb{R}^k)} \leq \|f\|_{L^p(\mathbb{R}^k)}$$

which concludes the proof of the first statement of the theorem. Before we continue the proof of the theorem we shall prove two auxiliary results.

Lemma 5.6. *For all compact $K \subset \Omega$ there exists an open neighborhood V of K such that $\overline{V} \subset \Omega$ and a continuous map $g : \Omega \to \mathbb{R}$ satisfying*

$$g(x) = 1 \quad \text{for all } x \in K,$$
$$g(x) = 0 \quad \text{for all } x \in \Omega \setminus V, \text{ and}$$
$$0 \leq g(x) \leq 1 \quad \text{for all } x \in \Omega.$$

Proof. Let $K \subset \Omega$ be a compact set. Consider $\delta > 0$ small and let V be δ-neighborhood of K whose closure $\overline{V}$ lies in Ω. Let $W = \Omega \setminus V$ and $\rho(x) = d(x, W) := \inf_{w \in W} \|x - w\|_2$ which is a continuous function.

Now let $\alpha = \inf_{x \in K} \rho(x) > 0$, and let $g(x) = \min\{1, \frac{1}{\alpha}\rho(x)\}$ which is also a continuous function. Clearly $g(x) = 1$ for $x \in K$, $g(x) = 0$ for $x \in W = \Omega \setminus V$, and $0 \le g(x) \le 1$ for $x \in V \setminus K$. $\qquad\square$

Lemma 5.7. $C_0(\Omega)$ *is dense in* $L^p(\Omega)$ *for all* $1 \le p < \infty$: $\overline{C_0(\Omega)}^{L^p(\Omega)} = L^p(\Omega)$ *(i.e., every* $L^p(\Omega)$ *function can be approximated by* $C_0(\Omega)$ *functions with respect to the usual norm of* $L^p(\Omega)$).

Proof. Let $u \in L^p(\Omega)$. We have $u = u^+ - u^-$, where both u^+ and u^- are nonnegative $L^p(\Omega)$ functions. So, it suffices to consider nonnegative $L^p(\Omega)$ functions u which we approximate by simple functions

$$s = \sum_{i=1}^{m} y_i \chi_{M_i},$$

where the sets $M_i \subset \Omega$ are mutually disjoint and measurable with $m(M_i) < \infty$, and the χ_{M_i} are their characteristic functions. Consider a sequence of simple functions (s_n), such that $0 \le s_n \le u$ and $s_n \to u$ as $n \to \infty$ for almost all $x \in \Omega$, so $s_n \to u$ in $L^p(\Omega)$. Thus u can be approximated by simple functions and so all reduces to approximating characteristic functions $u = \chi_M$ where $M \subset \Omega$ is a Lebesgue measurable set with $m(M) < \infty$. In fact, we only need to consider $K \subset M$ compact such that $m(M \setminus K) = m(M) - m(K)$ is small (see Exercise 3.2), so

$$\int_\Omega |\chi_K - \chi_M|^p \, dx = \int_{M \setminus K} 1 \, dx = m(M \setminus K)$$

is small. Now, choose V as in Lemma 5.6 such that $m(V \setminus K) < \varepsilon^p$, then there exists $g \in C_0(\Omega)$ such that

$$\int_\Omega |g - \chi_K|^p \, dx = \int_{V \setminus K} g^p \, dx \le \int_{V \setminus K} 1 \, dx = m(V \setminus K) < \varepsilon^p$$

so

$$\|g - \chi_K\|_{L^p(\Omega)} < \varepsilon.$$

Thus the characteristic functions $u = \chi_M$ can indeed be approximated by $C_0(\Omega)$ functions. $\qquad\square$

Proof of Theorem 5.5, continuation.
Consider $f \in L^p(\Omega)$ and approximate it using Lemma 5.7: for $\theta > 0$ there exists $g \in C_0(\Omega)$ such that

$$\|f - g\|_{L^p(\Omega)} < \frac{\theta}{3}. \qquad (5.2.5)$$

We have

$$\|f_\varepsilon - f\|_{L^p(\Omega)} \leq \|f_\varepsilon - g_\varepsilon\|_{L^p(\Omega)} + \|g_\varepsilon - g\|_{L^p(\Omega)} + \|g - f\|_{L^p(\Omega)}$$

which by the first statement of the theorem is

$$\leq 2\|f - g\|_{L^p(\Omega)} + \|g_\varepsilon - g\|_{L^p(\Omega)}$$

so by (5.2.5)

$$< \frac{2}{3}\theta + \|g_\varepsilon - g\|_{L^p(\Omega)}$$

which by Proposition 5.4

$$< \frac{2}{3}\theta + \underbrace{\text{constant} \cdot \|g_\varepsilon - g\|_{C(K')}}_{< \frac{\theta}{3}}$$

$$< \theta$$

for all $\varepsilon > 0$ small. Therefore,

$$\limsup_{\varepsilon \to 0^+} \|f_\varepsilon - f\|_{L^p(\Omega)} = 0 \implies \lim_{\varepsilon \to 0^+} \|f_\varepsilon - f\|_{L^p(\Omega)} = 0.$$

This completes the proof.

The following is a fundamental theorem.

Theorem 5.8. *Let $\Omega \subset \mathbb{R}^k$ be a nonempty open set. We have $\overline{C_0^\infty(\Omega)}^{L^p(\Omega)} = L^p(\Omega)$ for all $1 \leq p < \infty$ (i.e., every $L^p(\Omega)$ function can be approximated by test functions).*

Proof. Let $f \in L^p(\Omega)$. By Lemma 5.7 for all $\eta > 0$ there exists $g \in C_0(\Omega)$ such that $\|f - g\|_{L^p(\Omega)} < \eta/2$. On the other hand, there is a $g_\varepsilon \in C_0^\infty(\Omega)$ and by Theorem 5.5 $\|g_\varepsilon - g\|_{L^p(\Omega)} < \eta/2$ for $\varepsilon > 0$ small. Therefore,

$$\|f - g_\varepsilon\|_{L^p(\Omega)} \leq \|f - g\|_{L^p(\Omega)} + \|g_\varepsilon - g\|_{L^p(\Omega)} < \frac{\eta}{2} + \frac{\eta}{2} = \eta$$

for $\varepsilon > 0$ small. □

Theorem 5.9. *If $f \in L^1_{\text{loc}}(\Omega)$ is such that*

$$\int_\Omega f(x)\phi(x)\,dx = 0 \quad \forall \phi \in C_0^\infty(\Omega)\,, \tag{5.2.6}$$

then $f = 0$ a.e. on Ω.

Proof. First of all let us extend (5.2.6) to

$$\int_\Omega f(x)g(x)\,dx = 0 \tag{5.2.7}$$

for all $g \in L^\infty(\Omega)$ such that g vanishes almost everywhere on $\Omega \setminus K$, where $K \subset \Omega$ is a compact set. Obviously, such a function g belongs in particular to $L^1(\Omega)$ and (by Theorem 5.5)

$$\|g_\varepsilon - g\|_{L^1(\Omega)} \to 0 \ \text{ as } \varepsilon \to 0^+\,.$$

Hence, there exists a sequence $\varepsilon_j \to 0$ such that

$$g_{\varepsilon_j}(x) \to g(x) \ \text{ as } j \to \infty \ \text{ for a.a. } x \in \Omega\,. \tag{5.2.8}$$

Therefore by (5.2.6) we have

$$\int_\Omega f(x)g_{\varepsilon_j}(x)dx = 0\,, \tag{5.2.9}$$

for j large enough such that $\operatorname{supp} g_{\varepsilon_j} \subset K'$, where K' is a compact, $K \subset K' \subset \Omega$. We also have

$$|f(x)g_{\varepsilon_j}(x)| \le |f(x)| \cdot |g_{\varepsilon_j}(x)|$$
$$\le |f(x)| \int_\Omega |g(y)| \omega_{\varepsilon_j}(x-y)\,dy$$
$$\le |f(x)| \cdot \|g\|_{L^\infty(\Omega)}\,,$$

for j large enough and for almost all $x \in K'$. So we can apply the Lebesgue Dominated Convergence Theorem (see also (5.2.8) and (5.2.9)) to get (5.2.7) for all $g \in L^\infty(\Omega)$ such that g vanishes a.e. on $\Omega \setminus K$.

Now choose an arbitrary compact set $K \subset \Omega$ and let $g = \operatorname{sign} f \cdot \chi_K$. Then by (5.2.7) we have

$$\int_\Omega fg\,dx = \int_\Omega |f|\chi_K\,dx = \int_K |f|\,dx = 0,$$

which implies $f = 0$ for almost all $x \in K$. Since K is arbitrary, $f = 0$ a.e. on Ω. $\square$

5.3   Scalar Distributions

Let $\Omega \subset \mathbb{R}^k$ be a nonempty open set. Recall that $C_0^\infty(\Omega)$, topologized as the inductive limit of the $\mathcal{D}_K(\Omega)$'s, where K runs over all compact subsets of Ω, is denoted by $\mathcal{D}(\Omega)$ (see Sect. 5.1).

Definition 5.10. *A functional* $u : \mathcal{D}(\Omega) \to \mathbb{R}$ *is said to be a (scalar)* **distribution** *(on Ω) if u is linear and continuous, i.e.,*

- $u(\alpha_1 \phi_1 + \alpha_2 \phi_2) = \alpha_1 u(\phi_1) + \alpha_2 u(\phi_2)$ *for all* $\alpha_1, \alpha_2 \in \mathbb{R}$, *and all* $\phi_1, \phi_2 \in \mathcal{D}(\Omega)$;

- $\phi_n \to \phi$ *in* $\mathcal{D}(\Omega)$ *implies* $u(\phi_n) \to u(\phi)$ *in* $\mathbb{R}$.

In fact, it is enough to consider $\phi = 0$ for the second condition because of linearity. Let $\mathcal{D}'(\Omega)$ denote the set of all distributions on Ω.

It is easily seen that $\mathcal{D}'(\Omega)$ is a real linear space. Sometimes we shall write (u, ϕ) instead of $u(\phi)$.

Notice that, in general, a distribution is not defined point-wise on Ω, unless it is a regular distribution, i.e., a distribution defined by a usual function, as explained below.

Regular Distributions

Let $u \in L^1_{\mathrm{loc}}(\Omega)$ and define $\tilde{u} : \mathcal{D}(\Omega) \to \mathbb{R}$ by

$$\tilde{u}(\phi) = \int_\Omega u(x)\phi(x)\, dx \quad \forall \phi \in \mathcal{D}(\Omega) \, .$$

Since ϕ has compact support $u\phi \in L^1(\Omega)$ and so $\tilde{u}$ is well defined. Clearly, $\tilde{u}$ is linear and continuous and therefore a distribution. Note that the mapping $i : L^1_{\mathrm{loc}}(\Omega) \to \mathcal{D}'(\Omega)$, $i(u) = \tilde{u}$ is injective. Since i is linear, injectivity can be seen by showing the implication $\tilde{u} = i(u) = 0 \implies u = 0$ for a.a. $x \in \Omega$. This is indeed the case by Theorem 5.9. We now simply identify $\tilde{u}$ with u and write

$$u(\phi) = \int_\Omega u(x)\phi(x)\, dx \quad \forall \phi \in \mathcal{D}(\Omega).$$

A distribution which arises this way is called a **regular distribution**.

The Dirac Distribution[2]

Let $\Omega = \mathbb{R}^k$ and define $(\delta, \phi) = \delta(\phi) = \phi(0)$ for all $\phi \in \mathcal{D}(\Omega)$. It is linear and continuous, so $\delta \in \mathcal{D}'(\Omega)$. δ is called the *Dirac distribution* or *delta function*, to follow the original denomination, even though it is not in fact a function.

Claim: *The distribution δ is not a regular distribution.*

Proof. Suppose, by way of contradiction, that there exists a function $f \in L^1_{\text{loc}}(\mathbb{R}^k)$ such that

$$(\delta, \phi) = \int_{\mathbb{R}^k} f(x)\phi(x)\,dx \quad \forall \phi \in \mathcal{D}(\mathbb{R}^k).$$

This means

$$\int_{\mathbb{R}^k} f\phi\,dx = \phi(0) \quad \forall \phi \in \mathcal{D}(\mathbb{R}^k),$$

and, in particular,

$$\int_{\mathbb{R}^k} f\phi\,dx = 0 \quad \forall \phi \in \mathcal{D}(\mathbb{R}^k),\ \operatorname{supp}\phi \subset \mathbb{R}^k \setminus \{0\},$$

hence

$$\int_{\mathbb{R}^k \setminus \{0\}} f\phi\,dx = 0 \quad \forall \phi \in \mathcal{D}(\mathbb{R}^k \setminus \{0\}).$$

Then according to Theorem 5.9, $f = 0$ for almost all $x \in \mathbb{R}^k \setminus \{0\}$ so $f = 0$ for almost all $x \in \mathbb{R}^k$, thus $\phi(0) = (\delta, \phi) = 0$ for all $\phi \in \mathcal{D}(\mathbb{R}^k)$ which is false. $\qquad\qquad\square$

A physical interpretation of δ will be provided later.

For a given $x_0 \in \mathbb{R}^k$ one can define a similar Dirac distribution, denoted δ_{x_0}, by

$$(\delta_{x_0}, \phi) = \phi(x_0) \quad \forall \phi \in \mathcal{D}(\mathbb{R}^k).$$

The Dirac distribution associated with $x_0 = 0$ is precisely δ. Of course, linear combinations of Dirac distributions are also distributions. In fact, the space of distributions is a large one, as shown below.

---

[2]Paul Adrien Maurice Dirac, English theoretical physicist, 1902–1984.

5.3.1 Some Operations with Distributions

Besides addition and scalar multiplication there are some further operations we can perform on distributions.

- *Multiplication by a C^∞ function.*

For $u \in \mathcal{D}'(\Omega)$ and $a \in C^\infty(\Omega)$, define au by

$$(au, \phi) := (u, a\phi) \quad \forall \phi \in \mathcal{D}(\Omega).$$

Note that $a\phi$ is still a test function, and au is linear and continuous on $\mathcal{D}(\Omega)$, so $au \in \mathcal{D}'(\Omega)$.

This is a generalization of the usual multiplication of functions. Indeed, if $u \in L^1_{\mathrm{loc}}(\Omega)$ (i.e., u is a regular distribution), then

$$
\begin{aligned}
(au, \phi) &= (u, a\phi) \\
&= \int_\Omega u(a\phi)\, dx \\
&= \int_\Omega (au)\phi\, dx,
\end{aligned}
$$

so $(au)(x) = a(x)u(x)$ for almost all $x \in \Omega$.

- *Reflection about the origin.*

For the sake of simplicity, consider $\Omega = \mathbb{R}^k$. Let $u \in \mathcal{D}'(\mathbb{R}^k)$. Sometimes we write $u(x)$ instead of u even though it is not a function. For example, this helps denote the reflection of u

$$(u(-x), \phi(x)) := (u(x), \phi(-x)))\, dx \quad \forall \phi \in \mathcal{D}(\mathbb{R}^k).$$

Clearly $u(-x) \in \mathcal{D}'(\Omega)$. Notice that if $u \in L^1_{\mathrm{loc}}(\mathbb{R}^k)$, then $u(-x) \in \mathcal{D}'(\mathbb{R}^k)$ is precisely the regular distribution generated by the function $x \mapsto u(-x)$,

$$\int_{\mathbb{R}^k} u(-x)\phi(x)\, dx = \int_{\mathbb{R}^k} u(x)\phi(-x)\, dx \quad \forall \phi \in \mathcal{D}(\Omega),$$

and this explains the notation for the reflection of the distribution u.

- *Translation by a vector.*

For $u \in \mathcal{D}'(\mathbb{R}^k)$ and $h \in \mathbb{R}^k$, define $u(x + h)$ by

$$(u(x + h), \phi(x)) := (u(x), \phi(x - h)) \quad \forall \phi \in \mathcal{D}(\Omega).$$

It is clear that $u(x + h) \in \mathcal{D}'(\mathbb{R}^k)$. Again, the notation $u(x + h)$ is justified by the case when u is a locally integrable function.

Note that the Dirac distribution δ_{x_0} defined before is precisely $\delta(x - x_0)$ in terms of the above notation.

5.3.2 Convergence in Distributions

Let $(u_n)_{n \in \mathbb{N}}$ be a sequence in $\mathcal{D}'(\Omega)$. We say that (u_n) converges in $\mathcal{D}'(\Omega)$ if there exists $u \in \mathcal{D}'(\Omega)$ such that

$$\lim_{n \to \infty} (u_n, \phi) = (u, \phi) \quad \forall \phi \in \mathcal{D}(\Omega). \tag{5.3.10}$$

In fact, $\mathcal{D}'(\Omega)$ is sequentially complete, so if (5.3.10) holds then u is automatically in $\mathcal{D}'(\Omega)$. More precisely,

Claim: *If (u_n, ϕ) is convergent for all $\phi \in \mathcal{D}(\Omega)$, then the functional $u : \mathcal{D}(\Omega) \to \mathbb{R}$ defined by (5.3.10) is linear and continuous.*

Proof. While the linearity of u follows trivially from (5.3.10), its continuity is not immediate, see Gel'fand and Shilov [17].[3] Assume that u is not continuous, i.e., there exists a sequence $\phi_n \to 0$ in $\mathcal{D}(\Omega)$ such that, on a subsequence again denoted ϕ_n, we have

$$|u(\phi_n)| \geq \delta > 0 \quad \forall n. \tag{5.3.11}$$

Choosing another subsequence, we may assume that for all n

$$\sup_{x \in \Omega} |D^\alpha \phi_n(x)| < \frac{1}{2^{2n}} \quad \forall |\alpha| \leq n. \tag{5.3.12}$$

We consider $\psi_n = 2^n \phi_n$. By (5.3.12) we get

$$\psi_n \to 0 \quad \text{in } \mathcal{D}(\Omega). \tag{5.3.13}$$

On the other hand (see (5.3.11)),

$$|u(\psi_n)| \geq 2^n \delta \to \infty. \tag{5.3.14}$$

---

[3]Israel M. Gel'fand, Russian mathematician, 1913–2009; Georgiy E. Shilov, Russian mathematician, 1917–1975.

Let us now extract new subsequences, say $(\tilde{u}_n)$ and $(\tilde{\psi}_n)$. In view of (5.3.14) we can pick a $\tilde{\psi}_1$ such that $|(u, \tilde{\psi}_1)| > 1$. Thus, by virtue of (5.3.10), we can choose $\tilde{u}_1$ such that $|(\tilde{u}_1, \tilde{\psi}_1)| > 1$. Now, assuming that $\tilde{u}_j$ and $\tilde{\psi}_j$ have been chosen for $j = 1, 2, \ldots, n-1$, we can pick (by the continuity of the $\tilde{u}_j$'s and by (5.3.14)) a test function $\tilde{\psi}_n$ such that

$$|(\tilde{u}_k, \tilde{\psi}_n)| < \frac{1}{2^{n-k}}, \quad k = 1, 2, \ldots, n-1, \tag{5.3.15}$$

$$|(u, \tilde{\psi}_n)| > \sum_{j=1}^{n-1} |(u, \tilde{\psi}_j)| + n + 1. \tag{5.3.16}$$

Taking into account (5.3.10) and (5.3.16), we can pick $\tilde{u}_n$ such that

$$|(\tilde{u}_n, \tilde{\psi}_n)| > \sum_{j=1}^{n-1} |(\tilde{u}_n, \tilde{\psi}_j)| + n + 1. \tag{5.3.17}$$

So by induction we obtain $(\tilde{u}_n)$ and $(\tilde{\psi}_n)$ satisfying (5.3.15) and (5.3.17). Set

$$\psi = \sum_{n=1}^{\infty} \tilde{\psi}_n.$$

Since $(\tilde{\psi}_n)$ is a subsequence of (ψ_n) the above series is convergent in $\mathcal{D}(\Omega)$ (see (5.3.12)), hence $\psi \in \mathcal{D}(\Omega)$. Now, let us estimate $|(\tilde{u}_n, \psi)|$ by using the decomposition

$$(\tilde{u}_n, \psi) = \sum_{j \neq n} (\tilde{u}_n, \tilde{\psi}_j) + (\tilde{u}_n, \tilde{\psi}_n). \tag{5.3.18}$$

From (5.3.15) we get

$$\sum_{j=n+1}^{\infty} |(\tilde{u}_n, \tilde{\psi}_j)| < \sum_{j=n+1}^{\infty} \frac{1}{2^{j-n}} = 1. \tag{5.3.19}$$

Finally, from (5.3.17), (5.3.18), and (5.3.19) we see that

$$|(\tilde{u}_n, \psi)| > n,$$

which contradicts (5.3.10). $\qquad\qquad\qquad\qquad\qquad\qquad\qquad\square$

As an example, consider the sequence of Friedrichs mollifiers $u_n = \omega_{1/n}$, i.e., $u_n(x) = n^k \omega(nx)$ for $x \in \Omega = \mathbb{R}^k$, $n \in \mathbb{N}$, where ω is the test function defined before in (5.1.1). The graphs of the u_n's for $k = 1$ or $k = 2$ can be visualized in corresponding coordinate systems to observe the behavior of the u_n's as n gets larger and larger. The pointwise limit of (u_n) is as follows:

$$\lim_{n \to \infty} u_n(x) = \begin{cases} 0 & x \neq 0 \\ +\infty & x = 0 \end{cases}$$

which is not an $\mathbb{R}$-valued function. On the other hand, viewing the u_n's as distributions, we have $u_n \to \delta$ in $\mathcal{D}'(\mathbb{R}^k)$:

$$\begin{aligned} (u_n, \phi) &= \int_{\mathbb{R}^k} u_n(x)\phi(x)\, dx \\ &= \int_{B(0, \frac{1}{n})} u_n(x)\phi(x)\, dx \\ &\to \phi(0) = (\delta, \phi), \end{aligned}$$

for all $\phi \in \mathcal{D}(\mathbb{R}^k)$.

Physical Interpretation of the Dirac Distribution: The Dirac distribution represents, for instance, the density of a unit mass concentrated at some point. To explain that, let us suppose that a unit mass, which is concentrated at the origin of a coordinate system in $\mathbb{R}^3$, is distributed uniformly in $\overline{B(0, 1/n)} \subset \mathbb{R}^3$. Thus the corresponding mass density is given by

$$\delta_n(x) = \begin{cases} \frac{3n^3}{4\pi} & \|x\|_2 \leq \frac{1}{n}, \\ 0 & \text{otherwise}, \end{cases}$$

and obviously the total mass $\iiint \delta_n\, dx = 1$. For $n \to \infty$ the mass concentrates in $x = 0$. Obviously, $\delta_n(x) \to 0$ as $n \to \infty$ for all $x \neq 0$, and $\delta_n(0) \to +\infty$, so δ_n does not converge pointwise to a function. However, $\delta_n \to \delta$ in $\mathcal{D}'(\mathbb{R}^3)$ as $n \to \infty$:

$$(\delta_n, \phi) = \frac{3n^3}{4\pi} \iiint_{B(0, \frac{1}{n})} \phi(x)\, dx \to \phi(0) = (\delta, \phi) \quad \forall \phi \in \mathcal{D}'(\mathbb{R}^3),$$

so δ can indeed be interpreted as the density of an idealized point mass.

5.3.3   Differentiation of Distributions

For $u \in C^1(\Omega)$ and $\phi \in \mathcal{D}(\Omega)$ one can write

$$
\begin{aligned}
\left(\frac{\partial u}{\partial x_i}, \phi\right) &= \int_\Omega \frac{\partial u}{\partial x_i} \phi \, dx \\
&= \int_{\operatorname{supp}\phi} \frac{\partial u}{\partial x_i} \phi \, dx, \quad i = 1, \ldots, k
\end{aligned}
$$

and switching to a rectangular cell including $\operatorname{supp}\phi$ to ease computation

$$
= \int_{cell} \frac{\partial u}{\partial x_i} \phi \, dx
$$

and integrating by parts

$$
\begin{aligned}
&= -\int_{cell} u \frac{\partial \phi}{\partial x_i} \, dx \\
&= -\int_\Omega u \frac{\partial \phi}{\partial x_i} \, dx \\
&= -\left(u, \frac{\partial \phi}{\partial x_i}\right).
\end{aligned}
$$

Hence, if $u \in C^1(\Omega)$ we have

$$
\left(\frac{\partial u}{\partial x_i}, \phi\right) = -\left(u, \frac{\partial \phi}{\partial x_i}\right) \quad \forall \phi \in \mathcal{D}(\Omega), \ i = 1, \ldots, k. \tag{5.3.20}
$$

If u is an arbitrary distribution, then we use (5.3.20) as the definition of $\frac{\partial u}{\partial x_i}$ which is also an element of $\mathcal{D}'(\Omega)$. Whenever u is a smooth function, its distributional derivative defined by (5.3.20) coincides with the classical derivative of u.

Since $u \in \mathcal{D}'(\Omega)$ implies $\frac{\partial u}{\partial x_i} \in \mathcal{D}'(\Omega)$ for $i = 1, \ldots, k$, we deduce by induction that every distribution $u \in \mathcal{D}'(\Omega)$ is infinitely differentiable, and we have

$$
(D^\alpha u, \phi) = (-1)^{|\alpha|} (u, D^\alpha \phi) \quad \forall \phi \in \mathcal{D}(\Omega), \ \alpha = (\alpha_1, \ldots, \alpha_k) \in \mathbb{N}_0^k. \tag{5.3.21}
$$

By convention $D^{(0,0,\ldots,0)}u = u$. It is clear from (5.3.21) that mixed derivatives in the sense of distributions do not depend on the order of differentiation.

Let us now discuss some examples.

Example 1. Consider the Heaviside function H, defined on $\Omega = \mathbb{R}$ by

$$H(x) = \begin{cases} 1 & x \geq 0, \\ 0 & x < 0. \end{cases}$$

We use $\dot{H}$ to indicate the pointwise derivative, and H' for the derivative in $\mathcal{D}'(\mathbb{R})$. Obviously $\dot{H}(x) = 0$ for $x \neq 0$, hence $\dot{H} = 0$ a.e. On the other hand, $H' = \delta$: for all $\phi \in D(\mathbb{R})$ we have

$$\begin{aligned} (H', \phi) &= -(H, \dot{\phi}) \\ &= -\int_{-\infty}^{\infty} H(x)\dot{\phi}(x)\, dx \\ &= -\int_{0}^{\infty} \dot{\phi}(x)\, dx \\ &= -\phi(x)\big|_{x=0}^{x=\infty} \end{aligned}$$

and since ϕ has compact support

$$= \phi(0) = (\delta, \phi).$$

So, if u is not smooth, the distributional derivative may not coincide with the pointwise derivative.

Example 2. Consider $u = u(x_1, x_2) : \Omega = \mathbb{R}^2 \to \mathbb{R}$,

$$u = x_1 H(x_2) = \begin{cases} x_1 & x_2 \geq 0, \\ 0 & x_2 < 0. \end{cases}$$

By a straightforward computation we find that the distributional derivative $D^{(1,0)}u = H(x_2)$, which coincides with the classical partial derivative $\partial u / \partial x_1$. On the other hand,

$$\begin{aligned} (D^{(0,1)}u, \phi) &= -\iint_{\mathbb{R}^2} u \frac{\partial \phi}{\partial x_2}\, dx_1 dx_2 \\ &= -\int_{-\infty}^{+\infty} x_1 \left(\int_{0}^{+\infty} \frac{\partial \phi}{\partial x_2}\, dx_2 \right) dx_1 \\ &= \int_{-\infty}^{+\infty} x_1 \phi(x_1, 0)\, dx_1 \quad \forall \phi \in \mathcal{D}(\mathbb{R}^2). \end{aligned}$$

Note that $D^{(0,1)}u$ is not a regular distribution. Indeed, assuming the contrary, we obtain $D^{(0,1)}u = 0$ almost everywhere in $\mathbb{R}^2$ by using test functions with support in $\mathbb{R} \times (0, +\infty)$ and in $\mathbb{R} \times (-\infty, 0)$, while $D^{(0,1)}u$ cannot be zero. So $D^{(0,1)}u$ is different from the classical partial derivative $\partial u / \partial x_2$ (which is zero almost everywhere).

Example 3. Let $\Omega = \mathbb{R}^3$ and consider $u = 1/r$, where $r = \|x\|_2 = \sqrt{x_1^2 + x_2^2 + x_3^2}$. We want to calculate

$$\Delta u = \sum_{i=1}^{3} \frac{\partial^2 u}{\partial x_i^2},$$

where the derivatives are in the sense of distributions. The operator Δ is called the Laplace operator (or Laplacian).[4] Note that u is not an element of $L^1_{\text{loc}}(\mathbb{R}^3)$ (because of a singularity at the origin) so that we cannot define the distribution Δu directly. We replace u with

$$u_n = \begin{cases} \frac{1}{r} & r \geq \frac{1}{n}, \\ 0 & r < \frac{1}{n}, \end{cases}$$

which belongs to $L^1_{loc}(\mathbb{R}^3)$, for all $n \in \mathbb{N}$. For any test function $\phi \in \mathcal{D}(\mathbb{R}^3)$ we have

$$(\Delta u_n, \phi) = (u_n, \Delta \phi)$$

and since u_n is regular

$$= \iiint_{r \geq \frac{1}{n}} \frac{\Delta \phi}{r} \, dx.$$

We wish to accept

$$(\Delta u, \phi) = (\Delta \frac{1}{r}, \phi) = \lim_{n \to \infty} \iiint_{r \geq \frac{1}{n}} \frac{\Delta \phi}{r} \, dx$$

as the definition of Δu, but of course we must show that this limit exists. For a fixed $\phi \in \mathcal{D}(\mathbb{R}^3)$, define the spherical shell

$$S_n = \{x \in \mathbb{R}^3; \frac{1}{n} \leq r \leq a\}$$

---

[4]Pierre-Simon Laplace, French mathematician and astronomer, 1749–1827.

where a is large enough that $\operatorname{supp}\phi \subset B(0,a)$. We then use the second Green formula[5] (see, for example, [14, p. 628]) to deduce that

$$\iiint_{r \geq \frac{1}{n}} \left[(\Delta\phi)\frac{1}{r} - \phi\Delta\frac{1}{r} \right] dx = \iiint_{S_n} \left[(\Delta\phi)\frac{1}{r} - \phi\underbrace{\Delta\frac{1}{r}}_{=0} \right] dx$$

and changing the direction of the normal and consequently the sign in the double integral below (we can ignore the outer edge of the shell since ϕ vanishes there),

$$= -\iint_{r=\frac{1}{n}} \left[\frac{\partial\phi}{\partial r}\frac{1}{r} - \phi\frac{\partial}{\partial r}\left(\frac{1}{r}\right) \right] d\sigma$$

and since $r = \frac{1}{n}$ on the edge

$$= -n\iint_{r=\frac{1}{n}} \frac{\partial\phi}{\partial r} \, d\sigma - n^2 \iint_{r=\frac{1}{n}} \phi \, d\sigma$$

and because $\frac{\partial\phi}{\partial r}$ is bounded

$$= -n\mathcal{O}\left(\frac{1}{n^2}\right) - n^2 \iint_{r=\frac{1}{n}} \phi \, d\sigma$$

which as $n \to \infty$ becomes

$$= -4\pi\phi(0)$$
$$= -4\pi(\delta, \phi).$$

So, the limit exists and

$$\lim_{n\to\infty} \iiint_{r \geq \frac{1}{n}} \frac{\Delta\phi}{r} \, dx = -4\pi(\delta, \phi).$$

Hence

$$\left(\Delta\frac{1}{r}, \phi\right) = -4\pi(\delta, \phi) \quad \forall\phi \in C_0^\infty(\mathbb{R}^3),$$

that is to say, $\Delta\frac{1}{r} = -4\pi\delta$.

---

[5]George Green, British mathematical physicist, 1793–1841.

This result can be easily generalized to higher dimensions by showing that for $k \geq 3$,

$$\Delta\left(\frac{1}{r^{k-2}}\right) = -(k-2)a_k\delta$$

in $\mathcal{D}'(\mathbb{R}^k)$, where a_k is the "area" of the unit hyper-sphere in $\mathbb{R}^k$. Also, for $k = 2$ we have

$$\Delta\left(\ln\frac{1}{r}\right) = -2\pi\delta$$

in $\mathcal{D}'(\mathbb{R}^2)$, so that defining for $k \geq 2$

$$E(x) = \begin{cases} -\frac{1}{(k-2)a_k}\frac{1}{r^{k-2}} & k \geq 3, \\ -\frac{1}{2\pi}\ln\frac{1}{r} & k = 2, \end{cases}$$

we have

$$\Delta E = \delta \quad \text{in } \mathcal{D}'(\mathbb{R}^k). \tag{5.3.22}$$

E is called the fundamental solution of the Laplacian Δ. In particular, it can be used to find a solution to the **Poisson equation**[6]

$$\Delta u = f(x), \quad x \in \mathbb{R}^k. \tag{5.3.23}$$

Assume that $f \in L^\infty(\mathbb{R}^k)$ and vanishes almost everywhere outside a compact set. Then the function

$$u(x) = (E * f)(x) = \int_{\mathbb{R}^k} E(x-y)f(y)\,dy \tag{5.3.24}$$

is well defined (since E is locally summable) and satisfies Eq. (5.3.23). Indeed, we first notice that for all $y \in \mathbb{R}^n$ and $\phi \in C_0^\infty(\mathbb{R}^k)$

$$\int_{\mathbb{R}^k} E(x-y)\Delta\phi(x)\,dx = (E(x-y), \Delta\phi(x))$$

$$= (\Delta_x E(x-y), \phi(x))$$

and taking into account (5.3.22)

$$= (\delta(x-y), \phi(x))$$
$$= (\delta(x), \phi(x+y))$$
$$= \phi(y).$$

---

[6]Siméon Denis Poisson, French mathematician, engineer, and physicist, 1781–1840.

Now,

$$(\Delta u, \phi) = (u, \Delta \phi)$$

$$= \int_{\mathbb{R}^k} u(x) \Delta \phi(x)\, dx$$

$$= \int_{\mathbb{R}^k} \left(\int_{\mathbb{R}^k} E(x-y) f(y)\, dy \right) \Delta \phi(x)\, dx$$

$$= \int_{\mathbb{R}^k} f(y) \left(\int_{\mathbb{R}^k} E(x-y) \Delta \phi(x)\, dx \right) dy$$

and using the last relation above

$$= \int_{\mathbb{R}^k} f(y) \phi(y)\, dy$$

$$= (f, \phi) \quad \forall \phi \in \mathcal{D}(\mathbb{R}^k),$$

which implies $\Delta u = f$, as claimed.

Remark 5.11. We point out (without proof) the following result known as Weyl's regularity lemma[7] (see, e.g., [47]): *if* $\emptyset \neq \Omega \subset \mathbb{R}^k$ *is open,* $f \in C^\infty(\Omega)$, $u \in \mathcal{D}'(\Omega)$ *and* $\Delta u = f$ *in* $\mathcal{D}'(\Omega)$, *then* $u \in C^\infty(\Omega)$.

The following result says that differentiation in $\mathcal{D}'(\Omega)$ is a continuous operation.

Proposition 5.12. *Suppose that* $\emptyset \neq \Omega \subset \mathbb{R}^k$ *is open. If* $u_n \to u$ *in* $\mathcal{D}'(\Omega)$, *then* $\frac{\partial u_n}{\partial x_i} \to \frac{\partial u}{\partial x_i}$ *in* $\mathcal{D}'(\Omega)$ *for all* $i = 1, \ldots, k$.

Proof. Let $u_n \to u$ in $\mathcal{D}'(\Omega)$. We have for all $\phi \in \mathcal{D}(\Omega)$

$$\left(\frac{\partial u_n}{\partial x_i}, \phi \right) = -\left(u_n, \frac{\partial \phi}{\partial x_i} \right)$$

and since $\frac{\partial \phi}{\partial x_i}$ is a test function, as $n \to \infty$ the right-hand side converges to

$$= -\left(u, \frac{\partial \phi}{\partial x_i} \right)$$

$$= \left(\frac{\partial u}{\partial x_i}, \phi \right). \qquad \square$$

---

[7]Hermann Weyl, German mathematician, theoretical physicist, and philosopher, 1885–1955.

Remark 5.13. As an immediate consequence of Proposition 5.12, if $u_n \to u$ in $\mathcal{D}'(\Omega)$, then $D^\alpha u_n \to D^\alpha u$ in $\mathcal{D}'(\Omega)$ for all $\alpha = (\alpha_1, \ldots, \alpha_k) \in \mathbb{N}_0^k$ with $|\alpha| > 0$.

Series in $\mathcal{D}'(\Omega)$

Suppose $(u_n)_{n \in \mathbb{N}}$ is a sequence in $\mathcal{D}'(\Omega)$. Then we can associate with this sequence the series

$$u_1 + u_2 + \cdots + u_n + \cdots$$

and say that it converges in $\mathcal{D}'(\Omega)$ if the sequence of partial sums

$$s_n = u_1 + \cdots + u_n$$

converges, $s_n \to u$ in $\mathcal{D}'(\Omega)$, and write

$$u_1 + u_2 + \cdots + u_n + \cdots = u.$$

By Remark 5.13, $s_n \to u$ implies $D^\alpha s_n \to D^\alpha u$ in $\mathcal{D}'(\Omega)$ for all α, hence we can differentiate the series term by term as many times as we wish, i.e.,

$$D^\alpha u_1 + D^\alpha u_2 + \cdots + D^\alpha u_n + \cdots = D^\alpha u$$

in $\mathcal{D}'(\Omega)$. This is not the case in classical analysis. For example, with $\Omega = \mathbb{R}$, $u_n(x) = \frac{1}{n} \sin(nx)$ converges uniformly to 0 as $n \to \infty$ (and uniform convergence implies convergence in $\mathcal{D}'(\Omega)$), but $u_n'(x) = \cos(nx)$ which does not converge, even pointwise. However, it does converge in $\mathcal{D}'(\mathbb{R})$. In fact, $u_n^{(j)} \to 0$ as $n \to \infty$ in $\mathcal{D}'(\mathbb{R})$ for all $j = 1, 2, \ldots$.

5.3.4 Differential Equations for Distributions

Consider $\Omega = \mathbb{R}$, $u, b \in \mathcal{D}'(\mathbb{R})$ and smooth functions $a_1, a_2, \ldots, a_n \in C^\infty(\mathbb{R})$. Then, if $u^{(j)}$ indicates the j-th derivative of u in $\mathcal{D}'(\mathbb{R})$, the differential equations

$$u^{(n)} + a_1 u^{(n-1)} + \cdots + a_{n-1} u^{(1)} + a_n u = 0 \qquad (E_0)$$

$$u^{(n)} + a_1 u^{(n-1)} + \cdots + a_{n-1} u^{(1)} + a_n u = b \qquad (E)$$

make sense. Classically, there are nice solutions u to (E_0) and they are solutions in the sense of distributions as well. In fact, there are no other solutions to (E_0) in $\mathcal{D}'(\mathbb{R})$ as long as $a_i \in C^\infty(\mathbb{R})$, as proven below.

The equation $u' = 0$ has constant solutions C in $\mathcal{D}'(\mathbb{R})$ since for all $\phi \in \mathcal{D}(\mathbb{R})$

$$(C', \phi) = -(C, \phi') = -C \int_{-\infty}^{\infty} \phi'\, dt = 0,$$

But, are constant functions the only solutions to the equation $u' = 0$ in the sense of distributions? We answer this question in the following way. If $u \in \mathcal{D}'(\mathbb{R})$ and $u' = 0$ in $\mathcal{D}'(\mathbb{R})$, we have

$$0 = (u', \phi) = -(u, \phi') \quad \forall \phi \in \mathcal{D}(\mathbb{R}). \tag{5.3.25}$$

Given $\phi \in \mathcal{D}(\mathbb{R})$ define

$$\psi(t) = \phi(t) - \omega(t) \int_{-\infty}^{\infty} \phi(s)\, ds$$

for all $t \in \mathbb{R}$ with $\omega = \omega(t)$ defined as in (5.1.1) (where $k = 1$). Note that $\psi \in \mathcal{D}(\mathbb{R})$ and $\int_{-\infty}^{+\infty} \psi\, dt = 0$. Define

$$\phi_1(t) = \int_{-\infty}^{t} \psi(s)\, ds, \quad t \in \mathbb{R},$$

and notice that $\phi_1 \in \mathcal{D}(\mathbb{R})$ and $\phi_1' = \psi$. Now for all $\phi \in \mathcal{D}(\mathbb{R})$

$$(u, \phi) = (u, \psi) + \left(\int_{-\infty}^{\infty} \phi(s)\, ds \right) \underbrace{(u, \omega)}_{\text{constant}}$$

$$= (u, \phi_1') + \int_{-\infty}^{\infty} C\phi\, ds$$

and according to (5.3.25)

$$= (C, \phi),$$

thus $u = C$. Therefore, *any distributional solution of the equation $u' = 0$ is a constant distribution* (i.e., a distribution generated by a constant function).

Now consider the linear differential system

$$\begin{cases} u_1' = a_{11}u_1 + a_{12}u_2 + \cdots + a_{1n}u_n \\ u_2' = a_{21}u_1 + a_{22}u_2 + \cdots + a_{2n}u_n \\ \quad \vdots \\ u_n' = a_{n1}u_n + a_{n2}u_2 + \cdots + a_{nn}u_n \end{cases} \tag{5.3.26}$$

where the $a_{ij} \in C^\infty(\mathbb{R})$. Denoting by $A(x)$ the matrix $\big(a_{ij}(x)\big)$ and by u the column vector $(u_1, \ldots, u_n)^T$, we can rewrite (5.3.26) as

$$u' = Au. \tag{5.3.27}$$

Let $X = X(t)$ be a fundamental matrix of the system (5.3.26). We know from the classical theory of linear differential systems (see, e.g., [8, 11]) that X is invertible and $X' = AX$ for all $t \in \mathbb{R}$. Consider the transformation $u = Xz$ then in the sense of distributions

$$\begin{aligned} u' &= X'z + Xz' \\ &= AXz + Xz' \end{aligned}$$

and by (5.3.27)

$$= AXz.$$

Hence $Xz' = 0$, so having in mind the fact that X is invertible, we deduce that $z' = 0$. We have denoted by u' and z' the column vectors whose components are the distributional derivatives of $u_1, \ldots, u_n$ and $z_1, \ldots, z_n$, respectively. As $z' = 0$, z must be a constant vector $z = c \in \mathbb{R}^n$ and we find $u = Xc$. Therefore, there are no solutions in $\mathcal{D}'(\mathbb{R}^n)$ to system (5.3.26) other than the classical ones.

Finally, consider the homogeneous Eq. (E_0). Since it can be written in the vector form (5.3.27) which has only classical solutions, so does (E_0). The non-homogeneous case (E) has a general solution which is obtained by adding to the general solution of (E_0) a particular solution to (E) in the sense of distributions. Indeed, if $u_p \in \mathcal{D}'(\mathbb{R})$ is such a particular solution, and u is an arbitrary solution in $\mathcal{D}'(\mathbb{R})$ of (E), then $u - u_p$ is a (classical) solution of (E_0), hence a linear combination of the functions belonging to the fundamental system of solutions.

Example. Consider in $\mathcal{D}'(\mathbb{R})$ the differential equation

$$u'' - 2u' + u = 2\delta(t - 1), \quad t \in \mathbb{R}. \tag{5.3.28}$$

In order to solve this equation, we first notice that if u is a distributional solution of it, then

$$u'' - 2u' + u = 0 \quad \text{in } \mathcal{D}'(\Omega_i), \ i = 1, 2,$$

where $\Omega_1 = (-\infty, 1)$, $\Omega_2 = (1, +\infty)$. Therefore, u is a classical solution of the corresponding homogeneous equation within each of these two intervals, i.e., u is a function (regular distribution) of the form

$$u(t) = \begin{cases} (c_1 t + c_2)e^t, & t \in (-\infty, 1), \\ (c_3 t + c_4)e^t, & t \in (1, +\infty), \end{cases}$$

where c_1, c_2, c_3, c_4 are real constants. Not all these functions u are solutions of the given differential equation. The fact that such a function u is a solution means

$$\int_{-\infty}^{1} \left(u\phi'' + 2u\phi' + u\phi \right) dt$$

$$+ \int_{1}^{\infty} \left(u\phi'' + 2u\phi' + u\phi \right) dt = 2\phi(1) \quad \forall \phi \in \mathcal{D}(\mathbb{R}).$$

Integrating by parts and bearing in mind that u is a classical solution of the homogeneous equation in $(-\infty, 1)$ and also in $(1, \infty)$, plus the fact that $\phi(1)$ and $\phi'(1)$ can be any real numbers, we obtain

$$u(1+0) = u(1-0), \quad \dot{u}(1+0) - \dot{u}(1-0) = 2,$$

so

$$c_3 = c_1 + 2e^{-1}, \quad c_4 = c_2 - 2e^{-1}.$$

Thus the general solution of the given equation is

$$u(t) = (c_1 t + c_2)e^t + \begin{cases} 2(t-1)e^{t-1}, & t > 1, \\ 0, & t < 1. \end{cases}$$

i.e., $u(t) = (c_1 t + c_2)e^t + 2(t-1)e^{t-1}H(t-1)$.

It is worth pointing out that there is no classical solution of the given equation, more precisely there is a jump at $t = 1$ in the first derivative of any solution (which is caused by the Dirac distribution in the right-hand side of the equation).

Remark 5.14. Note that in equation (E) above the coefficient of $u^{(n)}$ is 1, i.e., we do not have any singularity in the coefficient of the leading term. Otherwise, some difficulties may occur. For example, consider the simple equation

$$tu' = 0 \quad \text{in } \mathcal{D}'(\mathbb{R}). \tag{5.3.29}$$

If u is a distributional solution of (5.3.29), then it must be constant in $(-\infty, 0)$ and in $(0, \infty)$ as well. So the general solution is

$$u(t) = c_1 + c_2 H(t), \quad t \in \mathbb{R},$$

where c_1, c_2 are real constants. Note that in this case there are two independent solutions (e.g., $u_1(t) = 1$, $u_2(t) = H(t)$), even if the given equation is of order one.

Now, in order to illustrate the need for distributions in solving problems associated with partial differential equations, consider the following examples:

Example 1.
Consider the equation of an infinite vibrating string with no external force acting on it

$$u_{tt} - u_{xx} = 0, \quad (t, x) \in \mathbb{R}^2, \tag{5.3.30}$$

with some conditions at $t = 0$, say,

$$u(0, x) = \psi(x), \quad u_t(0, x) = 0, \quad x \in \mathbb{R}, \tag{5.3.31}$$

where

$$u_t := \frac{\partial u}{\partial t}, \quad u_{tt} := \frac{\partial^2 u}{\partial t^2}, \quad u_{xx} := \frac{\partial^2 u}{\partial x^2}.$$

First assume that $\psi \in C^2(\mathbb{R})$. Recall that using the change of variables

$$\alpha = x + t, \quad \beta = x - t,$$

Eq. (5.3.30) can be reduced to the equation

$$u_{\alpha\beta} = 0.$$

So it is easily seen that any solution of the Eq. (5.3.30) has the form $u = g(x+t) + h(x-t)$, and so applying (5.3.31), we find the D'Alembert formula[8]

$$u = \frac{1}{2}[\psi(x + t) + \psi(x - t)] \quad \text{(D'Alembert's formula)}. \tag{5.3.32}$$

---

[8] Jean-Baptiste le Rond d'Alembert, French mathematician, mechanician, physicist, philosopher, and music theorist, 1717–1783.

Clearly, u is a C^2 function. It is the unique classical solution of problem (5.3.30), (5.3.31).

On the other hand, assuming that $\psi \in C^1(\mathbb{R})$, then u given by (5.3.32) is no longer a classical solution of Eq. (5.3.30). However, this u still satisfies conditions (5.3.31).

Now, assume that $\psi \in C(\mathbb{R})$. In this case, the function u given by (5.3.32) only satisfies classically the condition $u(0, x) = \psi(x)$, $x \in \mathbb{R}$. However, it should be some relation between this u and problem (5.3.30), (5.3.31). Indeed, we can show that this u satisfies (5.3.30) and the condition $u_t(0, x) = 0$ $(x \in \mathbb{R})$ in a weak sense, that is in the sense of distributions.

If ψ', ψ'' denote the first and second derivative of ψ in $\mathcal{D}'(\mathbb{R})$, then it is easily seen that

$$D^{(1,0)}\psi(x+t) = D^{(0,1)}\psi(x+t) = \psi'(x+t) \quad \text{in } \mathcal{D}'(\mathbb{R}^2),$$
$$D^{(2,0)}\psi(x+t) = D^{(0,2)}\psi(x+t) = \psi''(x+t) \quad \text{in } \mathcal{D}'(\mathbb{R}^2).$$

Similarly,

$$D^{(0,1)}\psi(x-t) \;\; = \psi'(x-t) = -D^{(1,0)}\psi(x-t) \quad \text{in } \mathcal{D}'(\mathbb{R}^2),$$
$$D^{(2,0)}\psi(x-t) \;\; = D^{(0,2)}\psi(x-t) = \psi''(x-t) \quad \text{in } \mathcal{D}'(\mathbb{R}^2).$$

Consequently, for all $\phi \in \mathcal{D}(\mathbb{R}^2)$, we have

$$\begin{aligned}
\left(D^{(2,0)}u(t,x), \phi(t,x)\right) &= \frac{1}{2}\left(\psi''(x+t) + \psi''(x-t), \phi(t,x)\right) \\
&= \left(D^{(0,2)}u(t,x), \phi(t,x)\right),
\end{aligned}$$

which shows that u given by (5.3.32) satisfies Eq. (5.3.30) in the sense of distributions:

$$D^{(2,0)}u - D^{(0,2)}u = 0 \quad \text{in } \mathcal{D}'(\mathbb{R}^2).$$

We also have

$$\begin{aligned}
D^{(1,0)}u(t,x) &= \frac{1}{2}\left[\psi'(x+t) - \psi'(x-t)\right] \\
&= 0, \quad \text{if } t = 0.
\end{aligned}$$

Example 2.

Here we discuss the boundary controllability of the 1-dimensional wave equation describing the vibrations of a finite string. Specifically, let us consider the following initial-boundary value problem:

$$\begin{cases} u_{tt} - u_{xx} = 0, & 0 < x < 1,\ t > 0, \\ u(t,0) = 0,\ u(t,1) = f(t), & t > 0, \\ u(0,x) = u^0(x),\ u_t(0,x) = u^1, & 0 < x < 1, \end{cases} \tag{5.3.33}$$

where $u^0,\ u^1 \in L^1(0,1)$.

We shall prove that

$$\exists T > 0,\ \forall (u^0,u^1) \in L^1(0,1)^2,\ \exists f \in L^1_{\text{loc}}[0,\infty),\ \forall t > T,\ u = 0,$$

where u is the corresponding solution of problem (5.3.33). In fact, we shall see that there exists a lowest time instant T with this property, precisely $T = 2$. Obviously, any $T > 2$ satisfies the same property. This result is in accordance with similar results previously obtained by other authors by using different arguments (see, e.g., [34, p. 57]). Our direct approach is more advantageous since it provides the solution u (in a generalized sense, under weak assumptions on the data) as a function of u^0, u^1, and f and allows us to determine the minimal time interval $(0,2)$ and an explicit control function f (depending on u^0 and u^1) which steers the solution u to zero. It may happen that this direct approach is known, but we could not find anything about it in the literature. Nevertheless, we present it here as a nice application and do not claim originality.

Existence of Solutions to Problem (5.3.33)

Denote $R = \{(t,x);\ t \geq 0,\ 0 \leq x \leq 1\}$. Consider in the first instance that $u = u(t,x)$ is a classical solution of problem (5.3.33) corresponding to regular u^0, u^1, and f. Obviously, the solution of the above wave equation has the general form

$$u(t,x) = g(x+t) + h(x-t). \tag{5.3.34}$$

From the initial and boundary conditions we can determine $g(x+t)$ and $h(x-t)$ (hence $u = u(t,x)$) within different subsets (triangles or squares) of R, as follows:

From the initial conditions we get

$$g(x) + h(x) = u^0(x),\ g(x) - h(x) = \int_0^x u^1 + c,\ 0 < x < 1,$$

where c is a real constant, hence

$$\begin{cases} g(x) = \frac{1}{2}\left(u^0(x) + \int_0^x u^1 + c\right), & 0 < x < 1, \\[2mm] h(x) = \frac{1}{2}\left(u^0(x) - \int_0^x u^1 - c\right), & 0 < x < 1. \end{cases} \tag{5.3.35}$$

From the boundary conditions we obtain

$$g(t) + h(-t) = 0, \ t > 0, \tag{5.3.36}$$

and

$$g(1+t) + h(1-t) = f(t), \ t > 0. \tag{5.3.37}$$

Now (5.3.36) yields (see also (5.3.35))

$$\begin{aligned} h(-t) &= -g(t) \\ &= -\frac{1}{2}\left(u^0(t) + \int_0^t u^1 + c\right), \ 0 < t < 1. \end{aligned} \tag{5.3.38}$$

From (5.3.37) we derive (see also (5.3.35))

$$\begin{aligned} g(1+t) &= f(t) - h(1-t) \\ &= f(t) - \frac{1}{2}\left(u^0(1-t) - \int_0^{1-t} u^1 - c\right), \ 0 < t < 1. \end{aligned} \tag{5.3.39}$$

We also have for $0 < t < 1$

$$\begin{aligned} h(-t-1) &= -g(1+t) \\ &= -f(t) + \frac{1}{2}\left(u^0(1-t) - \int_0^{1-t} u^1 - c\right), \end{aligned} \tag{5.3.40}$$

$$\begin{aligned} g(2+t) &= f(1+t) - h(-t) \\ &= f(1+t) + \frac{1}{2}\left(u^0(t) + \int_0^t u^1 + c\right), \end{aligned} \tag{5.3.41}$$

$$\begin{aligned} h(-t-2) &= -g(t+2) \\ &= -f(1+t) - \frac{1}{2}\left(u^0(t) + \int_0^t u^1 + c\right), \end{aligned} \tag{5.3.42}$$

$$g(3+t) = f(2+t) - h(-t-1)$$

$$= f(2+t) + f(t) - \frac{1}{2}\left(u^0(1-t) - \int_0^{1-t} u^1 - c\right), \quad (5.3.43)$$

and so on. By using the above formulas we can determine $u = u(t,x)$ in R. We decompose R into triangles and squares as in Fig. 5.1.

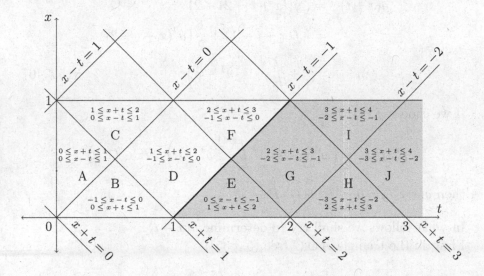

Figure 5.1: Regions of the plane

We first determine $g(x+t)$ in the triangle $A \cup B$ (i.e., the intersection of R and the strip $\{0 \le x+t \le 1\}$) (see (5.3.35)):

$$g(x+t) = \frac{1}{2}\left(u^0(x+t) + \int_0^{x+t} u^1 + c\right). \quad (5.3.44)$$

Now let us determine $g(x+t)$ in the parallelogram $C \cup D \cup E$ (i.e., the intersection of R and the strip $\{1 \le x+t \le 2\}$) (see (5.3.39)):

$$\begin{aligned}
g(x+t) &= g((x+t-1)+1) \\
&= f(x+t-1) - \frac{1}{2}\left(u^0(2-x-t)\right. \\
&\quad \left. - \int_0^{2-x-t} u^1 - c\right). \quad (5.3.45)
\end{aligned}$$

Note that choosing $f : (0, 1) \to \mathbb{R}$,

$$f(y) = \frac{1}{2}\left(u^0(1 - y) - \int_0^{1-y} u^1 - c\right) \quad \forall y \in (0, 1),$$

implies $g(x + t) = 0$ in $C \cup D \cup E$.

For $(t, x) \in F \cup G \cup H$ (i.e., $2 \le x + t \le 3$, $0 \le x \le 1$) we have (see (5.3.41))

$$
\begin{aligned}
g(x + t) &= g\big((x + t - 2) + 2\big) \\
&= f(x + t - 1) + \frac{1}{2}\big(u^0(x + t - 2) \\
&\quad + \int_0^{x+t-2} u^1 + c\big).
\end{aligned}
$$
(5.3.46)

If we choose $f : (1, 2) \to \mathbb{R}$,

$$f(z) = -\frac{1}{2}\left(u^0(z - 1) + \int_0^{z-1} u^1 + c\right) \quad \forall z \in (1, 2),$$

then $g(x + t) = 0$ in $F \cup G \cup H$.

In what follows we shall try to determine $h(x - t)$.
First, in the triangle $A \cup C$ (see (5.3.35))

$$h(x - t) = \frac{1}{2}\left(u^0(x - t) - \int_0^{x-t} u^1 - c\right).$$
(5.3.47)

Next, for $(t, x) \in B \cup D \cup F$ we have

$$
\begin{aligned}
h(x - t) &= h\big(-(t - x)\big) \\
&= -g(t - x) \\
&= -\frac{1}{2}\left(u^0(t - x) + \int_0^{t-x} u^1 + c\right).
\end{aligned}
$$
(5.3.48)

For $(t, x) \in E \cup G \cup I$ we have

$$
\begin{aligned}
h(x - t) &= h\big(-(t - x - 1) - 1\big) \\
&= -f(t - x - 1) + \frac{1}{2}\big(u^0(2 + x - t) \\
&\quad - \int_0^{2+x-t} u^1 - c\big).
\end{aligned}
$$
(5.3.49)

Observe that if $f : (0,1) \to \mathbb{R}$ is the function defined above then $h(x - t) = 0$ in $E \cup G \cup I$.

For $(t, x) \in H \cup J$ we have

$$
\begin{aligned}
h(x - t) &= h\big(-(t - x - 2) - 2\big) \\
&= -f(t - x - 1) - \frac{1}{2}\big(u^0(t - x - 2) \\
&\quad + \int_0^{t-x-2} u^1 + c\big).
\end{aligned}
\tag{5.3.50}
$$

With $f : (1,2) \to \mathbb{R}$ as defined before, we have $h(x - t) = 0$ in $H \cup J$. In fact all the above computations are valid for u^0, $u^1 \in L^1(0,1)$ and $f \in L^1_{\text{loc}}[0, \infty)$.

These calculations lead to the following theorem:

Theorem 5.15. *For any u^0, $u^1 \in L^1(0,1)$ and $f \in L^1_{\text{loc}}[0, \infty)$ (i.e., f is Lebesgue summable on $(0, m)$ for all $m > 0$) problem (5.3.33) has a unique weak solution u. If $u^0 \in C[0,1]$, $u^1 \in L^1(0,1)$, $f \in C[0, \infty)$, and the following compatibility conditions are satisfied:*

$$
u^0(0) = 0, \quad u^0(1) = f(0),
\tag{5.3.51}
$$

then $u = u(t, x) \in C([0, \infty) \times [0, 1])$.

Proof. Using the above computations we construct $u(t, x) = g(x+t) + h(x - t)$. Obviously, u satisfies the wave equation in the distribution sense on the interior of each of the sets A, B, C, D, E, F, G, and so on. In this sense, u is a weak solution of the wave equation. By construction, the initial and boundary conditions are also satisfied. It is easily seen that the constant c disappears when constructing $u = g(x+t) + h(x - t)$ in A, B, C, ... so the solution u is unique. If $u^0 \in C[0,1]$, $u^1 \in L^1(0,1)$, $f \in C[0, \infty)$ and u^0, f satisfy (5.3.51), then u is continuous on $[0, \infty) \times [0, 1]$. It suffices to observe that u is continuous on the characteristic lines $\{x - t = i\}$, $i = 0, 1, \ldots$, and $\{x + t = -j\}$, $j = 1, 2 \ldots$, restricted to the infinite strip R. $\qquad\square$

Remark 5.16. For higher regularity of u one needs to assume more regularity of the data and additional compatibility conditions.

Exact Boundary Controllability

A careful analysis of the above computations shows that there are pairs (u^0, u^1) for which there are no functions $f : (0, T) \to \mathbb{R}$, $T < 2$, making $u = 0$ in the trapezoid $A \cup B \cup C \cup D \cup F$. In other words, the waves cannot be controlled in $[0, T]$ if $T < 2$. On the other hand, we have

Theorem 5.17. *For any pair* $(u^0, u^1) \in L^1(0,1) \times L^1(0,1)$ *there exists a control function* $f : (0, +\infty) \to \mathbb{R}$ *defined by*

$$f(y) = \begin{cases} \frac{1}{2}\left(u^0(1-y) + \int_0^{1-y} u^1 - c \right), & y \in (0,1), \\[2mm] -\frac{1}{2}\left(u^0(y-1) + \int_0^{y-1} u^1 + c \right), & y \in (1,2), \\[2mm] 0, & y > 2, \end{cases} \qquad (5.3.52)$$

with $c = -u^0(1) - \int_0^1 u^1$, *which makes* $u = 0$ *in the infinite trapezoid* $\{x \leq t - 1\} \cap \{0 < x < 1\}$.

Proof. The proof follows easily from the computations performed above, including the remarks on the regions where $g(x+t) = 0$ or $h(x-t) = 0$ (based on the fact that f is that given in (5.3.52)). Of course, u vanishes in $\{x < t - 1\} \cap \{0 < x < 1\}$ since $f(t) = 0$ for $t > 2$. $\qquad \square$

Remark 5.18. Let us emphasize that, if f is chosen as in (5.3.52), then the corresponding (unique) solution u vanishes starting from the line segment defined as the intersection of R and the characteristic line $\{x - t = -1\}$ and remains zero everywhere on the right side of that segment, which can be interpreted as a threshold. So the waves can be controlled in the minimal time interval $(0, 2)$ and in fact in any interval $(0, T)$ with $T \geq 2$.

Remark 5.19. While the solution u is unique in Theorem 5.15, the control function f is not since the constant c in (5.3.52) can be chosen arbitrarily. Indeed, the restriction of f to the interval $(0, 2)$ is unique up to an additive constant, as follows from the computations above. We chose $c = -u^0(1) - \int_0^1 u^1$ in Theorem 5.17 in order to obtain a continuous control function f.

5.4 Sobolev Spaces

Let $\emptyset \neq \Omega \subset \mathbb{R}^k$ be an open set. For $m \in \mathbb{N}$, $1 \leq p \leq \infty$ define the **Sobolev space** of order m to be[9]

$$W^{m,p}(\Omega) = \{u \in L^p(\Omega);\, D^\alpha u \in L^p(\Omega)\ \forall \alpha \in \mathbb{N}_0^k,\, 0 < |\alpha| \leq m\},$$

where the derivatives $D^\alpha u$ are considered in the sense of distributions. Obviously, $W^{m,p}(\Omega)$ is a linear space with respect to the usual operations of addition and scalar multiplication. In particular,

$$W^{1,p}(\Omega) = \left\{ u \in L^p(\Omega);\, \frac{\partial u}{\partial x_i} \in L^p(\Omega)\ \forall i = 1, \ldots, k \right\},$$

where $\frac{\partial u}{\partial x_i}$ is the partial derivative of u with respect to x_i in the sense of distributions.

Theorem 5.20. *For all $m \in \mathbb{N}$, $1 \leq p \leq \infty$, $W^{m,p}(\Omega)$ is a real Banach space with respect to the norm*

$$\|u\|_{m,p} = \left(\sum_{|\alpha| \leq m} \|D^\alpha u\|_{L^p(\Omega)}^p \right)^{1/p}, \qquad 1 \leq p < \infty,$$

$$\|u\|_{m,\infty} = \max_{|\alpha| \leq m} \|D^\alpha u\|_{L^\infty(\Omega)}, \qquad p = \infty.$$

Proof. Obviously, $\|\cdot\|_{m,p}$ is a norm for all $1 \leq p \leq \infty$.
Let $(u_n)_{n \in \mathbb{N}}$ be a Cauchy sequence in $W^{m,p}(\Omega)$, i.e., for all $\varepsilon > 0$ there exists $N = N(\varepsilon) \in \mathbb{N}$ such that

$$\|u_n - u_m\|_{m,p} < \varepsilon \ \text{ for all } n, m > N.$$

It follows that $(D^\alpha u_n)$ is Cauchy in $L^p(\Omega)$ for all $\alpha \in \mathbb{N}_0^k$, $|\alpha| \leq m$. Since $L^p(\Omega)$ is a Banach space (with respect to $\|\cdot\|_{L^p(\Omega)}$), there exist $u, u_\alpha \in L^p(\Omega)$ such that

$$u_n \to u, \quad D^\alpha u_n \to u_\alpha \ \text{ in } L^p(\Omega)\ \forall \alpha \in \mathbb{N}_0^k,\, 0 < |\alpha| \leq m. \qquad (5.4.53)$$

On the other hand, we have the following.
Claim: In general, if $v_n \to v$ in $L^p(\Omega)$, $1 \leq p \leq \infty$, then $v_n \to v$ in $\mathcal{D}'(\Omega)$.

---

[9]Sergei, L. Sobolev, Russian mathematician, 1908–1989.

Indeed, for all $\phi \in \mathcal{D}(\Omega)$ we have

$$|(v_n - v, \phi)| = \left| \int_\Omega (v_n - v)\phi \, dx \right|$$

which for $p = 1$

$$\leq \|v_n - v\|_{L^1(\Omega)} \sup |\phi|$$

and for $1 < p < \infty$ by Hölder

$$\leq \|v_n - v\|_{L^p(\Omega)} \|\phi\|_{L^{p'}(\Omega)},$$

so $v_n \to v$ in $\mathcal{D}'(\Omega)$, and similarly for $p = \infty$.

By the above claim, it follows that the convergences in (5.4.53) also hold in $\mathcal{D}'(\Omega)$. Since D^α is a closed operation in $\mathcal{D}'(\Omega)$, it follows that

$$u_\alpha = D^\alpha u \quad \forall \alpha \in \mathbb{N}_0^k, \ 0 < |\alpha| \leq m.$$

Therefore, $u \in W^{m,p}(\Omega)$ and $\|u_n - u\|_{m,p} \to 0$ as $n \to \infty$. $\qquad\square$

Now, for $m \in \mathbb{N}$, $1 \leq p \leq \infty$, denote as usual by $W_0^{m,p}(\Omega)$ the closure of $C_0^\infty(\Omega)$ in $(W^{m,p}(\Omega), \|\cdot\|_{m,p})$. Obviously, $W_0^{m,p}(\Omega)$ is a Banach space with respect to $\|\cdot\|_{m,p}$ for all $m \in \mathbb{N}$, $1 \leq p \leq \infty$.

For $p = 2$ there are specific notations

$$H^m(\Omega) := W^{m,2}(\Omega), \quad H_0^m(\Omega) := W_0^{m,2}(\Omega),$$

and corresponding norm $\|\cdot\|_m := \|\cdot\|_{m,2}$. These are Hilbert spaces with the scalar product

$$(u, v)_m := \sum_{|\alpha| \leq m} \left(D^\alpha u, D^\alpha v \right)_{L^2(\Omega)}.$$

(A Banach space $(X, \|\cdot\|)$ is called Hilbert if $\|\cdot\|$ is given by a scalar product $(\cdot, \cdot)$, i.e., $\|x\| = \sqrt{(x,x)}$, $x \in X$; see Chap. 6 for more information on Hilbert spaces).

In particular, the scalar product of $H^1(\Omega)$ (and of $H_0^1(\Omega)$ as well) is

$$
\begin{aligned}
(u, v)_1 &= (u, v)_{L^2(\Omega)} + \sum_{j=1}^k \left(\frac{\partial u}{\partial x_j}, \frac{\partial v}{\partial x_j} \right)_{L^2(\Omega)} \\
&= \int_\Omega uv \, dx + \sum_{j=1}^k \int_\Omega \frac{\partial u}{\partial x_j} \frac{\partial v}{\partial x_j} \, dx,
\end{aligned}
$$

where the derivatives are in the sense of distributions, so that

$$\|u\|_1^2 = \|u\|_{L^2(\Omega)}^2 + \sum_{j=1}^{k} \int_\Omega \left(\frac{\partial u}{\partial x_j} \right)^2 dx\,.$$

It is well known that $W^{m,p}(\Omega)$ is separable if $1 \leq p < \infty$. See, e.g., [1, p. 47], where further results on Sobolev spaces can be found. See also [2, 6, 14].

Let us recall (without proof) the following approximation result (cf., e.g., [14, p. 252]).

Theorem 5.21. *Let $\emptyset \neq \Omega \subset \mathbb{R}^k$ be an open bounded set of class C^1, and let $1 \leq p < \infty$. Then for every $u \in W^{m,p}(\Omega)$ there exists a sequence (u_n) in $C^\infty(\overline{\Omega})$ such that $u_n \to u$ in $W^{m,p}(\Omega)$.*

For the definition of a C^1 open set see [6, p. 272]. Generally, in applications $\partial\Omega$ is smooth enough and consequently Ω is of class C^1.

Notice also that $W_0^{m,p}(\mathbb{R}^k) = W^{m,p}(\mathbb{R}^k)$, i.e., $C_0^\infty(\mathbb{R}^k)$ is dense in $W^{m,p}(\mathbb{R}^k)$ (see, e.g., [1, p. 56]). But, in general, $W_0^{m,p}(\Omega)$ is a proper subspace of $W^{m,p}(\Omega)$.

Let us also state (without proof) a unified version of some results due to Sobolev, Rellich & Kondrashov[10] (see, e.g., [2, pp. 3–4]).

Theorem 5.22. *If $\emptyset \neq \Omega \subset \mathbb{R}^k$ is an open set of class C^1 and $1 \leq p < \infty$, then there are the continuous embeddings*

(a) *if $m < \frac{k}{p}$, then $W^{m,p}(\Omega) \to L^q(\Omega)$ $\forall q \in [p, p^*]$, where $p^* = \frac{kp}{k-mp}$;*

(b) *if $m = \frac{k}{p}$, then $W^{m,p}(\Omega) \to L^q(\Omega)$ $\forall q \in [p, \infty)$;*

(c) *if $m > \frac{k}{p}$, then $W^{m,p}(\Omega) \to C^{0,\alpha}(\overline{\Omega})$ (which is the space of Hölder continuous functions defined on $\overline{\Omega}$ with exponent $\alpha \in (0,1)$, and with $\alpha = 1$ if $m - \frac{k}{p} > 1$).*

If, in addition, Ω is bounded, then all the above embeddings are compact except for the case $q = p^*$ in (a), and furthermore, if we replace $W^{m,p}(\Omega)$ by $W_0^{m,p}(\Omega)$, then all these embeddings (including the compact ones) hold without any regularity condition on $\partial\Omega$.

---

[10]Vladimir I. Kondrashov, Russian mathematician, 1909–1971; Franz Relich, Austrian-German mathematician, 1906–1955.

The above embeddings are the natural linear injective maps between the corresponding spaces. In particular, the embedding (c) above associates with every $u \in W^{m,p}(\Omega)$ (which is a class of functions with respect to the a.e. equality) its continuous representative. Continuity and compactness of the above embeddings are understood in the usual sense.

We continue with a few words on the *trace* of functions from $W^{m,p}(\Omega)$ on the boundary $\partial\Omega$ of Ω. The concept of trace is important for applications to boundary value problems for partial differential equations. We restrict our attention to $W^{1,p}(\Omega)$, $1 \le p < \infty$, since this case is sufficient for the applications that will be discussed later.

Clearly, for a function $u \in C(\overline{\Omega})$ its restriction to $\partial\Omega$, denoted $u|_{\partial\Omega}$, is well defined. But if $u \in W^{1,p}(\Omega)$ then u is only defined a.e. on Ω so it does not make sense to speak about the restriction of u to $\partial\Omega$ because the k-dimensional Lebesgue measure of $\partial\Omega$ is zero; however, there is a *trace* of u on $\partial\Omega$ which plays the role of the restriction $u|_{\partial\Omega}$. More precisely, we have the following theorem (cf. [14, pp. 258–259]):

Theorem 5.23. *Let $\emptyset \ne \Omega \subset \mathbb{R}^k$ be an open bounded set of class C^1, and let $1 \le p < \infty$. There exists a continuous linear operator γ : $W^{1,p}(\Omega) \to L^p(\partial\Omega)$ such that $\gamma(u) = u|_{\partial\Omega}$ for all $u \in W^{1,p}(\Omega) \cap C(\overline{\Omega})$. Moreover, $u \in W_0^{1,p}(\Omega)$ if and only if $u \in W^{1,p}(\Omega)$ and $\gamma(u) = 0$.*

In fact, the operator γ from the above statement is the extension by continuity of the classical restriction to $\partial\Omega$ from $W^{1,p}(\Omega) \cap C(\overline{\Omega})$ to $L^p(\partial\Omega)$. This extension is unique since $W^{1,p}(\Omega) \cap C(\overline{\Omega})$ is dense in $(W^{1,p}(\Omega), \|\cdot\|_{1,p})$ (see Theorem 5.21). If $u \in W_0^{1,p}(\Omega)$, hence $\gamma(u) = 0$, we say that $u = 0$ on $\partial\Omega$ in a generalized sense. For details on traces and $L^p(\partial\Omega)$, $1 \le p < \infty$, see [14].

The case $k = 1$

If $\Omega = (a, b) \subset \mathbb{R}$, $-\infty \le a < b \le +\infty$, we denote

$$L^p(a, b) := L^p\big((a, b)\big), \quad W^{m,p}(a, b) := W^{m,p}\big((a, b)\big),$$

$$W_0^{m,p}(a, b) := W_0^{m,p}\big((a, b)\big), H^m(a, b) := H^m\big((a, b)\big),$$
$$H_0^m(a, b) := H_0^m\big((a, b)\big).$$

The case $\Omega = (a, b)$ will be discussed later in Sect. 5.6 on vector distributions. In particular, we shall see that for $1 \le p < \infty$ and

$-\infty < a < b < +\infty$ every $u \in W^{1,p}(a,b)$ has a representative which is an absolutely continuous function on $[a,b]$, so identifying u with this representative, $u(a)$ and $u(b)$ make sense classically. According to Theorem 5.23, u is in $W_0^{1,p}(a,b)$ if and only if $u \in W^{1,p}(a,b)$ and $u(a) = 0 = u(b)$. This shows in particular that $W_0^{1,p}(a,b)$ is a proper subspace of $W^{1,p}(a,b)$.

Green's Identity

Let $\emptyset \neq \Omega \subset \mathbb{R}^k$ be an open and bounded set of class C^1. Recall the classical divergence (Gauss–Ostrogradski[11]) formula

$$\int_\Omega \nabla \cdot F \, dx = \int_{\partial\Omega} F \cdot n \, ds$$
$$\forall F = (f_1, \ldots, f_k), \ f_i \in C^1(\overline{\Omega}), \ i = 1, \ldots, k, \qquad (5.4.54)$$

where n is the outward pointing unit normal. Choosing in (5.4.54) $F = g\nabla f$, with $f \in C^2(\overline{\Omega})$ and $g \in C^1(\overline{\Omega})$, one obtains the classical Green identity

$$\int_\Omega g\Delta f \, dx + \int_\Omega \nabla f \cdot \nabla g \, dx = \int_{\partial\Omega} g\frac{\partial f}{\partial n} \, ds. \qquad (5.4.55)$$

Taking into account Theorems 5.21 and 5.23, the identity (5.4.55) can be easily extended by density to

$$\int_\Omega g\Delta f \, dx + \int_\Omega \nabla f \cdot \nabla g \, dx$$
$$= \int_{\partial\Omega} g\frac{\partial f}{\partial n} \, ds \quad \forall f \in W^{2,p}(\Omega), \ g \in W^{1,q}(\Omega), \qquad (5.4.56)$$

where $1 < p < \infty$ and q is the conjugate of p, i.e., $q = (p-1)/p$. Here, the functions in the right-hand side under $\int_{\partial\Omega}$ actually represent their traces on $\partial\Omega$.

Poincaré's Inequality[12]

Now we present an important inequality which holds in $W_0^{1,p}(\Omega)$ for $1 \leq p < \infty$ and Ω open and bounded.

---

[11]Mikhail V. Ostrogradski, Russian-Ukrainian mathematician, mechanician, and physicist, 1801–1862.

[12]Henri Poincaré, French mathematician, theoretical physicist, engineer, and philosopher of science, 1854–1912.

Theorem 5.24 (Poincaré). *Let $\emptyset \neq \Omega \subset \mathbb{R}^k$ be an open bounded set and let $1 \leq p < \infty$. Then*

$$\|u\|_{L^p(\Omega)} \leq C \|\nabla u\|_{L^p(\Omega)} \quad \forall u \in W_0^{1,p}(\Omega), \tag{5.4.57}$$

where C is a positive constant depending on Ω and

$$\|\nabla u\|_{L^p(\Omega)} := \left(\sum_{i=1}^{k} \|\partial u / \partial x_i\|_{L^p(\Omega)} \right)^{1/p}.$$

Proof. Taking into account the definition of $W_0^{1,p}(\Omega)$, it is enough to prove (5.4.57) for all $u \in C_0^\infty(\Omega)$.

Consider first the case $k = 1$, i.e., $\Omega = (a,b)$, $-\infty < a < b < \infty$. If $u \in C_0^\infty(a,b) := C_0^\infty((a,b))$, then

$$u(x) = \int_0^x u'(t)\, dt \implies |u(x)| \leq \int_a^b |u'(t)|\, dt \quad \forall x \in [a,b].$$

If $p = 1$ we obtain (5.4.57) with $C = b - a$ by integrating the last inequality over $[a,b]$.

If $1 < p < \infty$ then we can derive from the same inequality by using Hölder

$$|u(x)| \leq (b-a)^{\frac{1}{p'}} \|u'\|_{L^p(a,b)} \quad \forall x \in [a,b],$$

where $p' = p/(p-1)$. It follows that

$$\int_a^b |u(x)|^p\, dx \leq (b-a)^p \|u'\|_{L^p(a,b)}^p,$$

so (5.4.57) holds again with $C = b - a$.

Now, consider the case $k = 2$. Let $D = [a,b] \times [c,d]$ be a rectangle in the xy-plane such that $\Omega \subset D$. Take $u \in C_0^\infty(\Omega)$ and extend it as zero in $D \setminus \Omega$. We have

$$u(x,y) = \int_a^x \frac{\partial}{\partial s} u(s,y)\, ds \implies |u(x,y)|$$
$$\leq \int_a^b \left| \frac{\partial}{\partial s} u(s,y) \right| ds \quad \forall (x,y) \in D.$$

If $p = 1$ we obtain by integrating the last inequality over D

$$\|u\|_{L^1(D)} \leq (b-a) \|\frac{\partial u}{\partial x}\|_{L^1(D)} \implies \|u\|_{L^1(\Omega)} \leq (b-a) \|\frac{\partial u}{\partial x}\|_{L^1(\Omega)}.$$

If $1 < p < \infty$ we derive by using Hölder

$$\|u\|_{L^p(\Omega)} \le (b-a)\|\frac{\partial u}{\partial x}\|_{L^p(\Omega)}, \qquad (5.4.58)$$

so, in fact, (5.4.58) is valid for $p \in [1, \infty)$.
Similarly,

$$\|u\|_{L^p(\Omega)} \le (d-c)\|\frac{\partial u}{\partial y}\|_{L^p(\Omega)}. \qquad (5.4.59)$$

By (5.4.58) and (5.4.59) it follows that (5.4.57) holds with $C = 2\max$
$\{b-a, d-c\}$.
The proof is similar for $k \ge 3$. $\qquad\qquad\qquad\qquad\qquad\qquad\square$

Remark 5.25. An inspection of the above proof shows that the Poincaré inequality still holds if the Lebesgue measure of Ω is finite, and also if the projection of Ω on some coordinate plane is bounded.

Remark 5.26. If Ω is bounded or satisfies one of the conditions in the previous remark then, according to the Poincaré inequality, $W_0^{1,p}(\Omega)$ can be equipped with a new norm

$$\|u\|_{1,p}^* = \|\nabla u\|_{L^p(\Omega)},$$

which is equivalent to the usual norm $\|\cdot\|_{1,p}$.

5.5 Bochner's Integral

Let $\emptyset \ne \Omega \subset \mathbb{R}^k$ be a Lebesgue measurable set, and let $(X, \|\cdot\|)$ be a real Banach space.
As in the case of $\mathbb{R}$-valued functions, a function $g : \Omega \to X$ is a *simple function* if it is of the form

$$g(s) = \sum_{i=1}^{p} \chi_{M_i}(s) y_i$$

for some $y_i \in X$, $M_i \subset \Omega$ measurable with finite measure (i.e., $m(M_i) < \infty$), and $M_i \cap M_j = \emptyset$ if $i \ne j$. Here, we prefer to use s to denote a generic point in Ω (instead of x which could be used to designate points of X).
A function $f : \Omega \to X$ is called *strongly measurable* (or simply *measurable*) if there exists a sequence of simple functions $g_n : \Omega \to X$ such that

$$\lim_{n \to \infty} \|g_n(s) - f(s)\| = 0 \quad \text{for a.a. } s \in \Omega.$$

If g is a simple function as above, then it is clearly measurable. Define its integral over Ω to be

$$\int_\Omega g(s)\, ds := \sum_{i=1}^{p} m(A_i) y_i\,.$$

If g is a simple function, then $\|g\|$ (i.e., the function $s \mapsto \|g(s)\|$) is a simple function as well (hence Lebesgue integrable over Ω) and the following inequality holds:

$$\left\| \int_\Omega g(s)\, ds \right\| \le \int_\Omega \|g(s)\|\, ds\,.$$

Denote by $\mathcal{S}$ the set of all simple functions $g : \Omega \to X$. Clearly $\mathcal{S}$ is a real linear space with respect to the usual operations (addition of functions and scalar multiplication), and

$$\int_\Omega (\alpha_1 g_1 + \alpha_2 g_2)\, ds \;=\; \alpha_1 \int_\Omega g_1\, ds$$

$$+ \alpha_2 \int_\Omega g_2\, ds \quad \forall \alpha_1,\, \alpha_2 \in \mathbb{R},\; \forall g_1,\, g_2 \in \mathcal{S}\,.$$

Definition 5.27. $f : \Omega \to X$ *is said to be Bochner integrable (over* Ω*)*[13] *if there exists a sequence of simple functions* $g_n : \Omega \to X$ *converging strongly to* f *a.e. in* Ω *(so* f *is measurable) and*

$$\lim_{n,m\to\infty} \int_\Omega \|g_n(s) - g_m(s)\|\, ds = 0\,, \tag{5.5.60}$$

and the Bochner integral of f *is defined as*

$$\int_\Omega f(s)\, ds := \lim_{n\to\infty} \int_\Omega g_n(s)\, ds\,. \tag{5.5.61}$$

Let us justify the above definition. We have

$$\left\| \int_\Omega g_n\, ds - \int_\Omega g_m\, ds \right\| = \left\| \int_\Omega (g_n - g_m)\, ds \right\|$$

$$\le \int_\Omega \|g_n - g_m\|\, ds\,.$$

---

[13]Salomon Bochner, American mathematician, 1899–1982.

So (5.5.60) implies

$$\lim_{n,m\to\infty}\left\|\int_\Omega g_n\,ds - \int_\Omega g_m\,ds\right\| = 0\,,$$

i.e., the limit in (5.5.61) exists. To prove the limit does not depend on the choice of (g_n), consider another sequence $(\tilde{g}_n)$ satisfying the same properties. Then, by (5.5.60), we have for all $\varepsilon > 0$

$$\int_\Omega \|g_n - \tilde{g}_n - g_m + \tilde{g}_m\|\,ds \le \int_\Omega \|g_n - g_m\|\,ds + \int_\Omega \|\tilde{g}_n - \tilde{g}_m\|\,ds$$
$$\le \varepsilon \quad \forall n, m > N_\varepsilon\,.$$

Letting $m \to \infty$ it follows from Fatou's Lemma that

$$\int_\Omega \|g_n - \tilde{g}_n\|\,ds \le \varepsilon \quad \forall n > N_\varepsilon\,. \tag{5.5.62}$$

Now, since $g_n, \tilde{g}_n$ are simple functions, we have

$$\left\|\int_\Omega g_n\,ds - \int_\Omega \tilde{g}_n\,ds\right\| = \left\|\int_\Omega (g_n - \tilde{g}_n)\,ds\right\| \le \int_\Omega \|g_n - \tilde{g}_n\|\,ds\,. \tag{5.5.63}$$

From (5.5.62) and (5.5.63) we deduce

$$\lim_{n\to\infty}\int_\Omega \tilde{g}_n\,ds = \lim_{n\to\infty}\int_\Omega g_n\,ds = \int_\Omega f\,ds\,,$$

so the definition is correct.

Remark 5.28. Note that if $X = \mathbb{R}^N$, $N \in \mathbb{N}$, then $f = (f_1, \dots, f_N)$ is measurable in the sense above if and only if f_i is Lebesgue measurable for all $i = 1, \dots, N$, and integrability of f in the sense of Bochner means integrability of all f_i's in the sense of Lebesgue. If $(X, \|\cdot\|)$ is an infinite dimensional Banach space, then, in addition to the concept of strong measurability of a function from Ω to X as defined before, there is also a concept of weak measurability, namely $f : \Omega \to X$ is said to be weakly measurable if $s \mapsto x^*(f(s))$ is Lebesgue measurable for every continuous linear functional $x^* : (X, \|\cdot\|) \to \mathbb{R}$. If X is a separable Banach space, then the weak measurability of f is equivalent to its strong measurability. In fact, this equivalence holds if f is almost separably valued, that is $\{f(s); s \in \Omega \setminus M\}$ is a separable set, where $M \subset \Omega$ has zero Lebesgue measure. This result belongs to

Pettis,[14] see, e.g., [51, p. 131]. It is worth mentioning that, in all the applications discussed in this book, X will always stand for separable Banach spaces.

The next result says that Bochner integrability of any X-valued function f reduces to Lebesgue integrability of $\|f\|$.

Theorem 5.29 (Bochner). *Let $(X, \|\cdot\|)$ be a real Banach space and let $\Omega \subset \mathbb{R}^k$ be a measurable set. If $f : \Omega \to X$ is strongly measurable, then f is Bochner integrable if and only if $\|f\|$ is Lebesgue integrable, where $\|f\|(s) := \|f(s)\|$ for almost all $s \in \Omega$.*

Proof. Since f is strongly measurable, $\|f\|$ is also (Lebesgue) measurable because a sequence of simple functions gives a sequence of simple functions upon taking the norm.

To prove necessity, assume that f is Bochner integrable. If (g_n) is a sequence of simple functions as in Definition 5.27, we can write (see (5.5.60))

$$\int_\Omega \|g_n - g_m\| \, ds \leq \varepsilon \quad \forall n, m > N_\varepsilon .$$

Applying Fatou's Lemma, we get

$$\int_\Omega \|g_n - f\| \, ds \leq \varepsilon \quad \forall n > N_\varepsilon ,$$

i.e., $\|g_n - f\|$ is Lebesgue integrable for all $n > N_\varepsilon$. So integrating the obvious inequality

$$\|f\| \leq \|f - g_n\| + \|g_n\|$$

we obtain

$$\int_\Omega \|f\| \, ds \leq \int_\Omega \|f - g_n\| \, ds + \int_\Omega \|g_n\| \, ds < \infty \quad \forall n > N_\varepsilon ,$$

hence $\|f\|$ is Lebesgue integrable.

In order to prove sufficiency, assume that $\|f\|$ is Lebesgue integrable and consider a sequence of simple functions $h_n : \Omega \to X$ such that

$$\lim_{n \to \infty} \|h_n(s) - f(s)\| = 0 \quad \text{for almost all } s \in \Omega .$$

Define

$$g_n(s) = \begin{cases} h_n(s) & \text{if } \|h_n(s)\| \leq (1 + \delta)\|f(s)\|, \\ 0 & \text{otherwise,} \end{cases}$$

---

[14]Billy James Pettis, American mathematician, 1913–1979.

where δ is a positive constant. This is a sequence of simple functions and

$$\lim_{n\to\infty} \|g_n(s) - f(s)\| = 0 \quad \text{for a.a. } s \in \Omega. \tag{5.5.64}$$

We must show

$$\lim_{n,m\to\infty} \int_{\Omega} \|g_n - g_m\| \, ds = 0. \tag{5.5.65}$$

To do this, we shall apply the Lebesgue Dominated Convergence Theorem to the sequence $(\|g_n - f\|)$. The first condition of this theorem is satisfied (see (5.5.64)), and

$$\begin{aligned}
\|g_n(s) - f(s)\| &\leq \|g_n(s)\| + \|f(s)\| \\
&\leq (1+\delta)\|f(s)\| + \|f(s)\| \\
&= (2+\delta)\|f(s)\|,
\end{aligned}$$

so the second condition of the Lebesgue Dominated Convergence Theorem is also satisfied, hence

$$\lim_{n\to\infty} \int_{\Omega} \|g_n - f\| \, ds = 0.$$

This along with the obvious inequality

$$\int_{\Omega} \|g_n - g_m\| \, ds \leq \int_{\Omega} \|g_n - f\| \, ds + \int_{\Omega} \|g_m - f\| \, ds$$

implies (5.5.65). $\qquad\qquad\qquad\qquad\qquad\qquad\qquad\qquad\qquad\square$

Remark 5.30. It is worth pointing out that for every $f : \Omega \to X$ which is Bochner integrable, we have

$$\left\| \int_{\Omega} f \, ds \right\| \leq \int_{\Omega} \|f\| \, ds,$$

because this inequality holds for simple functions. In general, the usual properties of the Lebesgue integral are also satisfied by the Bochner integral.

Remark 5.31. Let $(X, \|\cdot\|)$ and $(Y, \|\cdot\|_*)$ be real Banach spaces. If $f : \Omega \to X$ is Bochner integrable over Ω and A is a continuous linear operator from $(X, \|\cdot\|)$ to $(Y, \|\cdot\|_*)$, then $A \circ f$ is also Bochner integrable and

$$\int_{\Omega} A \circ f \, ds = A \int_{\Omega} f \, ds.$$

Indeed, if (g_n) is a sequence of simple functions converging to f, then $(A \circ g_n)$ is also a sequence of simple functions which converges to $A \circ f$. Moreover,

$$\int_\Omega \|A \circ g_n - A \circ g_m\| \, ds \leq \|A\| \int_\Omega \|g_n - g_m\| \, ds \to 0 \quad \text{as } n, m \to \infty.$$

It follows that

$$\int_\Omega A \circ f \, ds = \lim_{n \to \infty} \int_\Omega A \circ g_n \, ds = \lim_{n \to \infty} A \int_\Omega g_n \, ds = A \int_\Omega f \, ds,$$

as claimed.

For X a real Banach space, $\Omega \subset \mathbb{R}^k$ measurable, and $1 \leq p < \infty$ define

$$\mathcal{L}^p(\Omega; X) = \{f : \Omega \to X; \ f \text{ is measurable and } \int_\Omega \|f\|^p ds < \infty\}.$$

We also define

$$\mathcal{L}^\infty(\Omega; X) = \{f : \Omega \to X; \ f \text{ is measurable}$$
$$\text{and } \operatorname*{ess\,sup}_{s \in \Omega} \|f(s)\| < \infty\},$$

where

$$\operatorname*{ess\,sup}_{s \in \Omega} \|f(s)\| := \inf\{C; \ \|f(s)\| \leq C \text{ a.e. on } \Omega\}.$$

Let $\sim$ denote equality a.e. and define the quotient space

$$L^p(\Omega; X) := \mathcal{L}^p(\Omega)/\sim.$$

This is a real Banach space for $1 \leq p \leq \infty$ with respect to the norm

$$\|f\|_{L^p(\Omega; X)} := \left(\int_\Omega \|f\|^p \, ds\right)^{1/p}, \quad 1 \leq p < \infty,$$
$$\|f\|_{L^\infty(\Omega; X)} := \operatorname*{ess\,sup}_{s \in \Omega} \|f(s)\|.$$

The proof follows by arguments similar to those from the proof of the classical theorem corresponding to the case $X = \mathbb{R}$ (Theorem 3.25), so we leave it to the reader as an exercise. The key condition is the completeness of X.

If $\Omega = (a, b)$ with $-\infty \leq a < b \leq \infty$ denote $L^p(a, b; X) := L^p((a, b); X)$.

5.6 Vector Distributions, $W^{m,p}(a,b;X)$ Spaces

Let X be a Banach space and let $-\infty \leq a < b \leq \infty$. Denote as before $\mathcal{D}(a,b) = C_0^\infty(a,b) := C_0^\infty((a,b))$ equipped with the inductive limit topology.

Definition 5.32. *An X-valued distribution over (a,b) is an operator $u : \mathcal{D}(a,b) \to X$ which is linear and continuous (in the sense that if $\phi_n \to 0$ in $\mathcal{D}(a,b)$ then $u(\phi_n) \to 0$). The set of all such vector distributions is denoted $\mathcal{D}'(a,b;X)$.*

As in the scalar case, a **regular distribution** is one which is generated by a locally integrable function $u \in L^1_{\text{loc}}(a,b;X)$, i.e., $u : (a,b) \to X$ is strongly measurable and $\|u\| \in L^1(K)$ for all $K \subset (a,b)$ compact. Define $\tilde{u} : \mathcal{D}(a,b) \to X$ by

$$\tilde{u}(\phi) := \int_a^b \phi(t)u(t)dt \quad \forall \phi \in \mathcal{D}(a,b).$$

The mapping $u \mapsto \tilde{u}$ is injective, as its null set is $\{0\}$. Indeed, for $\phi \in \mathcal{D}(a,b)$ and $v \in L^1_{\text{loc}}(a,b;X)$ satisfying

$$\int_a^b \phi(t)v(t)\,dt = 0,$$

we have (cf. Remark 5.31)

$$\int_a^b \phi(t)x^*(v(t))\,dt = 0 \quad \forall x^* \in X^*,$$

where X^* is the dual of X. Since $t \mapsto x^*(v(t))$ is a real, locally summable function, it follows by Theorem 5.9 that

$$x^*(v(t)) = 0 \quad \forall x^* \in X^*, \text{ and a.a. } t \in (a,b)$$

so $v(t) = 0$ for a.a. $t \in (a,b)$.

Consequently, one can identify the (regular) distribution $\tilde{u}$ with the locally summable function u, and write

$$u(\phi) := \int_a^b \phi(t)u(t)dt \quad \forall \phi \in \mathcal{D}(a,b).$$

Of course, as in the scalar case, not all vector distributions arise in this way, e.g., $u : \mathcal{D}(\mathbb{R}) \to X$ defined by $u(\phi) = \phi(0)x$ for all $\phi \in \mathcal{D}(\mathbb{R})$ and a fixed $x \in X \setminus \{0\}$.

For $u \in \mathcal{D}'(a, b; X)$ define the derivative

$$u'(\phi) := -u(\phi') \quad \forall \phi \in \mathcal{D}(a, b),$$

and inductively,

$$u^{(j)}(\phi) = (-1)^j u(\phi^{(j)}) \quad \forall \phi \in \mathcal{D}(a, b), \ j \in \mathbb{N},$$

and by convention

$$u^{(0)} = u.$$

In applications, intervals (a, b) are sufficient, though the theory extends to $\Omega \subset \mathbb{R}^k$.

For $m \in \mathbb{N}$, $1 \le p \le \infty$, we set

$$W^{m,p}(a, b; X) := \{u \in \mathcal{D}'(a, b; X);$$
$$u^{(j)} \in L^p(a, b; X), j = 0, 1, \ldots, m\},$$

so, in fact, u is a regular distribution because $j = 0$ is included. Also, all (distributional) derivatives above are regular as well.

Theorem 5.33. *If X is a Banach space then, for all $m \in \mathbb{N}$ and $1 \le p \le \infty$, $W^{m,p}(a, b; X)$ is a Banach space with respect to the norm*

$$\|u\|_{W^{m,p}(a,b;X)} := \left(\sum_{j=0}^{m} \|u^{(j)}\|_{L^p(a,b;X)}^p \right)^{1/p}, \qquad 1 \le p < \infty,$$

$$\|u\|_{W^{m,\infty}(a,b;X)} := \max_{0 \le j \le m} \|u^{(j)}\|_{L^\infty(a,b;X)}, \qquad p = \infty.$$

Proof. Similar to the proof of Theorem 5.20.   □

The notation $W_{\text{loc}}^{m,p}(a, b; X)$ indicates the set of all $u \in \mathcal{D}'(a, b; X)$ such that $u \in W^{m,p}(t_1, t_2; X)$ for every bounded interval $(t_1, t_2) \subset (a, b)$. For $p = 2$ denote $H^m(a, b; X) = W^{m,2}(a, b; X)$. If X is a Hilbert space, then so is $H^m(a, b; X)$ with respect to the inner product

$$(u, v)_{H^m(a,b; X)} = \sum_{j=0}^{m} \int_a^b \left(u^{(j)}(t), v^{(j)}(t) \right)_X dt \,.$$

Now for $-\infty < a < b < +\infty$ denote by $A^{m,p}(a, b; X)$ the space of all functions $f : [a, b] \to X$ which are absolutely continuous on $[a, b]$, the pointwise derivatives $d^j f / dt^j$ exist and are absolutely continuous on [a,b] for $j = 1, 2, \ldots, m - 1$, and $d^m f / dt^m \in L^p(a, b; X)$.

Remark 5.34. If X is reflexive, it follows by a well-known theorem due to Kōmura[15] (see [25]; see also [45, p. 105]) that

$$A^{1,1}(a, b; X) = AC([a, b]; X) \,,$$

where $AC([a, b]; X)$ is the space of all X-valued absolutely continuous functions on $[a, b]$.

Theorem 5.35. *For $m \in \mathbb{N}$, $1 \le p \le \infty$, $-\infty < a < b < \infty$, and $u \in L^p(a, b; X)$ then the following are equivalent:*

(j) $u \in W^{m,p}(a, b; X)$;

(jj) *there exists $u_1 \in A^{m,p}(a, b; X)$ such that $u_1(t) = u(t)$ for almost all $t \in (a, b)$.*

Proof. We shall prove the case $m = 1$, and then the result follows by induction.

To prove the implication $(j) \Rightarrow (jj)$ fix $u \in W^{1,p}(a, b; X)$ and extend it as zero in $\mathbb{R} \setminus (a, b)$. For $\varepsilon > 0$ small define u_ε as before, i.e.,

$$u_\varepsilon(t) = \int_{\mathbb{R}} \omega_\varepsilon(t - s) u(s) \, ds \,,$$

where

$$\omega_\varepsilon(t) = \frac{1}{\varepsilon} \omega(t/\varepsilon), \quad \text{and} \quad \omega(t) = \begin{cases} Ce^{-\frac{1}{1-t^2}}, & |t| < 1, \\ 0, & |t| \ge 1, \end{cases}$$

---

[15]Yukio Kōmura, Japanese mathematician, born 1931.

with $C > 0$ such that $\int_{\mathbb{R}} \omega(t) \, dt = 1$. We have

$$\dot{u}_\varepsilon(t) = \frac{d}{dt} u_\varepsilon(t)$$

$$= \int_{\mathbb{R}} \omega'_\varepsilon(t - s) u(s) \, ds, \quad \forall t \in \mathbb{R},$$

which is a function, but we understand it as a distribution and apply it to a test function $\phi \in C_0^\infty(\mathbb{R})$

$$(\dot{u}_\varepsilon, \phi) = \int_{\mathbb{R}} \phi(t) \dot{u}_\varepsilon(t) \, dt$$

$$= \int_{\mathbb{R}} \phi(t) \left(\int_{\mathbb{R}} \omega'_\varepsilon(t - s) u(s) \, ds \right) dt$$

and interchanging the order of integration

$$= \int_{\mathbb{R}} \left(\int_{\mathbb{R}} \omega'_\varepsilon(t - s) \phi(t) \, dt \right) u(s) \, ds$$

$$= - \int_{\mathbb{R}} \phi'_\varepsilon(s) u(s) \, ds$$

$$= -(u, \phi'_\varepsilon)$$

$$= u'(\phi_\varepsilon)$$

$$= \int_{\mathbb{R}} \phi_\varepsilon(t) u'(t) \, dt$$

$$= \int_{\mathbb{R}} \left(\int_{\mathbb{R}} \omega_\varepsilon(t - s) \phi(s) \, ds \right) u'(t) \, dt$$

and changing the order of integration again

$$= \int_{\mathbb{R}} \phi(s) \left(\int_{\mathbb{R}} \omega_\varepsilon(t - s) u'(t) \, dt \right) ds$$

$$= \int_{\mathbb{R}} \phi(s) (u')_\varepsilon(s) \, ds$$

so that

$$(\dot{u}_\varepsilon, \phi) = ((u')_\varepsilon, \phi), \quad \forall \phi \in C_0^\infty(\mathbb{R}). \tag{5.6.66}$$

In other words, the pointwise derivative $\dot{u}_\varepsilon$ is equal to $(u')_\varepsilon$.

Now, integrate to obtain

$$u_\varepsilon(t) - u_\varepsilon(s) = \int_s^t (u')_\varepsilon(\tau)\, d\tau\,. \qquad (5.6.67)$$

Note that $u_\varepsilon \to u$ and $(u')_\varepsilon \to u'$ in $L^p(a, b; X)$ as $\varepsilon \to 0^+$ (the proof is the same as in the scalar case).

Hence, there exists a function u_1 such that

$$u_1(t) - u_1(s) = \int_s^t u'(\tau)\, d\tau \quad \text{for a.a. } s, t \in (a, b)\,.$$

Therefore, $u_1 \in AC([a, b]; X)$ and $\dot{u}_1 = u'$ for almost all $t \in (a, b)$, i.e., the pointwise derivative $\dot{u}_1$ is a representative of the distributional derivative $u' \in L^p(a, b; X)$. So $\dot{u}_1 \in L^p(a, b; X)$, which together with absolute continuity implies that $u_1 \in A^{1,p}(a, b; X)$.

For the implication $(jj) \implies (j)$, assume there exists $u_1 \in A^{1,p}(a, b; X)$ an element of the class u. We must show that $u \in W^{1,p}(a, b; X)$. Since $u_1 \in AC[a, b]$, $u \in L^p(a, b; X)$, and we must show that $u' \in L^p(a, b; X)$. We start with $\dot{u}_1$ and interpret it as a distribution. For all $\phi \in \mathcal{D}(a, b)$, we have

$$(\dot{u}_1, \phi) = \int_a^b \phi \dot{u}_1\, dt$$

and, integrating by parts,

$$= -\int_a^b \dot{\phi} u_1\, dt$$

and, since changing u_1 to another element of its class won't affect the integral,

$$= -\int_a^b \dot{\phi} u\, dt$$
$$= -u(\dot{\phi})$$
$$= u'(\phi)\,.$$

Therefore, $\dot{u}_1 = u'$ as distributions, but since $\dot{u}_1$ is a function, so is u' and $\dot{u}_1 \in L^p(a, b; X)$ so $u' \in L^p(a, b; X)$. $\qquad \square$

Note that usually good representatives are preferred since their values at particular points make sense.

5.7   Exercises

1. Let $\Omega = \mathbb{R} \times (-1, +1) \subset \mathbb{R}^2$ and let $u : \Omega \to \mathbb{R}$ be defined by

$$u(x) = |x_1| x_1^2 (1 + x_1 x_2 + |x_2| x_2^2).$$

   Show that $u \in C^2(\Omega)$ and find $\operatorname{supp} u$.

2. Find a collection $\mathcal{F}$ of seminorms on $C[0,1] := C([0,1]; \mathbb{R})$ such that the topology generated by $\mathcal{F}$ coincide with the pointwise convergence topology.

3. Let $\Omega \subset \mathbb{R}^k$ be a nonempty open set. For any compact set $K \subset \Omega$ and $m \in \mathbb{N} \cup \{0\}$ define the seminorm $p : C^\infty(\Omega) \to \mathbb{R}$

$$p_{K,m}(f) = \sup_{x \in K, |\alpha| \leq m} |D^\alpha f(x)|, \quad f \in C^\infty(\Omega),$$

   where $\alpha = (\alpha_1, \ldots, \alpha_k)$ are multi-indices, $|\alpha| = \alpha_1 + \cdots \alpha_k$, and

$$D^\alpha f(x) = \frac{\partial^{|\alpha|}}{\partial x_1^{\alpha_1} \cdots \partial x_k^{\alpha_k}} f(x_1, \ldots, x_k).$$

   Consider a sequence of compact sets $K_1 \subset K_2 \subset \cdots \subset K_n \subset \cdots \subset \Omega$, such that $\Omega = \cup_{n=1}^\infty K_n$. Define for each $j \in \mathbb{N}$

$$d_j(f, g) = \sum_{m=0}^{j} \frac{1}{2^m} \cdot \frac{p_{K_j,m}(f-g)}{1 + p_{K_j,m}(f-g)}, \quad f, g \in C^j(\Omega),$$

   and

$$d(f, g) = \sum_{j=1}^{\infty} \frac{1}{2^j} \cdot \frac{d_j(f,g)}{1 + d_j(f,g)}, \quad f, g \in C^\infty(\Omega).$$

   Show that d is a metric on $C^\infty(\Omega)$.

4. Find a function $\phi \in C^\infty(\mathbb{R})$ with $\operatorname{supp} \phi = [0,4]$, $\phi \geq 0$ and $\max_{\mathbb{R}} \phi = 1$.

5. Let $\phi \in C_0^\infty(\mathbb{R}^k)$. Prove that there exists $\psi \in C_0^\infty(\mathbb{R}^k)$ such that $\phi = \frac{\partial^k \psi}{\partial x_1 \cdots \partial x_k}$ if and only if $\int_{\mathbb{R}^k} \phi(x)\, dx = 0$.

6. Let $(a_n)_{n \in \mathbb{N}}$ be a sequence of real numbers. Prove that there exists a function $\phi \in C_0^\infty(\mathbb{R})$ such that $\phi(n) = a_n \ \forall n \in \mathbb{N}$ if and only if there exists an $n_0 \in \mathbb{N}$ such that $a_n = 0 \ \forall n > n_0$.

7. Let $m \in \mathbb{N}$ and $\psi \in C_0^\infty(R^k)$. Define the sequence (ϕ_n) by

$$\phi_n(x) = 2^{-n} n^m \psi(nx), \quad x \in \mathbb{R}^k, \ n \in \mathbb{N}.$$

Show that $\phi_n \to 0$ in $D(\mathbb{R}^k)$ as $n \to \infty$.

8. Let h be a nonzero vector in $\mathbb{R}^k$ and let $\psi \in C_0^\infty(\mathbb{R}^k)$. Consider the sequence $(\phi_n)_{n\in\mathbb{N}}$, where

$$\phi_n(x) = n\Big(\psi\big(x + \frac{1}{n}h\big) - \psi(x)\Big), \quad x \in \mathbb{R}^k, \ n \in \mathbb{N}.$$

Prove that

$$\phi_n \to \sum_{j=1}^k h_j \frac{\partial \psi}{\partial x_j} \quad \text{in } D(\mathbb{R}^k).$$

Deduce from this result the convergence in $D(\mathbb{R}^k)$ to 0 of the sequence $(\gamma_n)_{n\in\mathbb{N}}$ defined by

$$\gamma_n(x) = n\Big(\psi\big(x + \frac{1}{n}h\big) - \psi\big(x - \frac{1}{n}h\big)\Big), \quad x \in \mathbb{R}^k, \ n \in \mathbb{N}.$$

9. Let $\Omega \subset \mathbb{R}^k$ be a nonempty open set. For $\phi \in C_0^\infty(\Omega)$ consider

$$\phi_n(x) = \int_\Omega \omega_{1/n}(x - y)\phi(y)\, dy, \quad x \in \Omega, \ n \in \mathbb{N} \text{ sufficiently large,}$$

where $\omega_{1/n}$ denotes the usual Friedrichs mollifier. Prove that ϕ_n converges to ϕ in $D(\Omega)$.

10. Let $\Omega \subset \mathbb{R}^k$ be a nonempty open set. For a given point $a \in \Omega$ and for a multi-index $\alpha \in \mathbb{N}_0^k$, define $u : D(\Omega) \to \mathbb{R}$ by

$$u(\phi) = D^\alpha \phi(a) \quad \forall \phi \in D(\Omega).$$

Here $\mathbb{N}_0 = \mathbb{N} \cup \{0\}$. Prove that $u \in D'(\Omega)$ and u is not a regular distribution.

11. Let $\Omega \subset \mathbb{R}^k$ be a nonempty open set. Show that, if $\phi \in D(\Omega)$ satisfies

$$u(\phi) = 0 \quad \forall u \in D'(\Omega),$$

then $\phi = 0$.

12. Let $u : D(\mathbb{R}) \to \mathbb{R}$,

$$u(\phi) = \sum_{i=1}^{\infty} \Big(\phi(1/i^2) - \phi(0) \Big), \quad \phi \in D(\mathbb{R}).$$

Prove that u is well defined, $u \in D'(\mathbb{R})$, and u is not a regular distribution.

13. Show that mixed derivatives of distributions do not depend on the order of differentiation.

14. Let $\emptyset \neq \Omega \subset \mathbb{R}^k$ be an open set, $u \in D'(\Omega)$, $a \in C^{\infty}(\Omega)$. Show that

$$\frac{\partial(au)}{\partial x_i} = \frac{\partial a}{\partial x_i} u + a \frac{\partial u}{\partial x_i}.$$

Extend this formula to $D^{\alpha}(au)$ for a general multi-index α.

15. Find the n-th derivatives ($n = 1, 2, 3$) in the sense of distributions of $f, g : \mathbb{R} \to \mathbb{R}$,

$$f(x) = \frac{1}{2} x |x|, \quad x \in \mathbb{R},$$
$$g(x) = H(x) \cdot \cos x, \quad x \in \mathbb{R},$$

where H denotes the usual Heaviside function.

16. Find a sequence $(H_n)_{n \in \mathbb{N}}$ in $C_0^{\infty}(\mathbb{R})$ such that $H_n \to H$ in $D'(\mathbb{R})$, where H is the Heaviside function.

17. Let $u : D(\mathbb{R}^2) \to \mathbb{R}$,

$$u(\phi) = \int_{-\infty}^{\infty} \phi(x_1, 0) \, dx_1 \ \ \forall \phi \in D(\mathbb{R}^2).$$

(i) Prove that $u \in D'(\mathbb{R}^2)$;

(ii) Show that u is not a regular distribution;

(iii) Check that $\frac{\partial u}{\partial x_1} = 0$.

18. Let $\Omega \subset \mathbb{R}^k$ be a nonempty open set and let $S \subset \Omega$ be a countably infinite set of isolated points, $S = \{x_1, x_2, \ldots, x_n, \ldots\}$. Show that for any sequence of real numbers $(a_n)_{n \in \mathbb{N}}$ the series $\sum_{n=1}^{\infty} a_n \delta_{x_n}$ converges in $D'(\Omega)$.

19. Let $(x_n)_{n \in \mathbb{N}}$ be a sequence in $\mathbb{R}^k$. Prove the following implication:

$$\delta_{x_n} \to 0 \ \text{in} \ D'(\mathbb{R}^k) \implies \|x_n\| \to \infty.$$

20. Solve the following equations in $D'(\mathbb{R})$:

   (a) $u' + tu = \chi_{[0,1]}(t)$ (where $\chi_{[0,1]}$ is the characteristic function of $[0,1]$);

   (b) $u'' + u = H + \delta'$ (where H denotes the Heaviside function and δ is the Dirac distribution);

   (c) $u'' - 2u' + u = 2\delta(t-1) + \delta(t-2)$;

   (d) $u'' - 4u = \delta' - \delta'' - 8$.

21. Solve the Cauchy problem

$$\begin{cases} u'' - u = \delta(t-1) + 2\delta(t-3) - 2t - 1 & \text{in} \ D'(\mathbb{R}), \\ u(0) = 1, \ u'(0) = 0. \end{cases}$$

22. Prove that the solution set of the equation

$$(\sin t) \cdot u' = 0 \ \text{in} \ D'(\mathbb{R})$$

is an infinite dimensional linear subspace of $D'(\mathbb{R})$.

23. Find u_1, u_2, $u_3 \in D'(\mathbb{R})$ satisfying the differential system

$$\begin{cases} u_1' = 4u_1 - u_2 + H, \\ u_2' = 3u_1 + u_2 - u_3 + \delta, \\ u_3' = u_1 + u_3 + H. \end{cases}$$

24. Let a be a given real number. If $u \in W_0^{1,1}(a, \infty) := W_0^{1,1}(a, \infty; \mathbb{R})$, prove that there exists a function $v \in C[a, \infty)$ which is a representative of the class u, and $v(a) = 0$.

25. Let $p \in (1, \infty)$. Show that $W^{2,p}(0,1)$ is compactly embedded into $C^1[0,1]$. The Sobolev space $W^{2,p}(0,1)$ is equipped with the usual norm, and $C^1[0,1]$ is equipped with the norm

$$\|f\|_{C^1} = \max_{0 \le t \le 1} |f(t)| + \max_{0 \le t \le 1} |f'(t)| \ \forall f \in C^1[0,1].$$

26. Let $\phi \in C_0^\infty(\mathbb{R}) \setminus \{0\}$ and let $1 \leq p \leq +\infty$. Define $u_n : \mathbb{R} \to \mathbb{R}$ by $u_n(t) = \phi(t + n)$, $t \in \mathbb{R}$, $n \in \mathbb{N}$. Prove that

    (i) $(u_n)_{n \in \mathbb{N}}$ is bounded in $W^{m,p}(\mathbb{R})$ for every $m \in \mathbb{N}$;

    (ii)   there exists no subsequence of (u_n) converging strongly in $L^q(\mathbb{R})$ for any $1 \leq q \leq \infty$.

27. Let $\emptyset \neq \Omega \subset \mathbb{R}^k$ be an open bounded set. If $u, v \in H^1(\Omega) = W^{1,2}(\Omega)$, show that $uv \in W^{1,1}(\Omega)$ and

$$\frac{\partial}{\partial x_i}(uv) = \frac{\partial u}{\partial x_i} \cdot v + u \cdot \frac{\partial v}{\partial x_i}, \quad i = 1, 2, \ldots, k,$$

in $D'(\Omega)$ and a.e. in Ω.

Chapter 6

Hilbert Spaces

Let X be a linear space over $\mathbb{K}$ equipped with a scalar (inner) product $(\cdot, \cdot)$ (i.e., X is an inner product space or a generalized Euclidean space, as defined in Chap. 1). As usual, throughout this chapter $\mathbb{K}$ is either $\mathbb{R}$ or $\mathbb{C}$. Define the norm

$$\|x\| = \sqrt{(x, x)}, \quad x \in X.$$

If $(X, \|\cdot\|)$ is a Banach space (i.e., (X, d) is a complete metric space, where $d(x, y) = \|x - y\|$, $x, y \in X$), then X is said to be a **Hilbert**[1] **space**. In other words, a Hilbert space is a Banach space $(X, \|\cdot\|)$ whose norm is given by a scalar product.

6.1 Examples

We have already met some Hilbert spaces, such as the Euclidean space $\mathbb{R}^k$, $\mathbb{C}^k$, $L^2(\Omega)$, $H^m(\Omega)$, $m \in \mathbb{N}$, these spaces being equipped with their usual scalar products, i.e.,

---

[1]David Hilbert, German mathematician, 1862–1943.

© Springer Nature Switzerland AG 2019

G. Moroşanu, *Functional Analysis for the Applied Sciences*,

Universitext, https://doi.org/10.1007/978-3-030-27153-4_6

$$(x, y) = \sum_{i=1}^{k} x_i y_i, \quad x = (x_1, \ldots, x_k), \ y = (y_1, \ldots, y_k) \in \mathbb{R}^k,$$

$$(x, y) = \sum_{i=1}^{k} x_i \overline{y_i}, \quad x = (x_1, \ldots, x_k), \ y = (y_1, \ldots, y_k) \in \mathbb{C}^k,$$

$$(u, v)_{L^2(\Omega)} = \int_{\Omega} uv \, dx, \quad u, v \in L^2(\Omega),$$

$$(u, v)_m = \sum_{|\alpha| \leq m} \left(D^\alpha u, D^\alpha v \right)_{L^2(\Omega)}, \quad u, v \in H^m(\Omega),$$

and the corresponding induced norms

$$\|x\|^2 = \sum_{i=1}^{k} x_i^2, \quad x = (x_1, \ldots, x_k) \in \mathbb{R}^k,$$

$$\|x\|^2 = \sum_{i=1}^{k} |x_i|^2, \quad x = (x_1, \ldots, x_k) \in \mathbb{C}^k,$$

$$\|u\|_{L^2(\Omega)}^2 = \int_{\Omega} u^2 \, dx, \quad u \in L^2(\Omega),$$

$$\|u\|_m^2 = \sum_{|\alpha| \leq m} \|D^\alpha u\|_{L^2(\Omega)}^2, \quad u \in H^m(\Omega),$$

where Ω is a measurable or open subset of $\mathbb{R}^k$ in the third and fourth cases, respectively.

Obviously, every Cauchy sequence in $\mathbb{R}^k$ is convergent since the corresponding coordinate sequences are Cauchy in $(\mathbb{R}, |\cdot|)$, hence convergent in that space. So the Euclidean space $\mathbb{R}^k$ equipped with the above scalar product and norm is a Hilbert space over $\mathbb{R}$. Similarly, $\mathbb{C}^k$ equipped with the above scalar product and norm is a Hilbert space over $\mathbb{C}$.

Note also that $L^p(\Omega)$ equipped with the usual norm is a Banach space for all $1 \leq p \leq \infty$ (see Theorem 3.25). So $(L^2(\Omega), \|\cdot\|_{L^2(\Omega)})$ is a real Hilbert space. Also, $H^m(\Omega)$ equipped with the above scalar product and norm is a real Hilbert space, and so is its closed subspace $H_0^m(\Omega)$, $m \in \mathbb{N}$.

It is worth pointing out that $H_0^1(\Omega)$ can be equipped with a different scalar product,

$$(u, v)_1^* = \int_{\Omega} \nabla u \cdot \nabla v \, dx, \quad u, v \in H_0^1(\Omega),$$

and the induced norm

$$\|u\|_1^* = \|\nabla u\|_{L^2(\Omega)}, \quad u \in H_0^1(\Omega),$$

whenever Ω is open and has finite measure, or its projection on a coordinate plane is bounded (see Theorem 5.24 and Remarks 5.25 and 5.26).

Note also that, for $-\infty \leq a < b \leq \infty$ and a Hilbert space X, $L^2(a, b; X)$ equipped with the scalar product

$$(u, v)_{L^2(a,b;X)} = \int_a^b \big(u(t), v(t)\big)_X \, dt, \quad u, v \in L^2(a, b; X),$$

and the induced norm

$$\|u\|_{L^2(a,b;X)}^2 = \int_a^b \|u(t)\|_{L^2(a,b;X)}^2 \, dt,$$

is a Hilbert space, too. Also, $H^m(a, b; X)$ is a Hilbert space for any $m \in \mathbb{N}$ with respect to the scalar product

$$(u, v)_m = \sum_{j=0}^m \int_a^b \big(u^{(j)}(t), v^{(j)}(t)\big)_X \, dt, \quad u, v \in H^m(a, b; X),$$

and the induced norm

$$\|u\|_m^2 = \sum_{j=0}^m \int_a^b \|u^{(j)}(t)\|_X^2 \, dt, \quad u \in H^m(a, b; X).$$

Let us point out that any inner product space can be extended (uniquely up to isomorphism) to a Hilbert space, by a completion procedure similar to that used in the proof of Theorem 2.8. To illustrate this consider the space $C[0, 2]$ endowed with the scalar product

$$\langle u, v \rangle = \int_0^2 u(t)v(t) \, dt, \quad u, v \in C[0, 2],$$

and the induced norm

$$\|u\|_{L^2}^2 = \langle u, u \rangle = \int_0^2 u(t)^2 \, dt, \quad u \in C[0, 2].$$

The space $(C[0,2], \| \cdot \|_{L^2})$ is not complete (i.e., it is not a Hilbert space), as can be seen by using the sequence $(u_n)_{n \geq 2}$ defined by

$$u_n(t) = \begin{cases} 0, & 0 \leq t \leq 1 - \frac{1}{n}, \\ nt - n + 1, & 1 - \frac{1}{n} < t < 1, \\ 1, & 1 \leq t \leq 2, \end{cases}$$

but it can be extended to the Hilbert space $(L^2(0,2), \| \cdot \|_{L^2})$ (each element $\in C[0,2]$ being identified with its L^2 equivalence class).

If X is a finite dimensional, inner product space, then it is a Hilbert space with respect to the norm induced by the corresponding inner product, so no extension is needed (in particular, $\mathbb{R}^k$ and $\mathbb{C}^k$ are Hilbert spaces).

6.2   Jordan–von Neumann Characterization Theorem

Our aim in this chapter is to present the main properties of Hilbert spaces which are of course common to all the particular spaces mentioned above. First of all, we state the following characterization result due to Jordan and von Neumann.[2]

Theorem 6.1 (Jordan–von Neumann). *Let $(H, \| \cdot \|)$ be a normed space. Then the norm $\| \cdot \|$ is given by a scalar product (i.e., there exists a scalar product $(\cdot, \cdot) : H \times H \to \mathbb{K}$ such that $\|x\| = \sqrt{(x,x)}$, $x \in H$) if and only if $\| \cdot \|$ satisfies the parallelogram law. (Hence, a Banach space $(H, \| \cdot \|)$ is Hilbert $\iff$ its norm $\| \cdot \|$ satisfies the parallelogram law).*

Proof. Necessity has already been proved in Chap. 1, though we repeat here the proof which is immediate. Assuming that $\| \cdot \|$ is generated by a scalar product $(\cdot, \cdot)$, we have for all $x, y \in H$

$$\begin{aligned} \|x + y\|^2 + \|x - y\|^2 &= (x + y, x + y) + (x - y, x - y) \\ &= 2(\|x\|^2 + \|y\|^2), \end{aligned} \tag{6.2.1}$$

i.e., the norm satisfies the parallelogram law.

---

[2] Pascual Jordan, German theoretical and mathematical physicist, 1902–1980; John von Neumann, Hungarian-American mathematician, physicist, and computer scientist, 1903–1957.

Now let us prove sufficiency. Assume that the norm $\|\cdot\|$ of H satisfies the parallelogram law (see (6.2.1)).
Consider first the case $\mathbb{K} = \mathbb{R}$. Define $f : H \times H \to \mathbb{R}$ by

$$f(x, y) = \frac{1}{4}\left(\|x + y\|^2 - \|x - y\|^2\right), \quad x, y \in H,$$

which we will show is a scalar product on H. Clearly,

$$f(x, x) = \frac{1}{4}\|2x\|^2 = \|x\|^2 \quad \forall x \in H, \tag{6.2.2}$$

$$f(x, y) = f(y, x) \quad \forall x, y \in H, \tag{6.2.3}$$

$$f(x, 0) = 0 \quad \forall x \in H. \tag{6.2.4}$$

Obviously, for any $x_1, x_2, y \in H$, we have

$$f(x_1 + x_2, y) = \frac{1}{4}\left(\|x_1 + x_2 + y\|^2 + \|x_1 + x_2 - y\|^2\right),$$

$$f(x_1 - x_2, y) = \frac{1}{4}\left(\|x_1 - x_2 + y\|^2 + \|x_1 - x_2 - y\|^2\right).$$

Add the two equations and apply the parallelogram law to get

$$
\begin{aligned}
f(x_1 + x_2, y) + f(x_1 - x_2, y) &= \frac{1}{2}\left(x_1 + y\|^2 + \|x_2\|^2 \right.\\
&\quad \left. - \|x_1 - y\|^2 - \|x_2\|^2\right) \\
&= \frac{1}{2}\left(\|x_1 + y\|^2 - \|x_1 - y\|^2\right) \\
&= 2f(x_1, y). \tag{6.2.5}
\end{aligned}
$$

In the special case $x_1 = x_2 = x$ we have (see also (6.2.4) and (6.2.3))

$$f(2x, y) = 2f(x, y) \quad \forall x, y \in H. \tag{6.2.6}$$

Now choose in (6.2.5) $x_1 + x_2 = x$ and $x_1 - x_2 = x'$ to obtain

$$f(x, y) + f(x', y) = 2f\left(\frac{x + x'}{2}, y\right),$$

which by (6.2.6) gives

$$f(x + x', y) = f(x, y) + f(x', y) \quad \forall x, x', y \in H. \tag{6.2.7}$$

From (6.2.7) we obtain $f(nx, y) = nf(x, y)$ for all $n \in \mathbb{N}$ which can be extended to

$$f(nx, y) = nf(x, y) \quad \forall x, y \in H, \forall n \in \mathbb{Z}, \tag{6.2.8}$$

since $f(-x, y) = -f(x, y)$ (by (6.2.7)). Now for a rational number $r = m/n$, $m, n \in \mathbb{Z}$, $n \neq 0$, we have (by (6.2.8))

$$f\left(\frac{m}{n}x, y\right) = mf\left(\frac{1}{n}x, y\right) = \frac{m}{n}f(x, y),$$

so

$$f(rx, y) = rf(x, y) \quad \forall x, y \in H, \ \forall r \in \mathbb{Q}.$$

Since f is continuous on $H \times H$, this extends to $r \in \mathbb{R}$, i.e.,

$$f(rx, y) = rf(x, y) \quad \forall x, y \in H, \ \forall r \in \mathbb{R}. \tag{6.2.9}$$

Summarizing, we see that f satisfies (6.2.2), (6.2.3), (6.2.7), and (6.2.9), so $f(\cdot, \cdot)$ is a scalar product and generates the given norm: $\|x\|^2 = f(x, x)$, $x \in H$.

Sufficiency in the complex case $\mathbb{K} = \mathbb{C}$ can be treated similarly, with $f : H \times H \to \mathbb{C}$ defined by

$$f(x, y) = \frac{1}{4} \sum_{m=0}^{3} i^m \|x + i^m y\|^2, \quad x, y \in H,$$

where i is the imaginary unit.   □

Remark 6.2. In fact, the scalar product generating a norm is unique. Indeed, if $(\cdot, \cdot)$ and $\langle \cdot, \cdot \rangle$ are two scalar products such that $(x, x) = \langle x, x \rangle = \|x\|^2$, $x \in H$, then we easily derive from

$$(x + y, x + y) = \langle x + y, x + y \rangle \quad \forall x, y \in H,$$

that

$$\mathrm{Re}(x, y) = \mathrm{Re}\langle x, y \rangle \quad \forall x, y \in H, \tag{6.2.10}$$

and this completes the proof in the real case. If $\mathbb{K} = \mathbb{C}$, then by replacing y by iy in (6.2.10), we also get

$$\mathrm{Im}(x, y) = \mathrm{Im}\langle x, y \rangle \quad \forall x, y \in H.$$

Remark 6.3. We have already noticed that $\mathbb{R}^k$ equipped with the usual Euclidean norm is a Hilbert space, but $\mathbb{R}^k$ is not Hilbert with respect to other norms, such as

$$\|u\|_1 = \sum_{i=1}^{k} |u_i|, \quad \text{or} \quad \|u\|_{\max} = \max_{1 \leq i \leq k} |u_i|, \quad u = (u_1, \ldots, u_k) \in \mathbb{R}^k.$$

Indeed, one can easily find pairs of vectors that do not satisfy the parallelogram law expressed in terms of these norms.

Similarly, $L^1(a, b)$, $-\infty \le a < b \le \infty$, equipped with its usual norm, is not a Hilbert space, as can be seen by finding a pair of functions $f, g \in L^1(a, b)$ that does not satisfy the parallelogram law (do it!).

6.3 Projections in Hilbert Spaces

A Hilbert space is similar in many respects to k-dimensional Euclidean space. That is why Hilbert spaces are more useful in applications than general Banach spaces.

Theorem 6.4. *Let H be a Hilbert space with scalar product $(\cdot, \cdot)$ and induced norm $\| \cdot \|$, and let C be a nonempty, convex, closed subset of H. Then for all $x \in H$ there exists a unique $y \in C$ such that*

$$\|x - y\| = d(x, C) := \inf_{v \in C} \|x - v\| . \qquad (6.3.11)$$

Proof. First we prove the existence of y. If $x \in C$ then $d(x, C) = 0$ so a good candidate is $y = x$.

Assume $x \in H \setminus C$. Denote $\rho = d(x, C)$. By the definition of inf, for all $n \in \mathbb{N}$ there exists $y_n \in C$ such that

$$\rho \le \|x - y_n\| < \rho + \frac{1}{n},$$

which gives

$$\lim_{n \to \infty} \|x - y_n\| = \rho . \qquad (6.3.12)$$

We have $\rho > 0$. Indeed if $\rho = 0$, then by (6.3.12) $y_n \to x$ and C is closed, so $x \in C$, contradiction.

Apply the parallelogram law (see (6.2.1)) to $x - y_n$ and $x - y_m$ to get

$$\|2x - (y_n + y_m)\|^2 + \|y_n - y_m\|^2 = 2 \left(\|x - y_n\|^2 + \|x - y_m\|^2 \right), \quad (6.3.13)$$

for all n, m. Consider the first term of the left-hand side of (6.3.13) and factor out a 4

$$4\|x - (1/2)(y_n + y_m)\|^2 \ge 4\rho^2. \qquad (6.3.14)$$

Note that $(1/2)(y_n + y_m)$ is a convex combination of elements of C and therefore is in C by convexity. Hence (see (6.3.13) and (6.3.14)),

$$\|y_n - y_m\|^2 \le 2 \left(\|x - y_n\|^2 + \|x - y_m\|^2 \right) - 4\rho^2. \qquad (6.3.15)$$

Using (6.3.12) we get that (y_n) is Cauchy because the right-hand side of (6.3.15) converges to 0 as $n, m \to \infty$. Therefore (y_n) converges strongly to some y, and $y \in C$ because C is closed. It follows from (6.3.12) that

$$\|x - y\| = \rho \,.$$

We now prove uniqueness. Suppose $\|x - y\| = \rho = \|x - y'\|$ for some $y, y' \in C$. We use the parallelogram law for $x - y, x - y'$ to obtain

$$\|2x - (y + y')\|^2 + \|y - y'\|^2 = 2\big(\|x - y\|^2 + \|x - y'\|^2\big)$$

which implies

$$4\|x - (1/2)(y + y')\|^2 + \|y - y'\|^2 = 4\rho^2 \,. \tag{6.3.16}$$

$(1/2)(y + y') \in C$ since it is a convex combination, therefore

$$4\|x - (1/2)(y + y')\|^2 \geq 4\rho^2$$

yielding (see (6.3.16))

$$\|y - y'\|^2 \leq 4\rho^2 - 4\rho^2 = 0 \,,$$

and thus $y = y'$. $\square$

Remark 6.5. Both assumptions (C closed and convex) are essential. For example, if C is an open disc in $\mathbb{R}^2$, then there is no y for $x \in \mathbb{R}^2 \backslash C$. On the other hand, if C is not convex there may exist more (possibly infinitely many) y's for the same x, as the reader can easily imagine.

Definition 6.6. *Let $\emptyset \neq C \subset H$ be a closed and convex set. A point y as above is called the* **projection** *of x on C and is denoted $y = P_C x$. Since a projection exists and is unique for any $x \in H$ we can define a* **projection operator** $P_C : H \to C : x \mapsto y = P_C x$.

Theorem 6.7. *Let H be a Hilbert space and let $\emptyset \neq C \subset H$ be a closed and convex set. For $x \in H$, $y \in C$ the following are equivalent:*

(a) $y = P_C x$;

(b) $\|x - y\| \leq \|x - v\|$ *for all $v \in C$;*

(c) $\mathrm{Re}(x - y, y - v) \geq 0$ *for all $v \in C$;*

(d) $\mathrm{Re}(x - v, y - v) \geq 0$ *for all $v \in C$.*

If H is a real Hilbert space, then the "Re" from (c) and (d) can be removed.

Proof.

$(a) \iff (b)$: Trivial.

$(b) \implies (c)$: $\|x - y\|^2 \leq \|x - v\|^2$ for all $v \in C$. Let $v = (1 - \lambda)y + \lambda w$ for $0 < \lambda < 1$, and $w \in C$. Since v is a convex combination, v is in C. We have

$$\|x - y\|^2 \leq \|x - y + \lambda(y - w)\|^2$$
$$\leq \|x - y\|^2 + 2\lambda \operatorname{Re}(x - y, y - w) + \lambda^2 \|y - w\|^2,$$

so that

$$0 \leq 2\operatorname{Re}(x - y, y - w) + \lambda \|y - w\|^2.$$

Let $\lambda \to 0^+$ to find

$$\operatorname{Re}(x - y, y - w) \geq 0 \quad \text{for all } w \in C.$$

$(c) \implies (b)$: Since $\operatorname{Re}(x - y, y - x + x - v) \geq 0$ we have

$$\|x - y\|^2 \leq \operatorname{Re}(x - y, x - v)$$
$$\leq |(x - y, x - v)|$$
$$\leq \|x - y\| \cdot \|x - v\| \quad \forall v \in C,$$

so if $\|x - y\| = 0$ then we are done; otherwise divide by it, and we get

$$\|x - y\| \leq \|x - v\|, \quad \forall v \in C.$$

$(c) \implies (d)$: $\operatorname{Re}(x - v + v - y, y - v) \geq 0$ for all $v \in C$ so that

$$\operatorname{Re}(x - v, y - v) \geq \|y - v\|^2 \geq 0, \quad \forall v \in C.$$

$(d) \implies (c)$: Replacing v in (d) by $(1 - \lambda)y + \lambda w$ for $\lambda \in (0, 1)$, $w \in C$, we get

$$\operatorname{Re}(x - y + \lambda(y - w), \lambda(y - w)) \geq 0,$$

and, as λ is strictly positive, this implies

$$\operatorname{Re}(x - y, y - w) + \lambda \|y - w\|^2 \geq 0.$$

Thus, letting $\lambda \to 0^+$ we obtain

$$\operatorname{Re}(x - y, y - w) \geq 0 \quad \forall w \in C. \qquad \square$$

Remark 6.8. The projection operator is Lipschitz.

Proof. Using condition (c) of Theorem 6.7 we have

$$\mathrm{Re}(x_1 - P_C x_1, P_C x_1 - P_C x_2) \geq 0,$$
$$\mathrm{Re}(x_2 - P_C x_2, P_C x_1 - P_C x_2) \geq 0.$$

Add the two to obtain

$$\mathrm{Re}(P_C x_1 - P_C x_2, x_1 - P_C x_1 - x_2 + P_C x_2) \geq 0,$$

which implies

$$\mathrm{Re}(P_C x_1 - P_C x_1, x_1 - x_2) \geq \|P_C x_1 - P_C x_2\|^2.$$

By the Bunyakovsky–Cauchy–Schwarz inequality, this leads to

$$\|P_C x_1 - P_C x_2\| \leq \|x_1 - x_2\| \quad \forall x_1, x_2 \in H.$$

Thus P_C is Lipschitz with constant $L = 1$. For this reason the operator P_C is also called **nonexpansive**.   □

Remark 6.9. Let $C \subset H$ be a closed linear subspace. By condition (c) of Theorem 6.7 we have for all $v \in C$, $\mathrm{Re}(x - y, y - v) \geq 0$, and in fact we can write it as $\mathrm{Re}(x - y, v) \geq 0$ for all $v \in C$ since C is a linear subspace. Both $v, -v \in C$ because of linearity and this gives equality $\mathrm{Re}(x - y, v) = 0$ for all $v \in C$. We can also replace v with iv, and so $\mathrm{Im}(x - y, v) = 0$, therefore

$$(x - y, v) = 0, \quad \forall v \in C. \tag{6.3.17}$$

In general, when two vectors $w_1, w_2 \in H$ satisfy $(w_1, w_2) = 0$ they are said to be **orthogonal** by analogy with orthogonality in Euclidean space, and we write $w_1 \perp w_2$. So, (6.3.17) can be expressed as $(x - y) \perp C$. The reader is invited to imagine what the orthogonality relation (6.3.17) looks like in the Euclidean space $\mathbb{R}^3$ equipped with the usual scalar product and norm.

6.4   The Riesz Representation Theorem

Let $(H, (\cdot, \cdot), \|\cdot\|)$ be a Hilbert space and let $M \subset H$ be a closed linear subspace. The **orthogonal complement** $M^\perp$ of M is defined as

$$M^\perp = \{u \in H; \ (u, v) = 0 \ \ \forall v \in M\}$$

and is a closed subset (subspace) because $(\cdot, \cdot) : H \times H \to \mathbb{K}$ is continuous.

Orthogonal Decomposition of H: We claim that *any vector $u \in H$ can be written as $u = u_1 + u_2$ with $u_1 \in M$ and $u_2 \in M^\perp$, and this decomposition is unique.* We write $H = M \oplus M^\perp$ and call it a **direct sum**.

Proof. Note that $u_1 = P_M u$ (which is unique) is the component in M, while $u_2 = u - u_1 = u - P_M u$ is in $M^\perp$ because $(u - P_M u, v) = 0$ for all $v \in M$ (see (6.3.17)). Let us now prove that this decomposition ($u = u_1 + u_2$) is unique.

Suppose that $u = u_1 + u_2 = u_1' + u_2'$ with $u_1, u_1' \in M$ and $u_2, u_2' \in M^\perp$. Then

$$0 = (u_1 - u_1' + u_2 - u_2', u_1 - u_1')$$
$$= \|u_1 - u_1'\|^2 + (u_2 - u_2', u_1 - u_1'),$$

where the second term is 0 because $u_1 - u_1' \in M$, $u_2 - u_2' \in M^\perp$. Thus $\|u_1 - u_1'\|^2 = 0$ so that $u_1 = u_1'$ which in turn implies $u_2 = u_2'$. $\square$

Theorem 6.10 (Riesz Representation Theorem). *Let $(H, (\cdot, \cdot), \|\cdot\|)$ be a Hilbert space. For all $f \in H^*$ (i.e., f is a continuous linear functional from H to $\mathbb{K}$) there exists a unique $v \in H$ such that*

$$f(u) = (u, v) \ \ \forall u \in H \quad and \quad \|v\| = \|f\|.$$

Proof.

Step 1.   We first show that such a v is unique. Suppose that $(u, v) = (u, v')$ for all $u \in H$, then $(u, v - v') = 0$ for all $u \in H$ and in particular $(v - v', v - v') = 0$ so $v = v'$.

Step 2. We now prove the existence of v. If $f = 0$ then clearly $v = 0$ works. If $f \neq 0$ consider the nullspace $N(f) = \{z \in H;\ f(z) = 0\}$. It is a closed linear subspace so $H = N(f) \oplus N(f)^{\perp}$. In fact $N(f) \neq H$ because f is not identically 0. Thus there exists $u_0 \in N(f)^{\perp} \setminus \{0\}$. We may assume $f(u_0) = 1$ by scaling. Let $u \in H$ be arbitrary and define

$$w = u - f(u)u_0 \,.$$

Now consider

$$f(w) = f(u) - f(u)f(u_0) = f(u) - f(u) = 0 \,,$$

showing that $w \in N(f)$. So

$$u = w + f(u)u_0 \quad \text{with} \quad w \in N(f),\ f(u)u_0 \in N(f)^{\perp} \,,$$

and this decomposition is unique. Thus

$$(u, u_0) = (w, u_0) + f(u)(u_0, u_0) = 0 + f(u)\|u_0\|^2 \,,$$

and solving for $f(u)$,

$$f(u) = \left(u, \frac{1}{\|u_0\|^2} u_0\right) ,$$

so $f(u)$ is of the given form with $v = \|u_0\|^{-2} u_0$, and v is unique by the previous step.

Step 3. We finally prove that $\|v\| = \|f\|$. For $f = 0$ this is obvious, so assume that $f \neq 0$, which implies $v \neq 0$. By Bunyakovsky–Cauchy–Schwarz

$$|f(u)| \leq \|v\| \cdot \|u\| \,,$$

and by considering those u with $\|u\| \leq 1$, we get

$$\|f\| \leq \|v\| \,. \tag{6.4.18}$$

Now $f(v) = (v, v) = \|v\|^2$ so that

$$f\left(\frac{1}{\|v\|} v\right) = \|v\| \,,$$

which combined with (6.4.18) shows that $\|f\| = \|v\|$.   □

Remark 6.11. Recall that in Sect. 4.4 of Chap. 4 we asked whether functionals f from the dual of $(C[a,b], \|\cdot\|_{L^2(a,b)})$, $-\infty < a < b < \infty$, can be expressed as $f(u) = (u,v)_{L^2(a,b)}$, $u \in C[a,b]$, with $v \in C[a,b]$. The answer is, in general, no. First of all, any $f \in (C[a,b], \|\cdot\|_{L^2(a,b)})^*$ can be extended by continuity to $(L^2(a,b), \|\cdot\|_{L^2(a,b)})$ which is a Hilbert space. By the Riesz Representation Theorem, for each such f (extended to $L^2(a,b)$) there exists a unique $v \in L^2(a,b)$ such that $f(u) = (u,v)_{L^2(a,b)}$, $\forall u \in L^2(a,b)$, but this v is not necessarily an element of $C[a,b]$ (i.e., v has no representative in $C[a,b]$). In fact, we can consider $f(u) = (u,v)_{L^2(a,b)}$, $u \in L^2(a,b)$, with $v \in L^2(a,b) \setminus C[a,b]$; this f is continuous on $(C[a,b], \|\cdot\|_{L^2(a,b)})$ and its representation as a scalar product, $f(u) = (u,v)_{L^2(a,b)}$, is unique (i.e., v is unique); but this v is not an element of $C[a,b]$, so the answer to the above question is negative.

Remark 6.12. In the proof of Theorem 6.10 we saw that for all $u \in H$, $0 \neq f \in H^*$ we have the decomposition $u = w + f(u)u_0$ with $w \in N(f)$, $u_0 \in N(f)^\perp$, $f(u_0) = 1$, so that $\dim N(f)^\perp = 1$. Another way to say this is that the **codimension** of $N(f)$ is 1. For such a functional f and for some $a \in \mathbb{K}$ we have an affine subspace of H,

$$Y := \{u \in H;\ f(u) = a\} = au_0 + N(f),$$

whose codimension is 1 (i.e., the codimension of $N(f)$ is 1), thus Y is a usual hyperplane if H is the Euclidean space.

Conversely, given a closed affine subspace Y of H of codimension 1, i.e., $Y = u_1 + Z$, for some $u_1 \in H$, $Z \subset H$ a closed linear subspace with codimension 1, there exists $u_0 \in H \setminus \{0\}$ which is orthogonal on Z, i.e., $(u, u_0) = 0$, $u \in Z$. Define $f : H \to \mathbb{K}$,

$$f(u) = (u, u_0), \quad \forall u \in H,$$

so that $f \in H^*$, $N(f) = Z$, $f \neq 0$ (since $f(u_0) = \|u_0\|^2 \neq 0$) and Y can be expressed by means of this f as follows:

$$Y = u_1 + N(f) = \{u \in H;\ f(u) = f(u_1)\}.$$

A simple example is $H = L^2(0,1)$, $Z = \{u \in H;\ \int_0^1 u(t)\,dt = 0\}$. Clearly, Z is a closed linear subspace of H with $\operatorname{codim} Z = 1$. Indeed, any $v \in H$ can be uniquely decomposed into

$$v(t) = \int_0^1 v(s)\,ds + \underbrace{\left[v(t) - \int_0^1 v(s)\,ds\right]}_{u(t)}$$

$$= C + u(t), \quad \text{for a.a. } t \in (0,1),$$

where $u \in Z$ and C is a constant, i.e., $H = \text{Span}\{1\} \oplus Z$. We can choose u_0 to be the constant function 1, so $f(u) = \int_0^1 u(t)\,dt$.

The Weak Topology of H

Taking into account the Riesz Representation Theorem, we see that the weak topology of H is generated by the neighborhood system

$$V_{v_1,v_2,\ldots,v_p;\varepsilon} = \{x \in H;\ |(x,v_j)| < \varepsilon,\ j = 1,\ldots,p\},$$

$$\varepsilon > 0,\ v_1,\ldots,v_p \in H,\ p \in \mathbb{N}.$$

So the fact that a sequence (x_n) in H converges weakly to some x means $(x_n, v) \to (x, v)$ for all $v \in H$.

If $\dim H = \infty$ then we can use the Gram–Schmidt method (see Chap. 1) to construct an infinite orthonormal sequence $(x_1, x_2, \ldots, x_n, \ldots)$. This sequence converges weakly to 0. Indeed, for $v \in H$ arbitrary, we have

$$\left\| \sum_{n=1}^N (v, x_n) x_n - v \right\|^2 = \sum_{n=1}^N |(x_n, v)|^2 - 2 \sum_{n=1}^N |(x_n, v)|^2 + \|v\|^2$$

$$= \|v\|^2 - \sum_{n=1}^N |(x_n, v)|^2$$

$$\geq 0,$$

so that

$$\sum_{n=1}^N |(x_n, v)|^2 \leq |v|^2, \quad \forall N \in \mathbb{N},$$

which is known as Bessel's inequality.[3] So the series $\sum_{n=1}^\infty |(x_n, v)|^2$ is convergent and consequently

---

[3]Friedrich Wilhelm Bessel, German astronomer, mathematician, physicist and geodesist, 1784–1846.

$$(x_n, v) \to 0, \quad \forall v \in H,$$

i.e., (x_n) converges weakly to 0. But (x_n) is not strongly convergent (to 0) since $\|x_n\| = 1$ for all $n \in \mathbb{N}$. Therefore, weak convergence in any infinite dimensional Hilbert space is different from strong convergence.

Based on the Riesz Representation Theorem, we can define the so-called **Riesz operator** $R : H \to H^*$ by $v \mapsto (\cdot, v)$ so that $(Rv)(u) = (u, v)$ for all $u, v \in H$ and $\|Rv\| = \|v\|$. As seen before, R is also bijective.

Theorem 6.13. *Every Hilbert space is reflexive.*

Proof. Let $\phi : H \to H^{**}$, $v \overset{\phi}{\mapsto} f_v \in H^{**}$ such that $f_v(x^*) = x^*(v)$ for all $x^* \in H^*$.
As we have already seen, ϕ is injective. For the convenience of the reader, let us prove this again in the present context. If $f_v = 0$, $x^*(v) = 0$ for all $x^* \in H^*$ which implies, by the Riesz Representation Theorem, that $(v, w) = 0$ for all $w \in H$ so that $v = 0$. Thus ϕ is injective.
We now prove that ϕ is surjective. Let $x^{**} \in H^{**}$ and define $u^* \in H^*$ by $u^*(v) := \overline{x^{**}(Rv)}$ for all $v \in H$. Denote $u = R^{-1}u^*$ and calculate

$$
\begin{aligned}
x^{**}(x^*) &= x^{**}(R(R^{-1}x^*)) \\
&= \overline{u^*(R^{-1}x^*)} \\
&= \overline{(R^{-1}x^*, u)} \\
&= (u, R^{-1}x^*) \\
&= x^*(u) \\
&= f_u(x^*),
\end{aligned}
$$

so that all functionals x^{**} are of the form $f_u(x^*)$, and ϕ is onto, i.e., for all $x^{**} \in H^{**}$ there exists $u \in H$ such that $x^{**} = f_u$. $\qquad \square$

Remark 6.14. The above proof is a direct one. In fact, Theorem 6.13 follows from the Milman–Pettis[4] general result we state without proof: *every uniformly convex Banach space is reflexive.*
Recall that a normed space $(H, \| \cdot \|)$ is said to be uniformly convex if $\forall \varepsilon \in (0, 2) \; \exists \delta > 0$ such that $\forall x, y \in H$, $\|x\| \le 1$, $\|y\| \le 1$, $\|x - y\| > \varepsilon$ we have $\|(1/2)(x + y)\| < 1 - \delta$.

---

[4]David P. Milman, Soviet and later Israeli mathematician, 1912–1982.

If H is a Hilbert space, it follows easily by using the parallelogram law that H is uniformly convex, hence reflexive (by Milman–Pettis).

6.5   Lax–Milgram Theorem

We begin this section with a preparatory lemma whose proof is based on the Banach Contraction Principle.

Lemma 6.15. *Let $(H, (\cdot, \cdot), \| \cdot \|)$ be a real Hilbert space and let $A : H \to H$ be a not necessarily linear operator satisfying*

(a) $(Au - Av, u - v) \geq c\|u - v\|^2$ *for all $u, v \in H$ (strong mono-tonicity);*

(b) $\|Au - Av\| \leq L\|u - v\|$ *for all $u, v \in H$ (Lipschitz condition),*

where c and L are given positive constants. Then for all $w \in H$ there exists a unique $u^* \in H$ such that $Au^* = w$, i.e., A is a bijection.

Proof. We first prove uniqueness: Suppose $u_1, u_2 \in H$ such that $Au_1 = w = Au_2$. Then by (a),

$$0 = (Au_1 - Au_2, u_1 - u_2) \geq c\|u_1 - u_2\|^2,$$

which implies $u_1 = u_2$.

We now prove existence: First we note that $c \leq L$ by using (a) and (b) together with Bunyakovsky–Cauchy–Schwarz. For a fixed $w \in H$, define $B : H \to H$ by

$$Bu = u - t(Au - w), \quad t > 0, \ u \in H.$$

Note that if there is a fixed point of B then it is u^* as desired. We wish to apply the Banach Contraction Principle in (H, d), where $d(u, v) = \|u - v\|$. We have for all $u, v \in H$

$$
\begin{aligned}
d(Bu, Bv)^2 &= \|Bu - Bv\|^2 \\
&= \|u - v\|^2 - 2t(u - v, Au - Av) + t^2\|Au - Av\|^2 \\
&\leq \|u - v\|^2 - \underbrace{2tc\|u - v\|^2}_{\text{from (a)}} + \underbrace{t^2 L^2\|u - v\|^2}_{\text{from (b)}} \\
&= \underbrace{(1 - 2tc + t^2 L^2)}_{\text{call this } m}\|u - v\|^2 \\
&= m\|u - v\|^2 \\
&= m\, d(u, v)^2.
\end{aligned}
$$

Obviously, $m \geq 0$. We choose t to minimize $m = m(t)$ and find that $t = \frac{c}{L^2}$. Thus the minimum value of m is

$$m = 1 - 2\frac{c^2}{L^2} + \frac{c^2}{L^2} = 1 - \frac{c^2}{L^2} \geq 0\,,$$

since $c \leq L$. If $c = L$, then $m = 0$, so B is constant, i.e., $Bu = w_0$, so that $w_0 = u - (c/L^2)(Au - w)$. In this case A is affine, namely

$$Au = \frac{L^2}{c}(u - w_0) + w\,,$$

so that $u^* = w_0$.

When $c < L$ then $0 < m < 1$ so that B is a contraction and hence by the Banach Contraction Principle (see Sect. 2.5) B has a unique fixed point u^*.   □

Theorem 6.16 (Nonlinear Lax–Milgram Theorem). [5] *Let H be a real Hilbert space and consider two functionals $a : H \times H \to \mathbb{R}$ and $b : H \to \mathbb{R}$ satisfying*

1. *For all $u \in H$ the map $v \mapsto a(u,v)$ is linear and continuous on H (i.e., it belongs to H^*);*

2. *$a(u, u - v) - a(v, u - v) \geq c\|u - v\|^2$ for all $u, v \in H$ and some $c > 0$;*

3. *$|a(u, w) - a(v, w)| \leq L\|u - v\| \cdot \|w\|$ for all $u, v, w \in H$ and some $L > 0$;*

4. *b is a continuous linear functional (i.e., $b \in H^*$).*

Then there exists a unique $u \in H$ such that

$$a(u, v) = b(v) \quad \forall v \in H\,. \tag{6.5.19}$$

Proof. By the first assumption and the Riesz Representation Theorem 6.10 for all $u \in H$ there exists a unique $z \in H$ such that $a(u, v) = (v, z)$ for all $v \in H$. So there exists an operator $A : H \to H$ defined by $Au := z$. We now rewrite the second condition

$$a(u, u - v) - a(v, u - v) = (u - v, Au) - (u - v, Av)$$
$$= (u - v, Au - Av)$$

---

[5]Peter D. Lax, Hungarian-born American mathematician, born 1926; Arthur N. Milgram, American mathematician, 1912–1961.

and since $\mathbb{K} = \mathbb{R}$

$$= (Au - Av, u - v)$$
$$\geq c\|u - v\|^2,$$

for all $u, v \in H$, so A satisfies condition (a) of the previous lemma. From the third assumption we have for all $u, v, z \in H$

$$|a(u, z) - a(v, z)| = |(z, Au) - (z, Av)|$$
$$= |(z, Au - Av)|$$
$$\leq L\|u - v\| \cdot \|z\|.$$

Choosing $z = Au - Av$ we see that operator A also satisfies condition (b) of Lemma 6.15.

On the other hand, by the fourth assumption and the Riesz Representation Theorem there exists a unique w such that $b(v) = (v, w)$ for all $v \in H$. Now (6.5.19) can be written as

$$[(v, Au) = (v, w), \quad \forall v \in H] \iff Au = w,$$

so the conclusion of the theorem follows by Lemma 6.15. $\square$

Theorem 6.17 (Classic Lax–Milgram Theorem). *Let H be a real Hilbert space and consider two functionals $a : H \times H \to \mathbb{R}$ and $b : H \to \mathbb{R}$ satisfying*

1. *a is bilinear;*

2. *a is bounded (continuous) on $H \times H$, namely $|a(u, v)| \leq L\|u\| \cdot \|v\|$ for all $u, v \in H$ for some $L > 0$;*

3. *a is strongly positive (or coercive), i.e., there exists $c > 0$ such that $a(v, v) \geq c\|v\|^2$ for all $v \in H$;*

4. *b is linear and continuous (i.e., $b \in H^*$).*

Then there exists a unique $u \in H$ satisfying

$$a(u, v) = b(v) \quad \forall v \in H. \tag{6.5.19'}$$

If, in addition, a is symmetric (i.e., $a(u, v) = a(v, u)$ for all $u, v \in H$) then u is a solution of (6.5.19') if and only if it is a solution (minimizer) of the quadratic minimization problem

$$\min_{v \in H} \left\{ \frac{1}{2} a(v, v) - b(v) \right\}. \tag{6.5.20}$$

Proof. Observe that the conditions of Theorem 6.16 are satisfied, so all that remains is to prove the final statement.
Define

$$F(v) = \frac{1}{2}a(v, v) - b(v), \quad v \in H.$$

If u is a solution of (6.5.20) then $F(u) \leq F(v)$ for all $v \in H$. Define $\phi(t) = F(u + tv)$ for $t \in \mathbb{R}$, $v \in H$. We have

$$\phi(t) = \frac{1}{2}a(u + tv, u + tv) - b(u + tv)$$

$$= \frac{1}{2}a(u, u) + ta(u, v) + \frac{1}{2}t^2 a(v, v) - b(u) - tb(v)$$

$$= F(u) + t[a(u, v) - b(v)] + \frac{1}{2}t^2 a(v, v).$$

Therefore,

$$\phi'(t) = a(u, v) - b(v) + ta(v, v),$$

hence

$$a(u, v) - b(v) = \phi'(0) = 0,$$

since $t = 0$ is a minimizer of ϕ, so that u satisfies (6.5.19') because v is arbitrary.
Conversely, suppose that u satisfies (6.5.19'). We must show $F(u) \leq F(v)$ for all $v \in H$. It is enough to prove $F(u+v) - F(u)$ is nonnegative:

$$F(u+v) - F(u) = \frac{1}{2}a(u + v, u + v) - b(u + v) - \frac{1}{2}a(u, u) + b(u)$$

$$= \frac{1}{2}a(u, u) + \underbrace{a(u, v)}_{\text{symmetric}} + \frac{1}{2}a(v, v) - b(v) - \frac{1}{2}a(u, u)$$

$$= \underbrace{a(u, v) - b(v)}_{=0} + \frac{1}{2}a(v, v)$$

$$= \frac{1}{2}a(v, v)$$

$$\geq 0.$$

So $F(u) \leq F(u + v)$ for all $v \in H$ which implies u is a solution to (6.5.20). $\square$

Next we illustrate the above results with some applications.

Dirichlet's Principle[6]: *Let $\emptyset \neq \Omega \subset \mathbb{R}^k$ be a bounded domain. For all $f \in L^2(\Omega)$ there exists a unique $u \in H_0^1(\Omega)$ which is a solution to the following minimization problem:*

$$\min_{v \in H_0^1(\Omega)} \left\{ \frac{1}{2} \int_\Omega \nabla v \cdot \nabla v \, dx - \int_\Omega fv \, dx \right\}, \qquad (6.5.21)$$

and equivalently u is a solution to

$$\begin{cases} u \in H_0^1(\Omega), \\ \int_\Omega \nabla u \cdot \nabla v \, dx = \int_\Omega fv \, dx \quad \forall v \in H_0^1(\Omega). \end{cases} \qquad (6.5.22)$$

Remark 6.18. In the sense of distributions we can rewrite (6.5.22) to be

$$u \in H_0^1(\Omega), \quad -\Delta u = f \ \text{ in } \ \Omega, \qquad (6.5.23)$$

which is known as the **Euler–Lagrange equation**[7] associated with the minimization problem (6.5.21) (being a Poisson equation in this example) and u being 0 on the boundary is interpreted as meaning the trace of u on the boundary $\partial\Omega$ is 0. Indeed, for every test function $\phi \in C_0^\infty(\Omega)$ we have

$$\int_\Omega \nabla u \cdot \nabla \phi \, dx = \int_\Omega f\phi \, dx \iff (-\Delta u, \phi) = (f, \phi),$$

i.e., $-\Delta u = f$ in $\mathcal{D}'(\Omega)$. Since f is in $L^2(\Omega)$, $-\Delta u$ is as well, so u satisfies the equation $-\Delta u = f$ for a.a. $x \in \Omega$. In fact, if $\partial\Omega$ is smooth enough, then $u \in H_0^1(\Omega) \cap H^2(\Omega)$ (see [39, Theorem 3.1, p. 212]). Moreover, if $f \in C^\infty(\Omega)$ then so is u. Actually, the following regularity result holds.

Lemma 6.19 (Weyl). *If $\emptyset \neq \Omega \subset \mathbb{R}^k$ is open, $f \in L^\infty(\Omega)$, and $u \in \mathcal{D}'(\Omega)$ satisfies the equation $-\Delta u = f$ in the sense of distributions, then $u \in C^\infty(\Omega)$.*

Proof of Dirichlet's Principle.
We wish to use the classical Lax–Milgram Theorem 6.17. Denote $H := H_0^1(\Omega)$. Recall that H is a real Hilbert space as a closed subspace of $H^1(\Omega)$. According to Remark 5.26, H can be equipped with the norm

---

[6]Johann Peter Gustav Lejeune Dirichlet, German mathematician, 1805–1859.
[7]Joseph-Louis Lagrange, Italian mathematician and astronomer, 1736–1813.

$$\|u\|_* = \left(\int_\Omega |\nabla u|^2 \, dx \right)^{1/2}, \quad u \in H = H_0^1(\Omega),$$

which is equivalent with the usual $H^1(\Omega)$ norm. Define $a : H \times H \to \mathbb{R}$ and $b : H \to \mathbb{R}$ by

$$a(u,v) := \int_\Omega \nabla u \cdot \nabla v dx, \quad b(v) := \int_\Omega fv \, dx.$$

Clearly a is bilinear and symmetric. Moreover, a is also continuous (bounded),

$$|a(u,v)| \le \|u\|_* \cdot \|v\|_* \quad \forall u, v \in H,$$

and coercive

$$a(v,v) = \int_\Omega \nabla v \cdot \nabla v \, dx = \|v\|_*^2 \quad \forall v \in H.$$

Obviously, b is linear and also continuous because

$$|b(v)| \le \|f\|_{L^2(\Omega)} \|v\|_{L^2(\Omega)}$$

so by Poincaré's inequality

$$\le C\|f\|_{L^2(\Omega)} \|v\|_* \quad \forall v \in H.$$

Thus all the conditions of Theorem 6.17 are fulfilled, so the proof of Dirichlet's Principle is complete.

Now let us consider the following nonlinear boundary value problem:

$$\begin{cases} -\Delta u(x) + \beta(u(x)) = f(x), & x \in \Omega, \\ u = 0, & x \in \partial\Omega, \end{cases} \tag{6.5.24}$$

where $\emptyset \ne \Omega \subset \mathbb{R}^k$ is a bounded domain, $f \in L^2(\Omega)$, and $\beta : \mathbb{R} \to \mathbb{R}$ is a nonlinear Lipschitz continuous, nondecreasing function. We wish to prove that problem (6.5.24) has a unique solution $u \in H_0^1(\Omega)$. To this purpose we can apply Theorem 6.16 with $H = H_0^1(\Omega)$ equipped with the norm $\|\cdot\|_*$ as above, and with $a : H \times H \to \mathbb{R}$, $b : H \to \mathbb{R}$ defined by

$$a(u,v) = \int_\Omega \nabla u \cdot \nabla v \, dx + \int_\Omega \beta(u)v \, dx, \quad b(v) = \int_\Omega fv \, dx.$$

It is a simple exercise to show that all the assumptions of Theorem 6.16 are fulfilled, so there is a unique $u \in H = H_0^1(\Omega)$ satisfying

$$u(u,v) = b(v), \quad \forall v \in H,$$

i.e., $-\Delta u + \beta \circ u = f$ in $\mathcal{D}'(\Omega)$. Note that $\beta \circ u \in L^2(\Omega)$ so $-\Delta u = f - \beta(u)$ is in $L^2(\Omega)$ as well, i.e., u satisfies the given equation for a.a. $x \in \Omega$. In fact, if $\partial\Omega$ is smooth enough, then $u \in H^2(\Omega)$ (cf. [39, Theorem 3.1, p. 212]).

6.6   Fourier Series Expansions

Let $(H, (\cdot,\cdot), \|\cdot\|)$ be a Hilbert space with $m := \dim H \geq 1$.
If $m < \infty$ then starting from a basis of H, say $B = \{e_1, \ldots, e_m\}$, one can construct by the Gram–Schmidt procedure (see Chap. 1) an orthonormal basis $B' = \{u_1, \ldots, u_m\}$, i.e., $(u_i, u_j) = \delta_{ij}$, $i, j = 1, \ldots, m$. So every $u \in H$ can be written as

$$u = \sum_{i=1}^m c_i u_i, \quad c_i \in \mathbb{K}, \ i = 1, \ldots, m.$$

This yields $c_i = (u, u_i)$, $i = 1, \ldots, m$, hence

$$u = \sum_{i=1}^m (u, u_i) u_i, \quad \forall u \in H. \tag{6.6.25}$$

(6.6.25) is called the **Fourier expansion** of u and (u, u_i) are called **Fourier coefficients**.[8]

In what follows we are interested in Fourier series expansions in the case $m = \infty$. A set $S \subset H$ is said to be an **orthonormal set** if for any pair $u, v \in S$, $u \neq v$, we have

$$\|u\| = \|v\| = 1, \quad \text{and} \quad (u, v) = 0.$$

An orthonormal set $S \subset H$ is called a **complete orthonormal system** in H if it is not properly included in any other orthonormal set in H.

---

[8]Jean-Baptiste Joseph Fourier, French mathematician and physicist,1768–1830.

Remark 6.20. Claim: any Hilbert space $H \neq \{0\}$ has a complete orthonormal system, and any orthonormal set can be extended to a complete orthonormal system. Indeed, choosing $x \in H \setminus \{0\}$ and denoting $u_1 = (1/\|x\|)x$ we see that $\{u_1\}$ is an orthonormal system in H. Consider the collection of all orthonormal systems in H which contain $\{u_1\}$. This collection is partially ordered with respect to the usual inclusion relation. By Zorn's Lemma there exists a maximal element for the collection, which is a complete orthonormal system in H. If $m = \infty$ then this system is infinite, be it countable or not (this issue will be clarified later).

Theorem 6.21. *Let $(H, (\cdot, \cdot), \| \cdot \|)$ be an infinite dimensional Hilbert space and let $S = \{u_n\}_{n \in \mathbb{N}} \subset H$ be a countably infinite orthonormal system. Then the following are equivalent:*

(a) *S is complete;*

(b) *$u = \sum_{n=1}^{\infty} (u, u_n)u_n \quad \forall u \in H$;*

(c) *$\sum_{n=1}^{\infty} |(u, u_n)|^2 = \|u\|^2 \quad \forall u \in H$ (Parseval's relation)[9];*

(d) *Span S is dense in H.*

Proof. First of all, using the orthogonality of system S, we have for all $u \in H$ and $N \in \mathbb{N}$

$$0 \leq \| \sum_{n=1}^{N} (u, u_n)u_n - u\|^2 = \|u\|^2 - \sum_{n=1}^{N} |(u, u_n)|^2. \qquad (6.6.26)$$

We deduce from (6.6.26) that (b) $\iff$ (c).

Let us prove that (b) $\implies$ (a). Assume by contradiction that (b) holds, but S is not complete, i.e., there exists a vector $\hat{u} \in H \setminus S$ such that $\|\hat{u}\| = 1$, and $(\hat{u}, u_n) = 0 \, \forall n \in \mathbb{N}$. From (b) with $u = \hat{u}$ it then follows $\hat{u} = 0$ which is a contradiction.

Now, we prove that (a) $\implies$ (b). Fix $u \in H$. By a standard computation we get

$$\left\| \sum_{n=m}^{m+p} (u, u_n)u_n \right\|^2 = \sum_{n=m}^{m+p} |(u, u_n)|^2. \qquad (6.6.27)$$

Since the numerical series $\sum_{n=1}^{\infty} |(u, u_n)|^2$ is convergent (see (6.6.26)), we deduce from (6.6.27) that the sequence of partial

---

[9] Marc-Antoine Parseval, French mathematician, 1755–1836.

sums of the series in (b) is Cauchy in H, hence convergent to some $\tilde{u} \in H$, so we can write

$$\tilde{u} = \sum_{n=1}^{\infty} (u, u_n) u_n. \tag{6.6.28}$$

We compute

$$(\tilde{u}, u_j) = \lim_{N \to \infty} \left(\sum_{n=1}^{N} (u, u_n) u_n, u_j \right) = (u, u_j) \quad \forall j \in \mathbb{N},$$

so

$$(\tilde{u} - u, u_j) = 0 \quad \forall j \in \mathbb{N},$$

which implies $\tilde{u} = u$ by the completeness of S. Therefore $\tilde{u}$ in (6.6.28) can be replaced by u.

It is clear that $(b) \implies (d)$. To complete the proof it suffices to show $(d) \implies (a)$. Assume by contradiction that (d) holds but S is not complete, i.e., there exists a vector $v \in H \setminus S$ such that $\|v\| = 1$, and $(v, u_n) = 0 \; \forall n \in \mathbb{N}$. According to (d), we obtain $(v, w) = 0 \; \forall w \in H$, hence $v = 0$, another contradiction. $\square$

Remark 6.22. According to Theorem 6.21, if $S = \{u_n\}_{n \in \mathbb{N}}$ is a complete orthonormal system in H, then every $u \in H$ is the sum of the Fourier series associated with it (see (b)), similar to the finite dimensional case $m < \infty$. That is why S is also called a countable **orthonormal basis** of H. The next result is a characterization of the Hilbert spaces possessing countable orthonormal bases.

Theorem 6.23. *A Hilbert space has a countable orthonormal basis if and only if it is separable.*

Proof. Let H be a Hilbert space. Denote $m := \dim H$.

If $m < \infty$, then the result is trivial, so let us assume $m = \infty$.

Let $S = \{u_n\}_{n \in \mathbb{N}}$ be a (countable) orthonormal basis in H. Then $\operatorname{Span} S$ is dense in H (cf. Theorem 6.21). On the other hand, using the fact that $\mathbb{Q}$ is dense in $\mathbb{R}$, we can show that there exists a countable subset of $\operatorname{Span} S$ which is dense in $\operatorname{Span} S$, hence in H. Indeed, for any $u \in \operatorname{Span} S$, say $u = \sum_{k=1}^{p} \alpha_k u_k$, and any $\varepsilon > 0$, there are numbers $r_k \in \mathbb{Q}$ if H is a real Hilbert space, or $r_k \in \mathbb{Q} + i\mathbb{Q}$ if H is a complex Hilbert space, such that

$$|r_k - \alpha_k| < \frac{\varepsilon}{p}, \; k = 1, \dots, p \implies \left\| u - \sum_{k=1}^{p} r_k u_k \right\| < \varepsilon.$$

Thus H is separable.

Conversely, assume H is separable, i.e., there exists a countably infinite set, say $M = \{x_1, x_2, \ldots, x_n, \ldots\}$ such that $\overline{M} = H$. Using Gram–Schmidt (see Chap. 1) we can construct with vectors from M an orthonormal system $S = \{u_1, u_2, \ldots, u_n, \ldots\}$ eliminating dependent vectors of M if any. An inspection of the Gram–Schmidt method shows that in fact $M \subset \mathrm{Span}\, S$ so that

$$H = \overline{M} \subset \overline{\mathrm{Span}\, S} \subset H \quad \Rightarrow \quad \overline{\mathrm{Span}\, S} = H \,,$$

so S is an orthonormal basis (cf. Theorem 6.21). $\qquad\qquad\qquad\square$

Remark 6.24. If H is not separable, then the existence of a complete orthonormal system $S = \{u_i\}_{i \in I}$ in H is still valid (cf. Remark 6.20). Obviously, the index set I is no longer countable. Surprisingly, in this case, for every $u \in H$ there is a sequence of indices $i_1, i_2, \ldots$ such that

$$u = \sum_{j=1}^{\infty} (u, u_{i_j}) u_{i_j} \,,$$

i.e., u has a Fourier series expansion as in the separable case. For the proof of this result, see [51, pp. 86–87].

A Classical Fourier Series Expansion

Let $H = L^2(-\pi, \pi)$ with the usual scalar product

$$(f, g) = \int_{-\pi}^{\pi} f(x) g(x) \, dx \,, \quad f, g \in H \,,$$

and $\|f\| = \sqrt{(f, f)}$ for all $f \in H$. Let $S = \{u_n\}_{n=0}^{\infty}$, where

$$u_0 = \frac{1}{\sqrt{2\pi}}, \quad u_{2k-1}(x) = \frac{1}{\sqrt{\pi}} \cos kx, \quad u_{2k}(x) = \frac{1}{\sqrt{\pi}} \sin kx, \quad k = 1, 2, \ldots.$$

By a straightforward computation we can see that S is an orthonormal system in H. Moreover, S is complete as stated in the following result.

Theorem 6.25 (Fischer[10]–Riesz). *The orthonormal system S as above is a basis in $H = L^2(-\pi, \pi)$.*

---

[10]Ernst Sigismund Fischer, Austrian mathematician, 1875–1954.

Proof. According to Theorem 6.21 it suffices to prove that $\overline{\text{Span}\,S} = H$. We know that $C_0^\infty(-\pi, \pi)$ is dense in $L^2(-\pi, \pi)$ (see Theorem 5.8). To conclude we can use Weierstrass' lemma below (cf. [52, p. 205]). This is an approximation result with respect to the sup-norm of $C[-\pi, \pi]$ which is obviously stronger than the norm of $H = L^2(-\pi, \pi)$. $\qquad\qquad\qquad\qquad\qquad\qquad\qquad\qquad\qquad\qquad$ $\square$

Lemma 6.26 (Weierstrass). Span S *is dense in the space*

$$X = \{f \in C[-\pi, \pi];\ f(-\pi) = f(\pi)\}$$

equipped with the sup-norm $\|\cdot\|_C$, *where* S *is the function system defined above.*

Proof. Let $f \in X$ be even, i.e., $f(-x) = f(x)$, $x \in [-\pi, \pi]$. Since the function $y \mapsto f(\arccos y)$ is continuous on $[-1, 1]$, for all $\varepsilon > 0$ there exists a Bernstein[11] polynomial p such that

$$\sup_{y \in [-1,1]} |f(\arccos y) - p(y)| < \varepsilon \iff \sup_{x \in [0,\pi]} |f(x) - p(\cos x)| < \varepsilon.$$
$$(6.6.29)$$

In fact, since both f and $x \mapsto p(\cos x)$ are even, we can extend (6.6.29) to $[-\pi, \pi]$,

$$\sup_{x \in [-\pi,\pi]} |f(x) - p(\cos x))| < \varepsilon. \qquad\qquad (6.6.30)$$

By elementary trigonometric formulas we see that $p(\cos x) \in \text{Span}\,S$, so (6.6.30) concludes the proof in the case when f is even.

Now, consider an odd function $f \in X$, so $f(-\pi) = f(\pi) = f(0) = 0$. Then $x \mapsto \frac{f(x)}{\sin x}$ is an even function, but has singularities at $x = 0, \pm\pi$. So we consider for $\delta > 0$ small

$$\tilde{f}(x) = \begin{cases} f\left(\frac{\pi(x-\delta)}{\pi - 2\delta}\right), & x \in (\delta, \pi - \delta), \\ 0, & x \in [0, \delta] \cup [\pi - \delta, \pi], \end{cases}$$

and $\tilde{f}(x) := -\tilde{f}(-x)$ for $x \in [-\pi, 0)$. Clearly, $\tilde{f}$ is a continuous odd function which approximates f uniformly. Now define

$$\psi(x) = \begin{cases} \frac{\tilde{f}(x)}{\sin x}, & x \in [-\pi, \pi] \setminus \{0, \pm\pi\}, \\ 0 & x \in \{0, \pm\pi\}. \end{cases}$$

---

[11]Sergei N. Bernstein, Russian mathematician, 1880–1968.

Thus, by the first part of the proof,

$$\forall \varepsilon > 0, \ \exists q \in \text{Span}\, S \text{ such that } \|\psi - q\|_C < \varepsilon \implies \|\tilde{f} - q \sin x\|_C \leq \varepsilon.$$

Obviously $q \sin x \in \text{Span}\, S$ and thus odd continuous functions can be approximated as well by elements in $\text{Span}\, S$.

To conclude the proof, it is enough to notice that any function f can be decomposed into $f = f_e + f_o$ where

$$f_e(x) = \frac{1}{2}[f(x) + f(-x)], \quad f_o(x) = \frac{1}{2}[f(x) - f(-x)],$$

are even and odd, respectively. $\square$

Some Comments

1. Since $L^2(-\pi, \pi)$ has a countable orthonormal basis S, it follows that $L^2(-\pi, \pi)$ is separable (by Theorem 6.23). Obviously, $L^2(a, b)$ is separable for any $a < b$. In fact, for any measurable set $\Omega \subset \mathbb{R}^k$, $L^p(\Omega)$ is separable for all $p \in [1, \infty)$ (see, e.g., [6, p. 95]).

2. By Theorems 6.21 and 6.25 it follows that every $u \in L^2(-\pi, \pi)$ is the sum of the Fourier series associated with it, i.e.,

$$u = \sum_{n=0}^{\infty} (u, u_n) u_n, \tag{6.6.31}$$

meaning that $s_n(u) = \sum_{k=0}^{n} (u, u_k) u_k$ converges strongly to u in $L^2(-\pi, \pi)$. Taking into account the structure of the basis S, (6.6.31) can be written as

$$u(x) = \frac{a_0}{2} + \sum_{n=1}^{\infty} (a_n \cos nx + b_n \sin nx), \tag{6.6.32}$$

where

$$a_0 = \frac{1}{\pi} \int_{-\pi}^{\pi} u(t)\, dt, \ a_k = \frac{1}{\pi} \int_{-\pi}^{\pi} u(t) \cos(kt)\, dt,$$

$$b_k = \frac{1}{\pi} \int_{-\pi}^{\pi} u(t) \sin(kt)\, dt, \tag{6.6.33}$$

for all $k \in \mathbb{N}$. Note that (6.6.32) is precisely the classical form of the Fourier series associated with u. For the moment, we know (by Fischer–Riesz) that for $u \in L^2(-\pi, \pi)$ the series expansion (6.6.32) is valid in $L^2(-\pi, \pi)$, i.e.,

$$s_n(u)(x) = \frac{a_0}{2} + \sum_{k=1}^{n}(a_k \cos kx + b_k \sin kx) \qquad (6.6.34)$$

converges to u in $L^2(-\pi, \pi)$. Then there is a subsequence of $(s_n(u))$ that converges to u for a.a. $x \in (-\pi, \pi)$. There is a question whether the sequence $(s_n(u))$ itself converges a.e., i.e., (6.6.32) holds for a.a. $x \in (-\pi, \pi)$. This question was posed in 1920 by Luzin.[12] In 1966, Carleson[13] proved that this is indeed the case. The proof is not trivial and is omitted. Later, Hunt[14] extended the result to L^p-functions, i.e., the series expansion (6.6.32) holds a.e. for every L^p-function u, for $1 < p < \infty$. On the other hand, in 1922 Kolmogorov[15] gave a counterexample showing that it does not hold for $p = 1$. However, the Fourier expansion (6.6.32) holds for L^1-functions in the sense of distributions, as explained below.

Fourier Series Expansions of L^1 Functions

Recall that in general L^1 functions do not admit Fourier series expansions in classical theory. However, the Fourier coefficients of u (see (6.6.33)) are still well defined if $u \in L^1(-\pi, \pi)$. Fix such a function $u \in L^1(-\pi, \pi)$ and associate with it the series

$$u(x) \approx \frac{a_0}{2} + \sum_{n=1}^{\infty}(a_n \cos nx + b_n \sin nx).$$

We can prove that

$$u(x) = \frac{a_0}{2} + \sum_{n=1}^{\infty}(a_n \cos nx + b_n \sin nx) \quad \text{in } \mathcal{D}'(-\pi, \pi), \qquad (6.6.35)$$

where a_k, b_k are the Fourier coefficients of u defined in (6.6.33).

---

[12]Nikolai N. Luzin, Russian mathematician, 1883–1950.

[13]Lennart Axel Edvard Carleson, Swedish mathematician, born 1928).

[14]Richard Allen Hunt, American mathematician, 1937–2009.

[15]Andrey N. Komogorov, Russian mathematician, 1903–1987.

Recall that distributions are not defined pointwise, and the appearance of x in (6.6.35) is simply for convenience.

In order to prove (6.6.35), consider the series

$$\frac{a_0}{4}x^2 + \sum_{n=1}^{\infty}\left(-\frac{a_n}{n^2}\cos nx - \frac{b_n}{n^2}\sin nx\right),$$

which is obtained by formally integrating twice in the right-hand side of (6.6.35). This series is uniformly and absolutely convergent since for all $n \geq 1$

$$\left|-\frac{a_n}{n^2}\cos nx - \frac{b_n}{n^2}\sin nx\right| \leq \frac{1}{n^2}(|a_n| + |b_n|)$$
$$\leq \frac{4}{n^2\pi}\int_{-\pi}^{\pi}|u(t)|\,dt$$
$$= C\frac{1}{n^2}.$$

Let

$$s(x) = \frac{a_0}{4}x^2 + \sum_{n=1}^{\infty}\left(-\frac{a_n}{n^2}\cos nx - \frac{b_n}{n^2}\sin nx\right). \qquad (6.6.36)$$

Of course, uniform convergence on $[-\pi, \pi]$ implies convergence in $\mathcal{D}'(-\pi, \pi)$, so (6.6.36) also holds in $\mathcal{D}'(-\pi, \pi)$. Differentiating (6.6.36) twice in the sense of distributions we get

$$\frac{a_0}{2} + \sum_{n=1}^{\infty}(a_n\cos nx + b_n\sin nx) = s'' \quad \text{in} \quad \mathcal{D}'(-\pi, \pi). \qquad (6.6.37)$$

Finally we must show that $s'' = u$, i.e., s'' is generated by the function u. We consider the partial sums

$$s_l(u)(x) = \frac{a_0}{2} + \sum_{n=1}^{l}(a_n\cos nx + b_n\sin nx).$$

For $\phi \in \mathcal{D}(-\pi, \pi) = C_0^\infty(-\pi, \pi)$ we have

$$
\begin{aligned}
(s_l(u), \phi) &= \int_{-\pi}^{\pi} s_l(u)(x)\phi(x)\, dx \\
&= \int_{-\pi}^{\pi} \phi(x) \left[\frac{a_0}{2} + \sum_{n=1}^{l} (a_n \cos nx + b_n \sin nx) \right] dx \\
&= \int_{-\pi}^{\pi} \phi(x) \left[\frac{1}{2\pi} \int_{-\pi}^{\pi} u(t)\, dt \right. \\
&\qquad + \sum_{n=1}^{l} \left(\frac{1}{\pi} \cos nx \int_{-\pi}^{\pi} u(t) \cos nt\, dt \right. \\
&\qquad \left. \left. + \frac{1}{\pi} \sin nx \int_{-\pi}^{\pi} f(t) \sin nt\, dt \right) \right] dx,
\end{aligned}
$$

and now we change the order of integration to get

$$
(s_l(u), \phi) = \int_{-\pi}^{\pi} u(t) s_l(\phi)(t)\, dt. \tag{6.6.38}
$$

On the other hand,

$$
\lim_{l \to \infty} s_l(\phi)(t) = \phi(t) \quad \text{uniformly for } t \in [-\pi, \pi]. \tag{6.6.39}
$$

Indeed, if we denote

$$
A_k = \frac{1}{\pi} \int_{-\pi}^{\pi} \phi(t) \cos kt\, dt \ (k \geq 0), \quad B_k = \frac{1}{\pi} \int_{-\pi}^{\pi} \phi(t) \sin kt\, dt \ (k \geq 1),
$$

we may integrate by parts twice (since ϕ is infinitely differentiable), so that for $k \geq 1$

$$
\begin{aligned}
A_k &= -\frac{1}{k\pi} \int_{-\pi}^{\pi} \phi'(t) \sin kt\, dt \\
&= -\frac{1}{k^2 \pi} \int_{-\pi}^{\pi} \phi''(t) \cos kt\, dt
\end{aligned}
$$

and similarly

$$
B_k = -\frac{1}{k^2 \pi} \int_{-\pi}^{\pi} \phi''(t) \sin kt\, dt.
$$

Therefore there exists a constant $C_1 > 0$ (depending on ϕ) such that

$$|A_k| \le \frac{C_1}{n^2}, \quad |B_k| \le \frac{C_1}{n^2}, \quad \forall k \ge 1. \tag{6.6.40}$$

As A_k, B_k are the Fourier coefficients of ϕ, we deduce from (6.6.40) that the Fourier series of ϕ is uniformly convergent (see Weierstrass' M Test) and its sum is ϕ (by the classical theory, or by Theorem 6.25), i.e., (6.6.39) holds. Finally, taking into account (6.6.39) and letting $l \to \infty$ in (6.6.38), we get

$$(s'', \phi) = \int_{-\pi}^{\pi} u(t)\phi(t)\, dt = (u, \phi).$$

As ϕ was arbitrarily chosen this implies $s'' = u$, as claimed.

6.7 Exercises

1. Let $\emptyset \neq \Omega \subset \mathbb{R}^k$ be an open set and let $p \in (1, \infty)$. It is well known that $L^p(\Omega)$ is a Banach space with respect to the usual norm

$$\|u\|_{L^p(\Omega)} = \left(\int_\Omega |u(x)|^p dx \right)^{1/p}, \quad u \in L^p(\Omega).$$

   Prove that $\big(L^p(\Omega), \| \cdot \|_{L^p(\Omega)} \big)$ is a Hilbert space if and only if $p = 2$.

2. Let H be a pre-Hilbert space, i.e., a linear space equipped with a scalar product $(\cdot, \cdot)$ and the induced norm $\| \cdot \|$. Show that for $x, y \in H$ we have $|(x, y)| = \|x\| \cdot \|y\|$ if and only if x and y are linearly dependent.

3. Let $-\infty < a < b < \infty$. Show that $C[a, b]$ with the sup-norm is not a Hilbert space.

4. Let n be a given natural number. Let C be the set of all polynomials with real coefficients of degree $\le n$. Show that for any $u \in L^2(0, 1)$ there exists a unique $p_u \in C$ such that

$$\|u - p_u\|_{L^2(0,1)} \le \|u - p\|_{L^2(0,1)} \quad \forall p \in C.$$

5. Let $(H, \| \cdot \|)$ be a Hilbert space. Define $P : H \to H$, by

$$Pu = \begin{cases} u & \text{if } \|u\| \le 1, \\ \|u\|^{-1}u & \text{if } \|u\| > 1. \end{cases}$$

(Operator P is called radial retraction). Prove that

(i) P is nonexpansive, i.e., Lipschitzian with Lipschitz constant $L = 1$;

(ii) if H is a general Banach space, then P is Lipschitzian with $L = 2$.

6. Let $\mathbb{R}^3$ be equipped with the usual scalar product and Euclidean norm. Set

$$M = \{x = (x_1, x_2, x_3)^T \in \mathbb{R}^3; \ 2x_1 - x_2 - 3x_3 = 0\}.$$

Show that M is a closed linear subspace of $\mathbb{R}^3$. Determine $M^\perp$ and for $x = (1, 2, -1)^T$ determine $P_M x$ and write x as a direct sum of vectors in M and $M^\perp$, i.e., $x = x_1 + x_2$, $x_1 \in M$, $x_2 \in M^\perp$.

7. Let $-\infty < a < b < +\infty$ and let $L^2(a, b) := L^2(a, b; \mathbb{R})$ be equipped with the usual scalar product and norm. Show that

$$M = \left\{u \in L^2(a, b); \ \int_a^b u(t)\, dt = 0\right\}$$

is a closed linear subspace of $L^2(a, b)$. Determine $M^\perp$ and write any $u \in L^2(a, b)$ as a direct sum of vectors in M and $M^\perp$, i.e., $u = u_1 + u_2$, $u_1 \in M$, $u_2 \in M^\perp$.

8. Same exercise for $L^2(-1, 1)$ and

$$M = \{u \in L^2(-1, 1); \ u(t) = u(-t) \text{ for a.a. } t \in (-1, 1)\}.$$

9. Show that any linear subspace Y of a Hilbert space $(H, (\cdot, \cdot))$ one has

$$(Y^\perp)^\perp = \text{Cl}\, Y.$$

10. Let $H = L^2(0, 1)$ be the real Hilbert space equipped with the usual scalar product and norm. Is the subspace $Y = \{u \in H; \ \int_0^1 \frac{u(t)}{t}\, dt = 0\}$ closed in H?

11. Prove that the dual of any Hilbert space is a Hilbert space, too.

12. Let $\{u_n\}_{n=1}^{\infty}$ be an orthonormal basis in a Hilbert space H and let $(a_n)_{n\in\mathbb{N}}$ be a bounded sequence in $\mathbb{R}$. Prove that

   (i) the sequence $(v_n)_{n\in\mathbb{N}}$ defined by

   $$v_n = \frac{1}{n}\sum_{i=1}^{n} a_i u_i, \quad n \in \mathbb{N},$$

   converges strongly to zero;

   (ii) the sequence $(\sqrt{n}v_n)_{n\in\mathbb{N}}$ converges weakly to zero.

13. Let $(H, \|\cdot\|)$ be a Hilbert space and $A \in L(H)$. Show that the following two conditions are equivalent:

   (i) there exists a constant $c > 0$ such that $c\|x\| \le \|Ax\| \ \forall x \in H$;

   (ii) there exists an operator $B \in L(H)$ such that $B \circ A = I$, where I is the identity operator on H.

14. Let $(H, \|\cdot\|, (\cdot,\cdot))$ be a real Hilbert space. For any $A \in L(H)$ satisfying $(Ax,x) \ge 0 \ \forall x \in H$, we have

   (i) $H = N(A) \oplus [\operatorname{Cl} R(A)]$;

   (ii) for all $t > 0$, $I + tA$ is bijective and

   $$\lim_{t\to\infty} (I + tA)^{-1}u = P_{N(A)}u \ \ \forall u \in H,$$

   where I denotes the identity operator.

15. Let $(u_n)_{n\in\mathbb{N}}$ be a sequence in a Hilbert space $(H, \|\cdot\|)$ which is weakly convergent to a point $u \in H$. If, in addition, $\limsup \|u_n\| \le \|u\|$ then show $\|u_n - u\| \to 0$.

16. Prove that for any $f \in L^1(0,1)$ there exists a unique $u \in H_0^1(0,1)$ satisfying

$$\int_0^1 u'(t)v'(t)\,dt + \int_0^1 u(t)v(t)\,dt = \int_0^1 f(t)v(t)\,dt \ \ \forall v \in H_0^1(0,1),$$

and, furthermore, $u \in W^{2,1}(0,1)$ and

$$\begin{cases} -u'' + u = f & \text{a.e. in } (0,1), \\ u(0) = 0, \ u(1) = 0. \end{cases}$$

17. Let $f \in L^2(0,1)$ and $\alpha > 0$.

   (*i*) Show that the following boundary value problem, denoted (P),

   $$\begin{cases} u \in H^2(0,1), \\ -u''(t) + \alpha u(t) = f(t) & \text{for a.a. } t \in (0,1), \\ u'(0) = 0, \ u'(1) = u(1), \end{cases}$$

   is equivalent to the variational formulation, denoted $(\tilde{P})$,

   $$u \in H^1(0,1), \ -u(1)v(1) + \int_0^1 u'v' + \alpha \int_0^1 uv$$
   $$= \int_0^1 fv \ \forall v \in H^1(0,1).$$

   (*ii*) Using Lax–Milgram prove that for α large enough there exists a unique solution u of problem (P).

   (*iii*) Show that the solution u can be expressed as the minimizer of a functional defined on $H^1(0,1)$.

18. Let $(H,(\cdot,\cdot))$ be a Hilbert space and let $Y \subset H$ be a closed subspace with an orthonormal basis $\{u_n\}_{n=1}^\infty$. Prove that $\forall y \in H$ the closest point to y in Y is $\sum_{i=1}^\infty (y,u_n)u_n$.

19. Let $(H, \|\cdot\|)$ be an infinite dimensional, separable Hilbert space. Show that for any $x \in H$, $\|x\| \leq 1$, there exists a sequence $(x_n)_{n \in \mathbb{N}}$ in H such that $\|x_n\| = 1$ for all $n \in \mathbb{N}$ and $x_n \to x$ weakly.

20. Find the Fourier expansions of the functions

$$f_1(x) = \cos x - |x|, \quad -\pi \le x \le \pi,$$

$$f_2(x) = -3x + \sin x, \quad -\pi \le x \le \pi,$$

$$f_3(x) = \begin{cases} -1 & -\pi \le x \le 0, \\ x+1 & 0 \le x \le \pi, \end{cases}$$

$$f_4(x) = \begin{cases} x+1 & -1 \le x \le 0, \\ x^2 - 1 & 0 \le x \le 1. \end{cases}$$

Chapter 7

Adjoint, Symmetric, and Self-adjoint Linear Operators

Here we first recall the definition of the adjoint of a linear operator and discuss some related results. Then we shall address the case of compact operators $A : H \to H$, where H is a Hilbert space, and present the Fredholm theorem as an application. The last section is devoted to symmetric operators and self-adjoint operators.

Throughout this chapter we consider linear operators between linear spaces over $\mathbb{K}$, where $\mathbb{K}$ is either $\mathbb{R}$ or $\mathbb{C}$, unless otherwise specified.

7.1 The Adjoint of a Linear Operator

Let X, Y be Banach spaces with duals X^* and Y^* and let $A : D(A) \subset X \to Y$ be a linear operator that is *densely defined*: $\overline{D(A)} = X$. The *adjoint of A* is an operator $A^* : D(A^*) \subset Y^* \to X^*$ defined as follows. The domain of A^* is the set

$$D(A^*) = \{y^* \in Y^*; \exists c > 0 \text{ such that } |y^*(Ax)| \le c\|x\| \, \forall x \in D(A)\},$$

which is a linear subspace of Y^*. Note that for $y^* \in D(A^*)$ the linear functional $f(x) = y^*(Ax)$ is continuous on $D(A)$ (equipped with the norm $\|\cdot\|$ of X), i.e., $|f(x)| \le c\|x\|$ for all $x \in D(A)$. According to the Hahn–Banach Theorem, f can be extended to a functional $g \in X^*$,

© Springer Nature Switzerland AG 2019
G. Moroşanu, *Functional Analysis for the Applied Sciences*,
Universitext, https://doi.org/10.1007/978-3-030-27153-4_7

such that $|g(x)| \leq c\|x\|$ for all $x \in X$. This extension is unique since $D(A)$ is dense in X. We now define

$$A^*y^* = g$$

and we can write

$$y^*(Ax) = (A^*y^*)(x) \quad \forall x \in D(A), \ y^* \in D(A^*). \tag{7.1.1}$$

Example.
Let $X = Y = l^1$ (for the definition of l^1 see Chap. 4). Let $A : D(A) \subset l^1 \to l^1$ be defined by

$$D(A) = \{(x_n)_{n \geq 1} \in l^1; \ (nu_n)_{n \geq 1} \in l^1\}, \quad A(u_n) = (nu_n).$$

Obviously, $D(A)$ is dense in l^1. It is also easily seen that

$$D(A^*) = \{(y_n)_{n \geq 1} \in l^\infty; \ (ny_n)_{n \geq 1} \in l^\infty\}, \quad A^*(y_n) = (ny_n).$$

Note that both A and A^* are closed operators, i.e., their graphs are closed in $l^1 \times l^1$ and $l^\infty \times l^\infty$, respectively. In fact, we have the following general result:

Theorem 7.1. *Let X and Y be Banach spaces and let $A : D(A) \subset X \to Y$ be a densely defined, linear operator. Then A^* is closed.*

Proof. Let (y_n^*) be a sequence in $D(A^*)$ such that $y_n^* \to y^*$ in Y^* and $A^*y_n^* \to x^*$ in X^*. We have

$$y_n^*(Ax) = (A^*y_n^*)(x) \quad \forall x \in D(A),$$

which yields (by letting $n \to \infty$)

$$y^*(Ax) = x^*(x) \quad \forall x \in D(A).$$

Therefore $y^* \in D(A^*)$ and $A^*y^* = x^*$. $\square$

We also have the following results about continuity.

Theorem 7.2. *Let X, Y be Banach spaces with duals X^* and Y^*. If $A \in L(X, Y)$, then $A^* \in L(Y^*, X^*)$, and $\|A\| = \|A^*\|$.*

Proof. Obviously, $D(A^*) = Y^*$. From (7.1.1) we deduce (using the same symbol $\| \cdot \|$ for different norms)

$$|(A^*y^*)(x)| \leq \|y^*\| \cdot \|A\| \cdot \|x\| \quad \forall x \in X, \ y^* \in Y^*.$$

Therefore,

$$\|A^*y^*\| \leq \|A\| \cdot \|y^*\| \quad \forall y^* \in Y^* \ \Rightarrow \ \|A^*\| \leq \|A\|.$$

On the other hand, using (7.1.1) again, we obtain

$$|y^*(Ax)| \leq \|A^*\| \cdot \|y^*\| \cdot \|x\| \quad \forall x \in X, \ y^* \in Y^*,$$

hence, by Corollary 4.18 in Chap. 4,

$$\|Ax\| \leq \|A^*\| \cdot \|x\| \quad \forall x \in X \ \Rightarrow \ \|A\| \leq \|A^*\|. \qquad \square$$

We continue with some simple properties of adjoint operators. Let X, Y, Z be three Banach spaces over $\mathbb{K}$, where $\mathbb{K}$ is the same (either $\mathbb{R}$ or $\mathbb{C}$) for all the three spaces. Then the following properties hold:

(a) If $A : D(A) \subset X \to Y$ is a densely defined, linear operator, and $B : D(B) \subset X \to Y$ is another linear operator, such that $A \subset B$ (i.e., $D(A) \subset D(B)$ and $Bx = Ax \ \forall x \in D(A)$), then $B^* \subset A^*$;

(b) For all $\alpha, \beta \in \mathbb{K}$ and $A, B \in L(X, Y)$,

$$(\alpha A + \beta B)^* = \alpha A^* + \beta B^*;$$

(c) If $A \in L(X, Y)$ and $B \in L(Y, Z)$, then $(B \circ A)^* = A^* \circ B^*$.

(d) If $A \in L(X, Y)$ is bijective, then A^* is bijective too, and

$$(A^*)^{-1} = (A^{-1})^*.$$

The proofs are left to the reader.

7.2   Adjoints of Operators on Hilbert Spaces

Let $(H, (\cdot,\cdot), \|\cdot\|)$ be a Hilbert space. Let $A : D(A) \subset H \to H$ be a densely defined, linear operator. Taking into account the Riesz Representation Theorem, the adjoint of A can be redefined as an operator from H into itself, as follows:

$$D(A^*) = \{y \in H; \exists c > 0 \text{ such that } |(Ax, y)| \leq c\|x\| \ \forall x \in D(A)\}.$$

Now, for $y \in D(A^*)$ the linear functional $x \mapsto (Ax, y)$ (which is continuous on $(D(A), \|\cdot\|)$) can be extended uniquely to a functional belonging to H^* so (by Riesz) there is a corresponding element in H, denoted A^*y. Thus we have a linear operator $A^* : D(A^*) \subset H \to H$, such that

$$(Ax, y) = (x, A^*y) \quad \forall x \in D(A), y \in D(A^*). \tag{7.2.2}$$

In fact, if $R : H \to H^*$ denotes the Riesz isomorphism, this adjoint is nothing else but the operator $R^{-1} \circ A^* \circ R$, with A^* being the adjoint defined in the previous section. Whenever we deal with a densely defined linear operator $A : D(A) \subset H \to H$, we shall associate with A the A^* defined in this section. It is easily seen that all the properties discussed in the previous section remain valid, except for (b) which now takes the form

(b') For all $\alpha, \beta \in \mathbb{K}$ and $A, B \in L(X, Y)$,

$$(\alpha A + \beta B)^* = \bar{\alpha}A^* + \bar{\beta}B^*,$$

where $\bar{\alpha}, \bar{\beta}$ denote the complex conjugates of α, β. In fact, for any $\alpha \in \mathbb{K}$ and any densely defined, linear operator $A : D(A) \subset X \to Y$, we have

$$(\alpha A)^* = \bar{\alpha}A^*.$$

If H is finite dimensional, then the matrix corresponding to A^* is the *transposed conjugate* of the matrix corresponding to A (while the matrix associated with the adjoint of A as defined in the previous section is just the *transpose* of the matrix corresponding to A. This shows the difference between the two notions of adjoint).

If $A \in L(H) := L(H, H)$, then $A^{**} := (A^*)^* = A$. Indeed, we have

$$
\begin{aligned}
(Ax, y) &= (x, A^* y) \\
&= \overline{(A^* y, x)} \\
&= \overline{(y, A^{**} x)} \\
&= (A^{**} x, y) \quad \forall x, y \in H \, ,
\end{aligned}
$$

which proves the assertion.

7.2.1 The Case of Compact Operators

Denote by $K(H) := K(H, H)$ the space of compact linear operators from H into itself. This is a closed subspace of $L(H) := L(H, H)$ with respect to the operator norm, hence $K(H)$ is a Banach space with respect to this norm (see Theorem 4.11).

Theorem 7.3. *If $(H, (\cdot, \cdot), \|\cdot\|)$ is a Hilbert space and $A \in K(H)$, then the nullspace of $I - A$, denoted $\mathcal{N} = N(I - A)$, is a finite dimensional subspace of H, where I denotes the identity operator of H.*

Proof. Obviously $\mathcal{N}$ is a (closed) linear subspace of $(H, \|\cdot\|)$. Let Q be a bounded subset of $\mathcal{N}$. Since A is compact and $Q = AQ$ we deduce that Q is relatively compact in $(\mathcal{N}, \|\cdot\|)$. According to Theorem 2.24, $\mathcal{N}$ is finite dimensional. $\qquad\square$

Theorem 7.4 (Schauder[1]). *If $(H, (\cdot, \cdot), \|\cdot\|)$ is a Hilbert space and $A \in K(H)$ then $A^* \in K(H)$, too.*

Proof. Let $r > 0$ be arbitrary but fixed. Since $A^* \in L(H)$, the set $A^* B(0, r)$ is bounded: $\|x\| < r \implies \|A^* x\| \leq r \|A^*\|$. As A is compact, it follows that for any sequence $(x_n)_{n \geq 1}$ in $B(0, r)$ the sequence $((A \circ A^*) x_n)_{n \geq 1}$ has a convergent subsequence, say $((A \circ A^*) x_{n_k})_{k \geq 1}$. We also have

$$
\begin{aligned}
\|A^* x_{n_k} - A^* x_{n_j}\|^2 &= \left(A^*(x_{n_k} - x_{n_j}), A^*(x_{n_k} - x_{n_j}) \right) \\
&= \left(x_{n_k} - x_{n_j}, A(A^*(x_{n_k} - x_{n_j})) \right) \\
&\leq 2r \|(A \circ A^*) x_{n_k} - (A \circ A^*) x_{n_j}\|,
\end{aligned}
$$

so $(A^* x_{n_k})_{k \geq 1}$ is convergent. $\qquad\square$

---

[1] Juliusz Pawel Schauder, Polish mathematician, 1899–1943.

Remark 7.5. Let $A \in L(H)$. Then A is compact if and only if A^* is compact. This follows from Schauder's Theorem above combined with $(A^*)^* = A$.

Remark 7.6. If $A, B \in L(H)$ and at least one is compact, then $A \circ B$ is compact as well.

We continue with an important result, essentially due to Fredholm,[2] that provides a necessary and sufficient condition for an operator equation involving a compact linear operator to have a solution.

Theorem 7.7 (Fredholm). *Let $(H, (\cdot, \cdot), \| \cdot \|)$ be a Hilbert space and let $A \in K(H)$. The equation $x - A^*x = f$ has a solution if and only if $f \in \mathcal{N}^\perp$, where $\mathcal{N} = N(I - A)$ (the nullspace of $I - A$).*

Corollary 7.8. *If $(H, (\cdot, \cdot), \| \cdot \|)$ is a Hilbert space and $A \in K(H)$, then the equation $x - Ax = f$ has a solution if and only if $f \in \big(N(I - A^*) \big)^\perp$.*

Proof. Use Theorem 7.7 with A^* instead of A. $\qquad\qquad\qquad\square$

In order to prove Fredholm's Theorem, we need the following lemma.

Lemma 7.9. *Let $(H, (\cdot, \cdot), \| \cdot \|)$ be a Hilbert space and let $A \in K(H)$. Then there exists a constant $C > 0$ such that*

$$C\|x\| \leq \|(I - A)x\| \quad \forall x \in \mathcal{N}^\perp, \qquad\qquad (7.2.3)$$

where $\mathcal{N} = N(I - A)$.

Proof. Assume by contradiction that (7.2.3) is not true, i.e., for all $n \in \mathbb{N}$ there exists an $x_n \in \mathcal{N}^\perp$ such that $\|x_n\| = 1$ and

$$\|(I - A)x_n\| < \frac{1}{n}.$$

Therefore,

$$x_n - Ax_n \to 0. \qquad\qquad\qquad (7.2.4)$$

As A is compact there is a subsequence of $(x_n)_{n \geq 1}$, say $(x_{n_k})_{k \geq 1}$, such that $(Ax_{n_k})_{k \geq 1}$ is convergent. By (7.2.4) we deduce that $(x_{n_k})_{k \geq 1}$ is also convergent, and its limit $x \in \mathcal{N}^\perp$ (since $\mathcal{N}^\perp$ is closed). Using again (7.2.4), we infer that $x - Ax = 0$, i.e., $x \in \mathcal{N}$. Since $\mathcal{N} \cap \mathcal{N}^\perp = \{0\}$, we have $x = 0$, which contradicts $\|x_n\| = 1 \ \forall n \geq 1$. $\qquad\square$

---

[2]Erik Ivar Fredholm, Swedish mathematician, 1866–1927.

Proof of Fredholm's Theorem.
Necessity. Assume that the equation $x - A^*x = f$ has a solution $x \in H$.
Then, for all $y \in \mathcal{N}$, we have

$$
\begin{aligned}
(f, y) &= (x, y) - (A^*x, y) \\
&= (x, y) - (x, Ay) \\
&= (x, \underbrace{(I - A)y}_{=0}) \\
&= 0.
\end{aligned}
$$

Therefore $f \in \mathcal{N}^\perp$.

Sufficiency. Assume $f \in \mathcal{N}^\perp$. Since $\mathcal{N}^\perp$ is a closed subspace of $(H, \|\cdot\|)$, $\mathcal{N}^\perp$ is a Hilbert space with the same scalar product and norm. According to Lemma 7.9, $\|\cdot\|$ is equivalent (on $\mathcal{N}^\perp$) with the norm defined by the scalar product

$$
\langle x, y \rangle = (Tx, Ty) \quad \forall x, y \in \mathcal{N}^\perp,
$$

where $T = I - A$. Since the functional $x \mapsto (x, f)$ is linear and continuous on $\mathcal{N}^\perp$, it follows by the Riesz Representation Theorem that there exists $x_f \in \mathcal{N}^\perp$ such that

$$
(x, f) = \underbrace{\langle x, x_f \rangle}_{=(Tx, Tx_f)} \quad \forall x \in \mathcal{N}^\perp. \tag{7.2.5}
$$

In fact, (7.2.5) holds for all $x \in H$, since $x = x' + x''$, with $x' \in \mathcal{N}$, $x'' \in \mathcal{N}^\perp$. Denoting $\tilde{x} = Tx_f$, we can write (see (7.2.5) extended to H)

$$
\underbrace{(Tx, \tilde{x})}_{=(x, \tilde{x} - A^*\tilde{x})} = (x, f) \quad \forall x \in H,
$$

so

$$
\tilde{x} - A^*\tilde{x} = f. \qquad \square
$$

The following result provides some information that supplements Theorem 7.7.

Theorem 7.10. *Let $(H, (\cdot, \cdot), \|\cdot\|)$ be a Hilbert space and let $A \in K(H)$. Then,*

$$
R(I - A) = H \iff \mathcal{N} = \{0\} \iff \mathcal{N}^* = \{0\} \iff R(I - A^*) = H,
$$

where $\mathcal{N} = N(I - A)$, $\mathcal{N}^* = N(I - A^*)$, and $R(I - A)$, $R(I - A^*)$ denote the ranges of $I - A$, $I - A^*$.

Proof. Keeping in mind Theorem 7.7 and Corollary 7.8, it suffices to prove that

$$R(I - A) = H \iff R(I - A^*) = H. \tag{7.2.6}$$

Assume $R(I - A) = H$. Let us prove that $\mathcal{N} = \{0\}$. Assume by way of contradiction that $\mathcal{N} \neq \{0\}$, i.e., there exists an $x_0 \in \mathcal{N}$, $x_0 \neq 0$. As $R(I - A) = H$ we can construct a sequence $(x_n)_{n \geq 1}$ in $D(A)$ such that

$$T x_n = x_{n-1} \quad \forall n \geq 1,$$

where $T := I - A$. We have

$$T^n x_n = x_0 \neq 0, \quad \text{and} \quad T^{n+1} x_n = 0,$$

where $T^k := T \circ T \circ \cdots \circ T$ (k factors). Hence, denoting $H_n = N(T^n)$, we have that H_n is a proper linear subspace of H_{n+1} for all $n \in \mathbb{N}$. According to Theorem 7.3, every H_n is a finite dimensional space, hence closed, since

$$T^n = (I - A)^n = I - \underbrace{\sum_{k=1}^{n} \binom{n}{k} (-1)^{k+1} A^k}_{\text{compact operator}}.$$

By Lemma 2.25 there exists a sequence $(u_n)_{n \geq 1}$ such that

$$u_n \in H_{n+1}, \ \|u_n\| = 1, \ \|u_n - u\| \geq \frac{1}{2} \quad \forall u \in H_n.$$

Since for $1 \leq m < n$

$$T^n(T u_n + A u_m) = T^{n+1} u_n + A T^n u_m = 0,$$

we have $T u_n + A u_m \in H_n$, and

$$\|A u_n - A u_m\| = \|u_n - (T u_n + A u_m)\| \geq \frac{1}{2}.$$

Thus the sequence $(A u_n)_{n \geq 1}$ cannot have Cauchy (hence convergent) subsequences. This contradicts the fact that A is compact combined with $\|u_n\| = 1$ for all $n \geq 1$. Therefore, $\mathcal{N} = \{0\}$, which (by Theorem 7.7) implies that $R(I - A^*) = H$. Thus we have proved the implication

$$R(I - A) = H \implies R(I - A^*) = H.$$

The converse implication follows by replacing A with A^*. $\qquad\square$

Remark 7.11. From Corollary 7.8 and Theorem 7.10 we deduce that if the equation $x - Ax = f$ has a solution u_f for all $f \in H$ then u_f is unique (since $N(I - A) = \{0\}$). So we can now state the so-called *Fredholm's alternative* regarding the equation $x - Ax = f$ with $A \in K(H)$, namely one of the following must hold:

- for every $f \in H$ the equation $x - Ax = f$ has a unique solution (equivalently, $N(I - A) = \{0\}$);

- $N(I - A) \neq \{0\}$, in which case the equation $x - Ax = f$ is solvable if and only if $f \perp N(I - A^*)$ (i.e., f satisfies m orthogonality relations, where $m = \dim N(I - A^*) = \dim N(I - A)$).

We shall later apply Fredholm's alternative to a class of integral equations that are named after him.

Remark 7.12. In fact, the above theory is valid in a general Banach space H (see, e.g., [6, Chapter 6] or [15, Chapter 5]).

7.3 Symmetric Operators and Self-adjoint Operators

We begin this section with the following definition.

Definition 7.13. *Let* $(H, (\cdot, \cdot), \| \cdot \|)$ *be a Hilbert space and let* $A : D(A) \subset H \to H$ *be a densely defined, linear operator.*
(a)   *A is called* **symmetric** *if* $A \subset A^*$, *i.e.,*

$$(Ax, y) = (x, Ay) \quad \forall x, y \in D(A) ;$$

(b)   *A is called* **self-adjoint** *if* $A = A^*$, *i.e.,* $A \subset A^*$ *and* $A^* \subset A$.

Obviously, if $D(A) = H$ then A is symmetric if and only if it is self-adjoint, and in this case A is closed (by Theorem 7.1), hence $A \in L(H)$ (by the Closed Graph Theorem).

Example 1. Let $X = L^2(a, b; \mathbb{K})$, where $-\infty < a < b < +\infty$ and let $A : X \to X$ be defined by

$$(Af)(t) = \int_a^b k(t, s) f(s) \, ds, \quad a \leq t \leq b,$$

where $k \in C([a, b] \times [a, b]; \mathbb{K})$. The space X equipped with the usual scalar product and norm is a Hilbert space and $A \in L(X)$. Moreover, it is easy to see (by using Arzelà–Ascoli's Criterion) that $A \in K(X)$. Note that for all $f, g \in X$ we have

$$
\begin{aligned}
(Af, g)_{L^2(a,b;\,\mathbb{K})} &= \int_a^b (Af)(t) \cdot \overline{g(t)} \, dt \\
&= \int_a^b \left(\int_a^b k(t,s) f(s) \, ds \right) \cdot \overline{g(t)} \, dt \\
&= \int_a^b f(s) \cdot \left(\int_a^b k(t,s) \overline{g(t)} \, dt \right) ds \\
&= \int_a^b f(t) \cdot \left(\overline{\int_a^b \overline{k(s,t)} g(s) \, ds} \right) dt \,,
\end{aligned}
$$

thus

$$
(A^* g)(t) = \int_a^b \overline{k(s,t)} \cdot g(s) \, ds \quad \forall g \in X \,.
$$

Obviously,

$$
A = A^* \iff k(t,s) = \overline{k(s,t)} \quad \forall t, s \in [a, b] \,.
$$

Example 2. Let $X = L^2(\mathbb{R}; \mathbb{K})$ with its usual scalar product and Hilbertian norm, and let $A : D(A) \subset X \to X$ be given by

$$
D(A) = \{ f \in X; \ t f(t) \in X \} \,,
$$

$$
(Af)(t) := t f(t) \quad \forall t \in \mathbb{K}, \ f \in D(A) \,.
$$

It is easily seen that A is self-adjoint.

Example 3. Let $H = L^2(\Omega)$ be equipped with the usual scalar product and norm, where $\emptyset \neq \Omega \subset \mathbb{R}^N$, $N \geq 2$, is a bounded domain with smooth boundary. Let $A : D(A) \subset H \to H$, where

$$
D(A) = C_0^\infty(\Omega), \quad Au = \Delta u \quad \forall u \in D(A) \,.
$$

Obviously, $D(A)$ is dense in H. By Green's identity, we have

$$
\int_\Omega v \Delta u \, dx = \int_\Omega u \Delta v \, dx \quad \forall u \in D(A) = C_0^\infty(\Omega), \ v \in H^2(\Omega) \,.
$$

Thus $H^2(\Omega) \subset D(A^*)$ and $A^* v = \Delta v$ for all $v \in D(A)$. Therefore A is symmetric but not self-adjoint because $D(A)$ is a proper subset of $D(A^*)$. If the domain of $A = \Delta$ is extended to $H_0^1(\Omega) \cap H^2(\Omega)$ then A becomes self-adjoint. More precisely, we have the following proposition.

Proposition 7.14. *Let $H = L^2(\Omega)$ be equipped with the usual scalar product $(\cdot, \cdot)$ and the induced norm $\| \cdot \|$, where $\emptyset \neq \Omega \subset \mathbb{R}^N$, $N \geq 2$, is a bounded domain with smooth boundary. Let $B : D(B) \subset H \to H$ be defined by $D(B) = H^2(\Omega) \cap H_0^1(\Omega)$, $Bu = \Delta u$ for all $u \in D(B)$. Then B is self-adjoint.*

Proof. Clearly, $D(B)$ is dense in H and, by Green's formula, we have for all $u, v \in D(B)$

$$(Bu, v) = \int_\Omega \Delta u \cdot v \, dx = \int_\Omega u \cdot \Delta v \, dx = (u, Bv),$$

hence $D(B) \subset D(B^*)$ and $B^*v = Bv$ for all $v \in D(B)$ (i.e., B is symmetric). Let us prove that $D(B^*) = D(B)$. Using the Lax–Milgram Theorem, we can see that $R(I + B) = H$. In addition, since B is positive, $I + B$ is invertible and $J := (I + B)^{-1} \in L(H)$. As B is symmetric, so is J. Now, let v be an arbitrary function in $D(B^*)$. Denoting $g = v + B^*v$, we have

$$(g, u) = (v, u + Bu) \quad \forall u \in D(B).$$

Therefore, for every $h \in H$, we have

$$(g, Jh) = (v, h) \implies (Jg, h) = (v, h),$$

so $v = Jg \in R(J) = D(B)$. $\qquad \square$

We know that, for every bijective $A \in L(H)$, A^* is also bijective and $(A^*)^{-1} = (A^{-1})^*$. In fact, the following more general result holds.

Theorem 7.15. *Let $(H, (\cdot, \cdot), \| \cdot \|)$ be a Hilbert space and let $A : D(A) \subset H \to H$ be a symmetric linear operator, with $\overline{R(A)} = H$. Then*

$$(A^{-1})^* = (A^*)^{-1},$$

where all operations are permitted. If, in addition, A is self-adjoint, then so is A^{-1}.

Proof. A is injective. Indeed, if $u \in D(A)$ and $Au = 0$ then

$$0 = (Au, v) = (u, Av) \quad \forall v \in D(A),$$

which implies $u = 0$ since $R(A)$ is dense in H.

A^* is also injective because if $v \in D(A^*)$ and $A^*v = 0$, then

$$(Au, v) = (u, A^*v) = 0 \quad \forall u \in D(A),$$

and thus $v = 0$ since $\overline{R(A)} = H$. Therefore, A^{-1} and $(A^*)^{-1}$ exist with $D(A^{-1}) = R(A)$, $D((A^*)^{-1}) = R(A^*)$. Since $D(A^{-1})$ is dense in H, $(A^{-1})^*$ exists. Denote $B := (A^{-1})^*$. We have

$$(u, v) = (A^{-1}(Au), v) = (Au, Bv) \quad \forall u \in D(A), \ v \in D(B), \quad (7.3.7)$$

and

$$\begin{aligned}(z, w) = (A(A^{-1}z), w) &= (A^{-1}z, A^*w) \quad \forall z \in D(A^{-1}) \\ &= R(A), \ w \in D(A^*). \end{aligned} \quad (7.3.8)$$

By (7.3.7) $Bv \in D(A^*)$ and

$$v = A^*(Bv) \quad \forall v \in D(B). \quad (7.3.9)$$

On the other hand, by (7.3.8), $A^*w \in D((A^*)^{-1}) = D(B)$ and

$$w = \underbrace{(A^{-1})^*}_{=B}(A^*w) \quad \forall w \in D(A^*). \quad (7.3.10)$$

From (7.3.9) and (7.3.10) we derive

$$B = (A^*)^{-1} \iff (A^{-1})^* = (A^*)^{-1}.$$

If $A = A^*$, then $(A^{-1})^* = A^{-1}$. $\square$

7.4   Exercises

1. Let X, Y be Banach spaces. Let $A : D(A) \subset X \to Y$ be a densely defined, closed linear operator and $B \in L(X, Y)$. Define $T : D(T) = D(A) \subset X \to Y$ by $Tx = Ax + Bx \ \forall x \in D(A)$. Prove that

   (i) T is a closed operator;

   (ii) $D(T^*) = D(A^*)$ and $T^* = A^* + B^*$.

2. Let X, Y be Banach spaces and let $A : D(A) \subset X \to Y$ be a densely defined linear operator. Show that A^* is injective if and only if $\mathrm{Cl}\, R(A) = Y$.

3. Let H be a Hilbert space. If $A : D(A) \subset H \to H$ is a symmetric linear operator with $R(A) = H$, then A is self-adjoint, i.e., $A = A^*$.

4. Let H be a Hilbert space, with the scalar product denoted $(\cdot, \cdot)$, and let A, $B \in L(H)$. Show that

$$A^*A = B^*B \iff (Ax, Ay) = (Bx, By) \ \forall x, y \in H.$$

5. Let H be a Hilbert space. For any $A \in L(H)$, show that $\|A^*A\| = \|A\|^2$.

6. Let $(H, (\cdot, \cdot))$ be a Hilbert space over $\mathbb{C}$ and let $A \in L(H)$. Prove that

A is symmetric (hence self-adjoint) $\iff (Ax, x) \in \mathbb{R} \ \forall x \in H$.

7. Let $(H, (\cdot, \cdot))$ be a Hilbert space over $\mathbb{R}$. Prove that for any $a > 0$ and any $A \in L(H)$ the operator $T = I + aA^*A$ is invertible and $T^{-1} \in L(H)$, where I denotes the identity operator on H.

8. Let H be a Hilbert space over $\mathbb{C}$ and let $A \in L(H)$ be a symmetric (hence self-adjoint) operator. Denote $T = A + iI$, where $i^2 = -1$ and I is the identity operator on H. Prove that

   (a) T is a normal operator (i.e., $T^*T = TT^*$);

   (b) T is invertible and $T^{-1} \in L(H)$.

9. Let H be a Hilbert space over $\mathbb{C}$. For $A \in L(H)$ and $a_0, a_1, \ldots, a_n \in \mathbb{C}$, denote by $P(A)$ the operator polynomial $a_0 I + a_1 A + \cdots + a_n A^n$, where I stands for the identity operator.

   (j) If A is symmetric (hence self-adjoint) and $a_0, a_1, \ldots, a_n \in \mathbb{R}$, then $P(A)$ is symmetric, too;

   (jj) If A is a normal operator (i.e., $A^*A = AA^*$), then so is $P(A)$.

10. Let H_1, H_2 be Hilbert spaces. Define $H = H_1 \times H_2$ to be the Hilbert space consisting of all pairs $(x_1, x_2)^T$, $x_1 \in H_1$ and $x_2 \in H_2$, with

$$\begin{pmatrix} x_1 \\ x_2 \end{pmatrix} + \begin{pmatrix} y_1 \\ y_2 \end{pmatrix} = \begin{pmatrix} x_1 + y_1 \\ x_2 + y_2 \end{pmatrix},$$

$$\alpha \begin{pmatrix} x_1 \\ x_2 \end{pmatrix} = \begin{pmatrix} \alpha x_1 \\ \alpha x_2 \end{pmatrix} \quad \forall \alpha \in \mathbb{K},$$

and a scalar product defined by

$$\left\langle \begin{pmatrix} x_1 \\ x_2 \end{pmatrix}, \begin{pmatrix} y_1 \\ y_2 \end{pmatrix} \right\rangle = (x_1, y_1)_{H_1} + (x_2, y_2)_{H_2} \; \forall \begin{pmatrix} x_1 \\ x_2 \end{pmatrix}, \begin{pmatrix} y_1 \\ y_2 \end{pmatrix} \in H.$$

Given $A_1 \in L(H_1)$ and $A_2 \in L(H_2)$, define the matrix operator

$$A = \begin{bmatrix} A_1 & 0 \\ 0 & A_2 \end{bmatrix}.$$

Prove that $A \in L(H)$ and $\|A\| = \max\{\|A_1\|, \|A_2\|\}$. Find A^*.

11. Let $A \in L(H)$, where H is a Hilbert space over $\mathbb{C}$. As in the previous exercise, define $Y = H \times H$ to be the Hilbert space consisting of all pairs $(x_1, x_2)^T$, $x_1 \in H$ and $x_2 \in H$, with the corresponding operations and scalar product. Define on Y the matrix operator B by

$$B = \begin{bmatrix} 0 & iA \\ -iA^* & 0 \end{bmatrix},$$

where $i = \sqrt{-1}$. Prove that $B \in L(Y)$, $\|B\| = \|A\|$, and that $B^* = B$.

Now, assume that $A : D(A) \subset H \to H$ is a linear, densely defined operator. Prove that $B : D(A^*) \times D(A) \subset Y \to Y$ is symmetric.

12. Let H be a Hilbert space and let $A \in L(H)$ satisfying $\|A\| \leq 1$. Prove that $Ax = x$ if and only if $A^*x = x$.

13. Let H be the real Hilbert space $L^2(0, 1)$ equipped with the usual scalar product and induced norm. Define $A : D(A) \subset H \to H$ by

$$D(A) = \{u \in H^1(0, 1); u(0) = 0\}, \quad Au = u'.$$

(a) Show that $D(A)$ is dense in H and that A is closed;

(b) Compute $N(A)$ and $R(A)$;

(c) Determine A^* and show that $D(A^*)$ is dense in H.

14. Let H be the real Hilbert space $L^2(0,1)$ equipped with the usual scalar product and induced norm. Let $A : D(A) \subset H \to H$ be the operator defined by $Au = u'$, where

(a) $D(A) = H_0^1(0,1)$;

(b) $D(A) = \{u \in H^1(0,1); u(0) = \alpha u(1)\}$ for some $\alpha \in \mathbb{R} \setminus \{0\}$.

Determine $N(A)$, $R(A)$, A^*, $N(A^*)$, $R(A^*)$ in each of these two cases.

15. Let H be the real Hilbert space $L^2(0,1)$ equipped with the usual scalar product and induced norm. Let $A : D(A) \subset H \to H$, $Au = u''$, where $D(A)$ is specified below. Determine A^* in each of the following cases:

(a) $D(A) = \{u \in H^2(0,1); u(0) = u(1) = 0\}$;

(b) $D(A) = \{u \in H^2(0,1); u(0) = u(1) = u'(0) = u'(1) = 0\}$;

(c) $D(A) = \{u \in H^2(0,1); u(0) = u'(1) = 0\}$;

(d) $D(A) = \{u \in H^2(0,1); u(0) = u(1)\}$;

16. Let $H = l^2(\mathbb{C})$ be the complex Hilbert space of all sequences of complex numbers $x = (x_n)_{n \in \mathbb{N}}$ satisfying $\sum_{n=1}^{\infty} |x_n|^2 < \infty$, with the usual scalar product

$$\langle x, y \rangle = \sum_{n=1}^{\infty} x_n \bar{y}_n \quad \forall x = (x_n), \, y = (y_n) \in H,$$

and the induced norm, denoted $\| \cdot \|$. Define the operators $A : H \to H$ and $B : D(B) \subset H \to H$ by

$$A(x_n) = (x_{p+1}, x_{p+2}, x_{p+3}, \dots), \quad \text{for a given } p \in \mathbb{N},$$

$$B(x_n) = \left(\frac{n^\alpha i^n}{1+n} x_n \right), \quad \text{for a given } \alpha \in \mathbb{R}.$$

(a) Show that $A \in L(H)$ and compute $\|A\|$ and A^*;

(b) Show that if $\alpha \leq 1$ then $D(B) = H$ and $B \in L(H)$; compute $\|B\|$;

(c) For $\alpha > 1$ find (the maximal domain) $D(B)$ and prove that $D(B)$ is dense in H;

(d) Compute B^* for all $\alpha \in \mathbb{R}$;

(e) Check whether A and B with $\alpha \leq 1$ are normal operators.

Chapter 8

Eigenvalues and Eigenvectors

In this chapter we present the main results regarding eigenvalues and eigenvectors of compact and/or symmetric operators. This includes the Hilbert–Schmidt Theorem and its applications to the main eigenvalue problems for the Laplacian.

Throughout this chapter we consider linear operators defined on linear spaces over $\mathbb{K}$, where $\mathbb{K}$ is either $\mathbb{R}$ or $\mathbb{C}$, unless otherwise specified.

8.1 Definition and Examples

We first introduce the concept of an eigenpair (i.e., eigenvector + the corresponding eigenvalue).

Definition 8.1. *Let X be a linear space. A vector $u \in X \setminus \{0\}$ is said to be an **eigenvector** of a linear operator $A : X \to X$ if there exists $\lambda \in \mathbb{K}$ such that $Au = \lambda u$. Such a λ is called an **eigenvalue** corresponding to u, and the pair (u, λ) is called an **eigenpair**.*

Remark 8.2. For a given eigenvector u of A the corresponding eigenvalue λ is unique. Indeed,

$$\lambda u = Au = \lambda_1 u \implies (\lambda - \lambda_1)u = 0 \implies \lambda - \lambda_1 = 0,$$

since $u \neq 0$.

© Springer Nature Switzerland AG 2019
G. Moroşanu, *Functional Analysis for the Applied Sciences*,
Universitext, https://doi.org/10.1007/978-3-030-27153-4_8

For a given eigenvalue λ of A, the set of the corresponding eigenvectors is $N(\lambda I - A) \setminus \{0\}$, where I is the identity operator of X.

Remark 8.3. Note also that a set of eigenvectors $u_1, u_2, \ldots, u_m$ of A corresponding to distinct eigenvalues $\lambda_1, \lambda_2, \ldots, \lambda_m$ ($m \in \mathbb{N}$) is a linearly independent system. The proof is by induction.

Example 1. Let $X = \mathbb{C}^n$, $A : X \to X$, $Au = Mu$ $\forall u = (u_1, \ldots, u_n)^T \in X$, where $M = (a_{ij})$ is an $n \times n$ matrix with entries $a_{ij} \in \mathbb{C}$. Then, λ is an eigenvalue of A if and only if $\det(\lambda I - M) = 0$, where I is the $n \times n$ identity matrix.

Example 2. Let $H = l^2(\mathbb{C})$ be the complex Hilbert space of all sequences of complex numbers $x = (x_n)_{n \in \mathbb{N}}$ satisfying $\sum_{n=1}^{\infty} |x_n|^2 < \infty$, with the usual scalar product

$$\langle x, y \rangle = \sum_{n=1}^{\infty} x_n \bar{y}_n \quad \forall x = (x_n), y = (y_n) \in H,$$

and the induced norm, denoted $\|\cdot\|$. Define the linear operator A by

$$A(x_n) = \left(\frac{2}{1} x_2, \frac{3}{2} x_3, \ldots, \frac{n}{n-1} x_n, \ldots \right).$$

We have for all $x = (x_n) \in H$

$$\begin{aligned}
\|Ax\|^2 &= \sum_{n=2}^{\infty} |n(n-1)^{-1} x_n|^2 \\
&\leq 4 \sum_{n=2}^{\infty} |x_n|^2 \\
&\leq 4 \|x\|^2,
\end{aligned}$$

so $A \in L(H)$ and $\|A\| \leq 2$. In fact, $\|A\| = 2$, since for $\tilde{x} = (0, 1, 0, 0 \ldots)$ we have $\|\tilde{x}\| = 1$ and $\|A\tilde{x}\| = 2$.

Consider the equation $Ax = \lambda x$, or, equivalently,

$$\frac{n+1}{n} x_{n+1} = \lambda x_n, \quad n = 1, 2, \ldots \tag{8.1.1}$$

Observe that $\lambda = 0$ is an eigenvalue of A with eigenvectors $(x_1, 0, 0, \ldots)$, $x_1 \in \mathbb{C} \setminus \{0\}$.

If $\lambda \neq 0$, then it follows easily from (8.1.1) that

$$x_n = \frac{1}{n} \lambda^{n-1} x_1, \quad n = 1, 2, \ldots$$

In order for (x_n) to be an eigenvector, we choose $x_1 \neq 0$. The condition $(x_n) \in H$ is equivalent to $|\lambda| \leq 1$. So the set $\{\lambda \in \mathbb{C}; |\lambda| \leq 1\}$ is the set of all eigenvalues of A.

8.2 Main Results

We begin this section with a general result about the eigenvalues of a compact linear operator.

Theorem 8.4. *Let $(X, \|\cdot\|)$ be a normed space and let $A \in K(X)$ (i.e., $A : X \to X$ is linear and sends bounded sets to relatively compact sets). Then A has a countable set of eigenvalues, and the only possible accumulation point of the set of eigenvalues is $\lambda = 0$. Moreover, for any eigenvalue $\lambda \neq 0$, $\dim N(\lambda I - A) < \infty$ (one says that λ has a finite rank or finite multiplicity).*

Proof. The proof is trivial if X is finite dimensional, so let us assume that X is infinite dimensional. To prove the first statement of the theorem, it suffices to show that for all $r > 0$ the set $\{\lambda \in \mathbb{K}; |\lambda| \geq r\}$ contains a finite number of eigenvalues. Suppose not, i.e., there exists $r_0 > 0$ and infinitely many distinct eigenvalues $\lambda_1, \lambda_2, \ldots$ such that $|\lambda_n| \geq r_0 \ \forall n \geq 1$. Then there exists a sequence $u_n \in X \setminus \{0\}$ such that $Au_n = \lambda_n u_n \ \forall n \geq 1$, and we may assume that $\|u_n\| = 1 \ \forall n \geq 1$. Because the λ_n's are distinct, $B_n = \{u_1, u_2, \ldots, u_n\}$ are independent systems. Set $X_n = \text{Span} \, B_n$, $n = 1, 2, \ldots$ By Lemma 2.25, there exists $y_n \in X_n \setminus X_{n-1}$ such that $\|y_n\| = 1 \ \forall n \geq 2$ and

$$\|y_n - v\| \geq \frac{1}{2} \ \forall v \in X_{n-1}, \ n \geq 2 \implies \|y_n - y_m\| \geq \frac{1}{2} \ \forall n \neq m.$$

Thus (y_n) has no Cauchy (hence no convergent) subsequence. On the other hand, assuming that $1 \leq m < n$, we have

$$Ay_n - Ay_m = \underbrace{\lambda_n y_n}_{\in X_n \setminus X_{n-1}} - \underbrace{\lambda_n y_m}_{\in X_m} + \underbrace{(Ay_n - \lambda_n y_n)}_{\in X_{n-1}} - \underbrace{(Ay_m - \lambda_m y_m)}_{\in X_m \subset X_{n-1}}$$

$$= \lambda_n y_n - v_{mn}$$

with $v_{mn} \in X_{n-1}$, because

$$
Ay_n - \lambda_n y_n = A\Big(\sum_{i=1}^{n} \alpha_i^n u_i\Big) - \lambda_n \Big(\sum_{i=1}^{n} \alpha_i^n u_i\Big)
$$

$$
= \Big(\sum_{i=1}^{n} \alpha_i^n \lambda_i u_i\Big) - \lambda_n \Big(\sum_{i=1}^{n} \alpha_i^n u_i\Big)
$$

$$
= \sum_{i=1}^{n} \alpha_i^n (\lambda_i - \lambda_n) u_i
$$

$$
= \sum_{i=1}^{n-1} \alpha_i^n (\lambda_i - \lambda_n) u_i
$$

which is in X_{n-1}. Hence we have

$$
\begin{aligned}
\|Ay_n - Ay_m\| &= \|\lambda_n y_n - v_{nm}\| \\
&= |\lambda_n| \cdot \|y_n - \lambda_n^{-1} v_{nm}\| \\
&\geq r_0 \|y_n - \lambda_n^{-1} v_{nm}\| \\
&\geq \frac{r_0}{2},
\end{aligned}
$$

so (Ay_n) has no Cauchy (hence no convergent) subsequence. But A is compact and $\|y_n\| = 1$ $\forall n \geq 1$ so (Ay_n) must have a convergent subsequence. This contradiction shows that $\{\lambda \in \mathbb{K}; |\lambda| \geq r\}$ contains a finite number of eigenvalues of A for all $r > 0$, as claimed.

The proof of the latter statement of the theorem is similar to the proof of Theorem 7.3.   □

Proposition 8.5. *Let* $(H, (\cdot, \cdot), \|\cdot\|)$ *be a Hilbert space and let* $A :$ $H \to H$ *be a symmetric (hence self-adjoint) operator. Then,*
(i)   *every eigenvalue of A is real, even if $\mathbb{K} = \mathbb{C}$;*
(ii)   *every two eigenvectors of A corresponding to distinct eigenvalues are orthogonal.*

Proof. To prove (i) suppose λ is an eigenvalue of A. Let $u \in H \setminus \{0\}$ be a corresponding eigenvector, i.e., $Au = \lambda u$. Then

$$
\begin{aligned}
(Au, u) &= (\lambda u, u) = \lambda \|u\|^2, \\
(u, Au) &= (u, \lambda u) = \overline{\lambda} \|u\|^2.
\end{aligned}
$$

As A is symmetric and $\|u\| \neq 0$, we infer that $\lambda = \overline{\lambda}$.

To prove (ii), consider two eigenpairs of A, (u_1, λ_1), (u_2, λ_2), where $\lambda_1, \lambda_2 \in \mathbb{R}$ (from (i)) and $\lambda_1 \neq \lambda_2$. We have

$$\lambda_1(u_1, u_2) = (Au_1, u_2) = (u_2, Au_1) = \lambda_2(u_1, u_2),$$

thus

$$\underbrace{(\lambda_1 - \lambda_2)}_{\neq 0}(u_1, u_2) = 0,$$

so $(u_1, u_2) = 0$. $\square$

Proposition 8.6. *Let $(H, (\cdot, \cdot), \|\cdot\|)$ be a Hilbert space, $H \neq \{0\}$, and let $A \in L(H)$ be a symmetric operator. Then,*

$$\|A\| = \sup\left\{\, |(Ax, x)|;\ x \in H,\ \|x\| = 1 \,\right\}.$$

Proof. Trivial if $A = 0$ (equivalently $\|A\| = 0$). Assume $A \neq 0$ ($\|A\| > 0$) and set

$$a = \sup\left\{\, |(Ax, x)|;\ x \in H,\ \|x\| = 1 \,\right\}.$$

Since

$$|(Ax, x)| \leq \|Ax\| \cdot \|x\| \leq \|A\| \cdot \|x\|^2 \quad \forall x \in H,$$

we infer that

$$a \leq \|A\|. \tag{8.2.2}$$

Now, for given $b > 0$ and $x \in H$ such that $\|x\| = 1$ and $\|Ax\| > 0$, we have

$$\|Ax\|^2 = \frac{1}{4}\left[\left(A(bx + \frac{1}{b}Ax), bx + \frac{1}{b}Ax\right) - \left(A(bx - \frac{1}{b}Ax), bx - \frac{1}{b}Ax\right)\right]. \tag{8.2.3}$$

We also have

$$|(Av, v)| \leq a\|v\|^2 \quad \forall v \in H. \tag{8.2.4}$$

Combining (8.2.3) and (8.2.4) we obtain

$$\|Ax\|^2 \leq \frac{a}{4}\left(\|bx + \frac{1}{b}Ax\|^2 + \|bx - \frac{1}{b}Ax\|^2\right)$$

$$\leq \frac{a}{2}\left(b^2\|x\|^2 + \frac{1}{b^2}\|Ax\|^2\right),$$

so for $\|x\| = 1$ and $b = \|Ax\| > 0$ we have

$$\|Ax\|^2 \leq a\|Ax\|.$$

Therefore,

$$\|Ax\| \leq a \quad \forall x \in H,\ \|x\| = 1 \implies \|A\| \leq a.$$

This together with (8.2.2) implies $\|A\| = a$. $\square$

Note that the assumption that A is symmetric in the above theorem is essential.

We have the following central theorem.

Theorem 8.7 (Hilbert–Schmidt). *Let $(H, (\cdot, \cdot), \| \cdot \|)$ be an infinite dimensional, separable Hilbert space and let $A : H \to H$ be a symmetric (equivalently, self-adjoint), compact linear operator, with $N(A) = \{0\}$. Then there exist a sequence of eigenvalues of A, $(\lambda_1, \lambda_2, \ldots, \lambda_n, \ldots)$, such that $(|\lambda_n|)$ is a decreasing sequence of positive numbers converging to 0 and a complete orthonormal system (basis) in H of corresponding eigenvectors $\{u_n\}_{n=1}^{\infty}$ (i.e., $Au_n = \lambda_n u_n$ for $n = 1, 2, \ldots$).*

Proof. We first observe that $\|A\| > 0 \iff A \neq 0$ (since $N(A) = \{0\}$). Let us prove that either $\|A\|$ or $-\|A\|$ is an eigenvalue of A. By Proposition 8.6 there exists $(v_n)_{n \geq 1}$, with $\|v_n\| = 1 \ \forall n \geq 1$, such that $|(Av_n, v_n)| \to \|A\|$. In fact, one can extract from (v_n) a subsequence, again denoted (v_n), such that (Av_n, v_n) converges to either $\|A\|$ or $-\|A\|$, say

$$(Av_n, v_n) \to \lambda_1 := \|A\|. \tag{8.2.5}$$

Since A is compact we can now take another subsequence, also denoted (v_n), such that

$$Av_n \to u_1, \tag{8.2.6}$$

and this is the subsequence we keep. Now, passing to the limit in

$$0 \leq \|Av_n - \lambda_1 v_n\|^2 = \|Av_n\|^2 - 2\lambda_1(Av_n, v_n) + \lambda_1^2, \tag{8.2.7}$$

we get (see (8.2.5) and (8.2.6))

$$0 \leq \|u_1\|^2 - \lambda_1^2 \implies |\lambda_1| \leq \|u_1\|.$$

Hence, in particular, $u_1 \neq 0$. The converse is also true since we have

$$\|Av_n\| \leq \|A\| \cdot \|v_n\| = \|A\|,$$

so by (8.2.6)

$$\|u_1\| \leq \|A\| = |\lambda_1|.$$

Therefore,

$$\|u_1\| = |\lambda_1| = \|A\|. \tag{8.2.8}$$

From (8.2.7) (see also (8.2.5), (8.2.6) and (8.2.8)) we derive

$$\|Av_n - \lambda_1 v_n\| \to 0. \tag{8.2.9}$$

So, in view of (8.2.6), $(\lambda_1 v_n)$ converges to u_1 and thus by (8.2.9) and continuity of A we get

$$Au_1 = \lambda_1 u_1,$$

i.e., (u_1, λ_1) is an eigenpair of A. We normalize without changing notation, $u_1 := |\lambda_1|^{-1} u_1$, since we want an orthonormal system of eigenvectors.

It is worth pointing out that any other eigenvalue λ satisfies $|\lambda| \le |\lambda_1|$. Indeed, if we assume by contradiction the existence of an eigenpair (u, λ), with $|\lambda| > |\lambda_1|$ and $\|u\| = 1$, then $|(Au, u)| = |\lambda|$ which contradicts $|\lambda_1| = \|A\|$ being the supremum from Proposition 8.6.

We now use induction to prove the existence of eigenpairs (u_n, λ_n) for $n = 2, 3, \ldots$

Denote by Y the orthogonal complement of $\mathrm{Span}\{u_1\}$, i.e.,

$$Y = \{\, u \in H;\ (u, u_1) = 0 \,\}.$$

Since H is infinite dimensional, so is Y. Moreover Y is a Hilbert space (with the scalar product and norm of H), and is invariant to A in the sense that $AY \subset Y$ because for $y \in Y$,

$$(Ay, u_1) = (y, Au_1),$$

since A is symmetric, and

$$\begin{aligned} &= (y, \lambda_1 u_1) \\ &= \lambda_1 (y, u_1) \\ &= 0. \end{aligned}$$

The restriction $A|_Y$ is not 0 since then $N(A) = Y$. In fact, all the properties are inherited (symmetric, compact, and $N(A|_Y) = \{0\}$) and by the previous step we have an eigenvalue

$$\lambda_2 = \pm \sup \{\, |(Av, v)|;\ v \in Y, \|v\| = 1 \,\},$$

and a corresponding eigenvector

$$u_2 \in Y, \quad \|u_2\| = 1, \quad Au_2 = \lambda_2 u_2.$$

Moreover, $|\lambda_2| \le \lambda_1$.

Next, take

$$Z = \{ u \in Y;\ (u, u_2) = 0 \} = \Big(\operatorname{Span}\{u_1, u_2\} \Big)^{\perp},$$

which is an infinite dimensional (Hilbert) subspace of H, and obtain a new eigenpair (u_3, λ_3), with $\|u_3\| = 1$, $|\lambda_3| \le |\lambda_2|$. We may continue doing this, each time obtaining an infinite dimensional subspace. We thus construct a sequence of eigenvalues (λ_n) such that

$$|\lambda_1| \ge |\lambda_2| \ge \cdots \ge |\lambda_n| \ge \cdots, \qquad (8.2.10)$$

and the corresponding sequence of eigenvectors (u_n),

$$A u_n = \lambda_n u_n, \quad \|u_n\| = 1,\ n \ge 1,$$

forms an orthonormal system by construction.

Next, we prove that

$$Au = \sum_{n=1}^{\infty} \lambda_n (u, u_n) u_n \quad \forall u \in H. \qquad (8.2.11)$$

Define the space

$$V_m := \{ u \in H;\ (u, u_j) = 0,\ j = 1, \ldots, m \} = \Big(\operatorname{Span}\{u_1, \ldots, u_m\} \Big)^{\perp},$$

which is an infinite dimensional Hilbert space (with respect to $(\cdot, \cdot)$, $\|\cdot\|$), invariant under A (i.e., $Av \in V_m \ \forall v \in V_m$). By the previous step of our proof, there is an eigenpair (u_{m+1}, λ_{m+1}) of A such that

$$|\lambda_{m+1}| = \|A|_{V_m}\| = \sup \{ |(Av, v)|;\ v \in V_{m+1},\ \|v\| = 1 \}.$$

In particular,

$$\|Av\| \le |\lambda_{m+1}| \cdot \|v\| \quad \forall v \in V_{m+1}. \qquad (8.2.12)$$

Now, choose a particular

$$w_m = u - \sum_{n=1}^{m} (u, u_n) u_n$$

and notice that $w_m \in V_m$ because $(v_m, u_j) = (u, u_j) - (u, u_j) = 0 \ \forall j = 1, \ldots, m$. Calculate

$$\|w_m\|^2 = \|u\|^2 - \sum_{n=1}^{m} |(u, u_n)|^2 \le \|u\|^2. \qquad (8.2.13)$$

Combining (8.2.12) and (8.2.13) we get

$$Aw_m = Au - \sum_{n=1}^{m} (u, u_n) Au_n$$

$$= Au - \sum_{n=1}^{m} \lambda_n (u, u_n) u_n,$$

so that

$$\|Aw_m\| \le \|A|_{V_m}\| \cdot \|w_m\|$$
$$= |\lambda_{m+1}| \cdot \|w_m\|$$
$$\le |\lambda_{m+1}| \cdot \|u\|. \qquad (8.2.14)$$

On the other hand, $\lambda_n \to 0$. Indeed, since $(|\lambda_n|)$ is decreasing (see (8.2.10)), there exists

$$\lim_{n \to \infty} |\lambda_n| = \alpha \ge 0.$$

Suppose by way of contradiction that $\alpha > 0$. Obviously, $|\lambda_n| \ge \alpha$ for all $n \ge 1$ and so

$$\|\lambda_n^{-1} u_n\| = \frac{\|u_n\|}{|\lambda_n|} = \frac{1}{|\lambda_n|} \le \frac{1}{\alpha} \quad \forall n \ge 1.$$

Since A is compact $u_n = A(\lambda_n^{-1} u_n)$ has a convergent subsequence. But this is impossible because

$$\|u_n - u_m\|^2 = \|u_n\|^2 + \|u_m\|^2 = 2 \quad \forall n \ne m.$$

So $\alpha = 0$, i.e., $\lambda_n \to 0$, as claimed.

Consequently, we have by (8.2.14) that $\|Aw_m\| \to 0$ as $m \to \infty$, i.e., (8.2.11) holds true.

Finally, let us prove that $\{u_n\}_{n=1}^{\infty}$ is a basis in H.

We know from the proof of Theorem 6.21 that for all $u \in H$ the series $\sum_{n=1}^{\infty} (u, u_n) u_n$ converges (as $\{u_n\}_{n=1}^{\infty}$ is an orthonormal system), so we can write

$$v = \sum_{n=1}^{\infty} (u, u_n) u_n$$

and we simply need to check that $u = v$. Consider the sequence of partial sums $s_m = \sum_{n=1}^m (u, u_n) u_n$ which converges strongly to v as $m \to \infty$, so $A s_m \to A v$. On the other hand, by (8.2.11) we have that

$$A s_m = \sum_{n=1}^m \lambda_n (u, u_n) u_n \to A u \quad \text{as } m \to \infty.$$

Hence,

$$A v = A u \implies A(v - u) = 0 \implies v = u,$$

since $\ker A = \{0\}$. Thus the system $\{u_n\}_{n=1}^\infty$ is complete, i.e., a basis in H (cf. Theorem 6.21). $\square$

Remark 8.8. If we assume in addition that A is positive (i.e., $(Av, v) > 0$ for all $v \in H \setminus \{0\}$), then it has eigenvalues $\lambda_1 \geq \lambda_2 \geq \cdots \geq \lambda_n \geq \cdots$, with $\lambda_n > 0 \ \forall n \geq 1$. This follows from

$$(A u_n, u_n) = \lambda_n \|u_n\|^2 = \lambda_n, \ n \geq 1.$$

Note also that

$$\lambda_1 = \|A\| = \sup\{(Av, v); \ v \in H, \ \|v\| = 1\} \text{ and}$$

$$\lambda_{n+1} = \|A|_{V_n}\| = \sup\{(Av, v); \ v \in V_n, \ \|v\| = 1\} \ \forall n \geq 1,$$

where $V_n = \left(\text{Span}\{u_1, u_2, \ldots, u_n\} \right)^\perp, \ n \geq 1$.

8.3   Eigenvalues of $-\Delta$ Under the Dirichlet Boundary Condition

In what follows we apply the Hilbert–Schmidt Theorem to an eigenvalue problem for the Laplace operator. Specifically, let $\emptyset \neq \Omega \subset \mathbb{R}^N$, $N \geq 2$, be a bounded domain with smooth boundary $\partial \Omega$. Consider the Dirichlet eigenvalue problem

$$\begin{cases} -\Delta u = \lambda u & \text{in } \Omega, \\ u = 0 & \text{on } \partial \Omega. \end{cases} \tag{8.3.15}$$

Definition 8.9. *A real number λ is said to be an eigenvalue of the Dirichlet problem (8.3.15) if there is a function $u \in H_0^1(\Omega) \setminus \{0\}$ such that the problem is satisfied in the sense that*

$$\int_\Omega \nabla u \cdot \nabla v \, dx = \lambda \int_\Omega u v \, dx \quad \forall v \in H_0^1(\Omega), \tag{8.3.16}$$

or, equivalently,

$$-\Delta u = \lambda u \ \text{in } D'(\Omega).$$

Remark 8.10. As $\partial\Omega$ is assumed to be smooth, the eigenfunction u is in fact more regular (see [6, Theorem 9.25, p. 298]).

Theorem 8.11. *Let $\emptyset \neq \Omega \subset \mathbb{R}^N$ be a bounded domain with smooth boundary $\partial\Omega$. Then there exist an increasing sequence of positive eigenvalues λ_n for (8.3.15) such that $\lambda_n \to +\infty$ and a complete orthonormal system (in $H = L^2(\Omega)$) of eigenfunctions u_n satisfying problem (8.3.15) with $\lambda = \lambda_n$, $n = 1, 2, \ldots$*

Proof. Let $H = L^2(\Omega)$ equipped with the usual inner product and norm. H is an infinite dimensional, separable Hilbert space (over $\mathbb{R}$). We know that for every $f \in H = L^2(\Omega)$ the problem

$$\begin{cases} -\Delta u = f & \text{in } \Omega, \\ u = 0 & \text{on } \partial\Omega, \end{cases}$$

has a unique solution $u \in H_0^1(\Omega)$ (by Dirichlet's Principle, Chap. 6). Define an operator $A : H \to H$ by assigning $f \mapsto u$. Note that A is linear and $N(A) = \{0\}$. Moreover, A is symmetric (hence self-adjoint since $D(A) = H$). Indeed, if $v = Ag$ with $g \in H$, i.e.,

$$\begin{cases} -\Delta v = g & \text{in } \Omega, \\ v = 0 & \text{on } \partial\Omega, \end{cases}$$

then, by Green, we can write

$$\int_\Omega \nabla u \cdot \nabla v \, dx = \int_\Omega f v \, dx = \int_\Omega f A g \, dx,$$

$$\int_\Omega \nabla v \cdot \nabla u \, dx = \int_\Omega g u \, dx = \int_\Omega g A f \, dx,$$

so $\int_\Omega f A g \, dx = \int_\Omega g A f \, dx$ as desired.

Let us show that operator A is also compact, i.e., for every constant $M > 0$, the set

$$S_M := \{Af; \ f \in L^2(\Omega), \ \|f\|_{L^2(\Omega)} \leq M\}$$

is relatively compact in $H = L^2(\Omega)$. Indeed, if $u = Af \in S_M$ it follows from (8.3.16) with $v = u$ that

$$\|\nabla u\|_{L^2(\Omega)}^2 = \int_\Omega f u \, dx$$
$$\leq \|f\|_{L^2(\Omega)} \cdot \|u\|_{L^2(\Omega)}$$

and by the Poincaré inequality

$$\leq C\|f\|_{L^2(\Omega)} \cdot \|\nabla u\|_{L^2(\Omega)}.$$

Finally,

$$\|\nabla u\|_{L^2(\Omega)} \leq C\|f\|_{L^2(\Omega)} \leq CM,$$

so that $\|Af\|_{H_0^1(\Omega)}$ is less than or equal to some constant. We know that bounded sets in $H_0^1(\Omega)$ are relatively compact in $L^2(\Omega)$ so that S_M is relatively compact in this space.

We can apply the Hilbert–Schmidt Theorem which guarantees the existence of a sequence of eigenpairs for A, $\{(u_n, \mu_n)\}_{n=1}^\infty$, such that $|\mu_n|$ decreases to zero and $\{u_n\}_{n=1}^\infty$ is a complete orthonormal system (basis) in $H = L^2(\Omega)$. Note that $Au_n = \mu_n u_n$ says that u_n satisfies the problem

$$\begin{cases} -\Delta u_n = \lambda_n u_n & \text{in } \Omega, \\ u_n = 0 & \text{on } \partial\Omega, \end{cases}$$

where $\lambda_n = 1/\mu_n$, i.e., (u_n, λ_n) is an eigenpair of problem (8.3.15). Note also that

$$\lambda_n = \lambda_n \int_\Omega \|u_n\|^2 dx = \int_\Omega |\nabla u_n|^2 dx > 0 \quad \forall n \geq 1,$$

so $(\lambda_n)_{n \geq 1}$ is an increasing sequence of positive numbers, and $\lambda_n \to +\infty$ (since $|\mu_n| = \mu_n$ decreases to 0). $\qquad \square$

8.4 Eigenvalues of $-\Delta$ Under the Robin Boundary Condition

Let again $\emptyset \neq \Omega \subset \mathbb{R}^N$, $N \geq 2$, be a bounded domain with smooth boundary $\partial\Omega$. Consider the classical Robin eigenvalue problem

$$\begin{cases} -\Delta u = \lambda u & \text{in } \Omega, \\ \frac{\partial u}{\partial \nu} + \alpha u = 0 & \text{on } \partial\Omega, \end{cases} \qquad (8.4.17)$$

where α is a positive constant and $\partial u/\partial \nu$ denotes the outward unit normal to $\partial\Omega$. In this case we have the following natural definition:

Definition 8.12. *A real number λ is said to be an eigenvalue of the Robin problem (8.4.17) if there is a function $u \in H^1(\Omega) \setminus \{0\}$ such that*

$$\int_\Omega \nabla u \cdot \nabla v \, dx + \alpha \int_{\partial\Omega} uv \, ds = \lambda \int_\Omega uv \, dx \quad \forall v \in H^1(\Omega). \quad (8.4.18)$$

Remark 8.13. Again, as $\partial\Omega$ was assumed to be smooth enough, the eigenfunction u is, in fact, more regular.

Theorem 8.14. *Assume $\emptyset \neq \Omega \subset \mathbb{R}^N$ is a bounded domain with smooth boundary $\partial\Omega$ and α is a positive constant. Then there exists an increasing sequence of positive eigenvalues λ_n for (8.4.17) such that $\lambda_n \to +\infty$ and a complete orthonormal system (in $H = L^2(\Omega)$) of eigenfunctions u_n satisfying problem (8.4.17) with $\lambda = \lambda_n$, $n = 1, 2, \dots$*

Proof. Again, let $H = L^2(\Omega)$ equipped with the usual inner product and norm. By the Lax–Milgram Theorem (see Chap. 6) we easily infer that for every $f \in H = L^2(\Omega)$ the problem

$$\begin{cases} -\Delta u + u = f & \text{in } \Omega, \\ \frac{\partial u}{\partial \nu} + \alpha u = 0 & \text{on } \partial\Omega, \end{cases}$$

has a unique solution $u \in H^1(\Omega)$. Now define $A : H \to H$ by assigning $f \mapsto u$. It is an easy exercise to check that A is positive and satisfies all the conditions of the Hilbert–Schmidt Theorem. In contrast with the previous Dirichlet case, we have replaced $-\Delta$ by $-\Delta + I$ in order to ensure the strong positivity (coercivity) of the corresponding bilinear form as well as the compactness of A (based on Theorem 5.22). Therefore there exists a sequence of eigenpairs for A, $\{(u_n, \mu_n)\}_{n=1}^\infty$, such that $|\mu_n| = \mu_n$ decreases to 0 and $\{u_n\}_{n=1}^\infty$ is an orthonormal basis in H. The fact that $Au_n = \mu_n u_n$ can be written as

$$\begin{cases} -\Delta u_n = \lambda_n u_n & \text{in } \Omega, \\ \frac{\partial u}{\partial \nu} + \alpha u_n = 0 & \text{on } \partial\Omega, \end{cases}$$

where $\lambda_n = -1 + 1/\mu_n$, i.e., (u_n, λ_n) is an eigenpair of problem (8.4.17). Note that

$$\lambda_n = \lambda_n \int_\Omega \|u_n\|^2 dx = \int_\Omega \|\nabla u_n\|^2 dx + \alpha \int_{\partial\Omega} u_n^2 \, ds > 0 \quad \forall n \geq 1, \quad (8.4.19)$$

so $(\lambda_n)_{n \geq 1}$ is an increasing sequence of positive numbers converging to ∞ (since $|\mu_n| = \mu_n$ decreases to 0). $\qquad \square$

8.5   Eigenvalues of $-\Delta$ Under the Neumann Boundary Condition

Under the same conditions on Ω we consider the Neumann eigenvalue problem

$$\begin{cases} -\Delta u = \lambda u & \text{in } \Omega, \\ \frac{\partial u}{\partial \nu} = 0 & \text{on } \partial\Omega, \end{cases} \tag{8.5.20}$$

i.e., $\alpha > 0$ in the Robin eigenvalue problem is replaced by $\alpha = 0$. The definition of an eigenvalue is the same as before (see Definition 8.12) with $\alpha = 0$ in (8.4.18). We have a result similar to Theorem 8.14 which we explain in what follows. One can again consider $H = L^2(\Omega)$ with its usual scalar product and norm, and $A : H \to H$ the operator which associates with each $f \in H$ the unique solution $u \in H^1(\Omega)$ of the problem

$$\begin{cases} -\Delta u + u = f & \text{in } \Omega, \\ \frac{\partial u}{\partial \nu} = 0 & \text{on } \partial\Omega. \end{cases}$$

The Hilbert–Schmidt Theorem is again applicable (see also Remark 8.8), thus there exist a decreasing sequence of positive eigenvalues of operator A, say $(\mu_n)_{n\geq0}$, $\mu_n \to 0$, and a corresponding complete orthonormal system $\{u_n\}_{n=0}^{\infty}$, i.e., $Au_n = \mu_n u_n$ for $n = 0, 1, 2, \ldots$ So denoting $\lambda_n = -1 + 1/\mu_n$ we have

$$\begin{cases} -\Delta u_n = \lambda_n u_n & \text{in } \Omega, \\ \frac{\partial u_n}{\partial \nu} = 0 & \text{on } \partial\Omega, \end{cases}$$

for $n = 0, 1, 2, \ldots$ and (λ_n) is an increasing sequence converging to ∞. We also have (8.4.19) with $\alpha = 0$, hence $\lambda_n \geq 0$ for all $n \geq 0$. Note that $\lambda_0 = 0$ is the first eigenvalue of problem (8.5.20), the corresponding eigenfunctions being the nonzero constant functions. Thus $\lambda_0 = 0$ has multiplicity one (so $\lambda_0 = 0$ is said to be a simple eigenvalue) and the corresponding normalized eigenfunction is $u_0 = \pm 1/\sqrt{m(\Omega)}$, where $m(\Omega)$ denotes the Lebesgue measure of Ω. Consequently, a result similar to Theorem 8.14 holds, with the only difference that the first eigenvalue is no longer a positive number (it is $\lambda_0 = 0$).

In fact, the proof can also be done as in the Dirichlet case, as explained below. Denote by V_0 the one-dimensional space generated by $u_0 =$

$1/\sqrt{m(\Omega)}$: $V_0 = \mathrm{Span}\{u_0\} = \mathrm{Span}\{1\}$. Obviously, the space $H = L^2(\Omega)$ can be written as a direct sum

$$H = V_0 \oplus V_1, \quad V_1 = V_0^{\perp} = \{v \in H; \int_{\Omega} v\, dx = 0\}.$$

The space V_1 is a closed linear subspace of H, so it is a real Hilbert space with respect to the same scalar product and norm. We can use V_1 as a basic space to show the existence of (λ_n, u_n) for $n = 1, 2, \ldots$ Note that $W = V_1 \cap H^1(\Omega)$ is a real Hilbert space with respect to the scalar product (see (8.5.21) below)

$$\langle v, w \rangle = \int_{\Omega} \nabla v \cdot \nabla w\, dx \quad \forall v, w \in W,$$

and the corresponding induced norm. Indeed, we can show that

$$\beta = \inf\{\int_{\Omega} \|\nabla v\|^2 dx;\ v \in W,\ \int_{\Omega} v^2\, dx = 1\}$$

$$= \inf_{v \in W \setminus \{0\}} \underbrace{\frac{\int_{\Omega} \|\nabla v\|^2 dx}{\int_{\Omega} \|v\|^2 dx}}_{\text{Rayleigh quotient}},$$

is a positive number. If we assume by way of contradiction that $\beta = 0$ then there exists a minimizing sequence $(v_k)_{k \geq 1}$ in W, $\|v_k\|_{L^2(\Omega)} = 1\ \forall k \geq 1$, such that (v_k) converges to some $\hat{v}$ weakly in $H^1(\Omega)$ and strongly in V_1. From

$$\|\nabla \hat{v}\|^2_{L^2(\Omega)} = \int_{\Omega} \nabla \hat{v} \cdot \nabla(\hat{v} - v_k)\, dx + \int_{\Omega} \nabla \hat{v} \cdot \nabla v_k\, dx$$

$$\leq \int_{\Omega} \nabla \hat{v} \cdot \nabla(\hat{v} - v_k)\, dx + \|\nabla \hat{v}\|_{L^2(\Omega)} \|\nabla v_k\|_{L^2(\Omega)}$$

we derive

$$\int_{\Omega} \|\nabla \hat{v}\|^2 dx \leq \liminf \int_{\Omega} \|\nabla v_k\|^2 dx = 0,$$

which implies

$$\int_{\Omega} \|\nabla \hat{v}\|^2 dx = 0,$$

and so $\hat{v}$ is a constant function. Since $\hat{v} \in V_1$ it follows that $\hat{v} = 0$. On the other hand, one can derive from $\|v_k\|_{L^2(\Omega)} = 1$, $k \geq 1$, that

$\|\hat{v}\|_{L^2(\Omega)} = 1$, a contradiction. Thus $\beta > 0$ as claimed. In particular, this implies the following Poincaré-type inequality:

$$\beta\|v\|_{L^2(\Omega)}^2 \leq \|\nabla v\|_{L^2(\Omega)}^2 \quad \forall v \in W. \tag{8.5.21}$$

Now, according to the Lax–Milgram Theorem, for each $f \in V_1$ the problem

$$\begin{cases} -\Delta u = f & \text{in } \Omega, \\ \frac{\partial u}{\partial \nu} = 0 & \text{on } \partial\Omega, \end{cases}$$

has a unique solution $u \in W$. Moreover the operator $A : V_1 \to V_1$ defined by $Af = u$, $f \in V_1$ (i.e., $A = (-\Delta)^{-1}$), is positive and satisfies the conditions of the Hilbert–Schmidt Theorem. Therefore the existence of $\{(\lambda_n, u_n)\}_{n=1}^{\infty}$ is again guaranteed.

Summarizing what we have done so far, we obtain the following result.

Theorem 8.15. *Assume $\emptyset \neq \Omega \subset \mathbb{R}^N$ is a bounded domain with smooth boundary $\partial\Omega$. Then there exist a sequence of eigenvalues for (8.5.20), $0 = \lambda_0 < \lambda_1 \leq \lambda_2 \leq \cdots \leq \lambda_n \leq \cdots$, such that $\lambda_n \to \infty$ and a complete orthonormal system (in $H = L^2(\Omega)$) of eigenfunctions u_n verifying problem (8.5.20) with $\lambda = \lambda_n$, $n = 0, 1, 2, \ldots$; in addition $\lambda_0 = 0$ is simple and $u_0 = \pm 1/\sqrt{m(\Omega)}$.*

8.6 Some Comments

1. Let $f \in L^2(\Omega)$. The Neumann problem

$$\begin{cases} -\Delta u = f & \text{in } \Omega, \\ \frac{\partial u}{\partial \nu} = 0 & \text{on } \partial\Omega, \end{cases}$$

has a solution (that is unique up to additive constant) if and only if $f \in V_1$ (i.e., $\int_\Omega f \, dx = 0$). Sufficiency follows by the Lax–Milgram Theorem, as noticed before, while the converse implication follows by Green's Identity.

2. Define

$$\lambda_1^D = \inf\left\{ \int_\Omega \|\nabla v\|^2 dx; \, v \in H_0^1(\Omega), \int_\Omega v^2 \, dx = 1 \right\}$$

$$= \inf_{v \in H_0^1(\Omega)\setminus\{0\}} \underbrace{\frac{\int_\Omega \|\nabla v\|^2 dx}{\int_\Omega \|v\|^2 dx}}_{\text{Rayleigh quotient}}.$$

It is easily seen that λ_1^D is positive and is attained for a function $u_1^D \in W_D = H_0^1(\Omega)$, $\|u_1^D\|_{L^2(\Omega)} = 1$ which is an eigenfunction corresponding to λ_1^D. Moreover, λ_1^D is the first eigenvalue (or principal eigenvalue), i.e., $\lambda_1^D = \lambda_1$ given by Theorem 8.11, λ_1^D is simple, and u_1^D is positive within Ω (see [14, Theorem 2, p. 336]). If we define

$$W_1^D = \{v \in H_0^1(\Omega); \int_\Omega u_1^D v \, dx = 0\},$$

then

$$\lambda_2^D = \inf \{\int_\Omega \|\nabla v\|^2 dx; \ v \in W_1^D, \ \int_\Omega v^2 \, dx = 1\},$$

is attained at some $u_2^D \in W_1^D$, $\|u_2^D\|_{L^2(\Omega)} = 1$, which is an eigenfunction corresponding to λ_2^D, $u_2^D \perp u_1^D$. In general, setting

$$W_{n-1}^D = \{v \in H_0^1(\Omega); \int_\Omega u_j^D v \, dx = 0, \ j = 1, \ldots, n-1\}, \ n \geq 2,$$

$$\lambda_n^D = \inf \{\int_\Omega \|\nabla v\|^2 dx; \ v \in W_{n-1}^D, \ \int_\Omega v^2 \, dx = 1\},$$

we obtain a sequence of eigenpairs (λ_n^D, u_n^D), such that

$$\lambda_1^D < \lambda_2^D \leq \cdots \leq \lambda_n^D \leq \cdots, \quad \lambda_n^D = \lambda_n \to \infty,$$

and $\{u_n^D\}_{n=1}^\infty$ is an orthonormal basis in $L^2(\Omega)$.

This method is an alternative to that described in the proof of the Hilbert–Schmidt Theorem.

Similar arguments work for the Robin and Neumann eigenvalue problems. We just recall that the lowest positive eigenvalues are given by

$$\begin{aligned}
\lambda_1^R &= \inf \{\int_\Omega \|\nabla v\|^2 dx + \alpha \int_{\partial\Omega} v^2 \, ds; \ v \in H^1(\Omega), \ \int_\Omega v^2 \, dx = 1\} \\
&= \inf_{v \in H^1(\Omega)\setminus\{0\}} \frac{\int_\Omega \|\nabla v\|^2 dx + \alpha \int_{\partial\Omega} v^2 \, ds}{\int_\Omega v^2 \, dx}, \quad\quad (8.6.22)
\end{aligned}$$

$$\begin{aligned}
\lambda_1^N &= \inf \{\int_\Omega \|\nabla v\|^2 dx; \ v \in W = V_1 \cap H^1(\Omega), \ \int_\Omega v^2 \, dx = 1\} \\
&= \inf_{v \in W\setminus\{0\}} \frac{\int_\Omega \|\nabla v\|^2 dx}{\int_\Omega v^2 \, dx}.
\end{aligned}$$

It is readily seen that both λ_1^R and λ_1^N (which is equal to β defined before) are positive numbers and are attained for functions $u_1^R \in H^1(\Omega)$ and $u_2^N \in W$ which are the corresponding eigenfunctions. It is also well known that both λ_1^R and λ_1^N are simple and $u_1^R \in H^1(\Omega)$, $u_2^N \in W$ do not change sign within Ω.

3. For all $f \in L^2(\Omega)$ the Robin problem

$$\begin{cases} -\Delta u = f & \text{in } \Omega, \\ \frac{\partial u}{\partial \nu} + \alpha u = 0 & \text{on } \partial\Omega, \end{cases}$$

where α is a given positive constant, has a unique solution $u \in H^1(\Omega)$. Indeed, by (8.6.22) we have the inequality

$$\lambda_1^R \int_\Omega v^2 \, dx \leq \int_\Omega \|\nabla v\|^2 dx + \alpha \int_{\partial\Omega} v^2 \, ds \quad \forall v \in H^1(\Omega), \qquad (8.6.23)$$

which (along with the continuity of the canonical injection of $H^1(\Omega)$ into $L^2(\partial\Omega)$) shows that its right-hand side defines a norm equivalent to the usual norm in $H^1(\Omega)$. So the claim follows from Lax–Milgram applied to the bilinear form

$$(u, v) \mapsto \int_\Omega \nabla u \cdot \nabla v \, dx + \alpha \int_{\partial\Omega} uv \, ds.$$

4. For some particular sets $\Omega \subset \mathbb{R}^N$ the eigenpairs (λ_n, u_n) can be calculated. In the one-dimensional case ($N = 1$), if $\Omega = (0, 1)$, the three eigenvalue problems look as follows:

$$\begin{cases} -u'' = \lambda u, & 0 < x < 1, \\ u(0) = u(1) = 0, \end{cases}$$

$$\begin{cases} -u'' = \lambda u, & 0 < x < 1, \\ u'(0) = u'(1) = 0, \end{cases}$$

$$\begin{cases} -u'' = \lambda u, & 0 < x < 1, \\ -u'(0) + \alpha u(0) = 0 = u'(1) + \alpha u(1), \end{cases}$$

where α is a given positive constant. In the first two cases (Dirichlet and Neumann) we obtain by easy computations

$$\lambda_n^D = \pi^2 n^2, \quad u_n^D(x) = \sqrt{2}\sin(n\pi x), \quad n = 1, 2, \ldots;$$

$\lambda_0^N = 0, u_0^N(x) = 1; \quad \lambda_n^N = \pi^2 n^2, \quad u_n^N(x) = \sqrt{2}\cos(n\pi x), \ n = 1, 2, \ldots$

In the Robin case we cannot calculate by elementary methods the corresponding eigenpairs (u_n, λ_n), $n \geq 1$.

5. In the Dirichlet case above, the system $\{w_n = \lambda_n^{-1/2} u_n\}_{n=1}^{\infty}$ is an orthonormal basis in $W_D = H_0^1(\Omega)$. Indeed, we can deduce from

$$\begin{cases} -\Delta u_n = \lambda_n u_n & \text{in } \Omega, \\ u_n = 0 & \text{on } \partial\Omega, \end{cases} \qquad (8.6.24)$$

that

$$\int_\Omega \nabla w_n \cdot \nabla w_k \, dx = \int_\Omega u_n u_k \, dx = \delta_{nk} \quad \forall n, k \geq 1,$$

which shows that $\{w_n\}_{n=1}^{\infty}$ is an orthonormal system in W_D. Now, since $\{u_n\}_{n=1}^{\infty}$ is complete in $H = L^2(\Omega)$, any $u \in H$ can be written as (see Theorem 6.21)

$$u = \sum_{n=1}^{\infty} (u, u_n)_{L^2(\Omega)} u_n = \sum_{n=1}^{\infty} \left(\int_\Omega u u_n \, dx \right) u_n,$$

so, according to (8.6.24),

$$u = \sum_{n=1}^{\infty} \left(\int_\Omega \nabla u \cdot \nabla w_n \, dx \right) w_n.$$

Thus $\{w_n\}_{n=1}^{\infty}$ is complete in W_D.

Similar statements hold true for the other two cases (Neumann and Robin) within $W_N = V_1 \cap H^1(\Omega)$ and $W_R = H^1(\Omega)$ equipped with the scalar products

$$(w_1, w_2)_N = \int_\Omega \nabla w_1 \cdot \nabla w_2 \, dx,$$

$$(w_1, w_2)_R = \int_\Omega \nabla w_1 \cdot \nabla w_2 \, dx + \alpha \int_{\partial\Omega} w_1 w_2 \, dx.$$

In fact, these statements on the negative Laplacian with Dirichlet, Neumann or Robin boundary conditions can be derived from the abstract framework we describe below, related to the so-called *energetic extension* of a linear operator Q satisfying the following assumptions:

(a) $Q : D(Q) \subset H \to H$ is a linear, densely defined, self-adjoint, strongly positive operator, where $(H, (\cdot, \cdot), \|\cdot\|)$ is a real, infinite dimensional, separable Hilbert space.

Define on the vector subspace $D(Q)$ the so-called *energetic scalar product*

$$(u, v)_E = (Qu, v) \quad \forall u, v \in D(Q).$$

It induces the *energetic norm* on $D(Q)$: $\|u\|_E^2 = (u, u)_E, \ u \in D(Q)$. Denote by H_E the completion of $(D(Q), \|\cdot\|_E)$. Then H_E is a Hilbert space with respect to the scalar product

$$(u, v)_E := \lim_{k \to \infty} (u_k, v_k)_E,$$

where (u_k) and (v_k) are sequences in $D(Q)$ converging to u and v, respectively. Since Q is strongly positive, i.e., there exists a constant $c > 0$ such that

$$(Qu, u) \geq c\|u\|^2 \quad \forall u \in D(Q), \tag{8.6.25}$$

we have

$$\|u\| \leq \frac{1}{\sqrt{c}} \|u\|_E \quad \forall u \in H_E,$$

so the identity map from H_E to H is continuous (i.e., H_E is continuously embedded in H). Denote by Q_E the Riesz isomorphism from $(H_E, \|\cdot\|_E)$ onto its dual H_E^*, namely,

$$(Q_E u)(v) = (u, v)_E \quad \forall u, v \in H_E.$$

Identifying H with its dual, we have

$$D(Q) \subset H_E \subset H \subset H_E^*.$$

Since $D(Q)$ is dense in H, we see that

$$Q_E u = Qu \quad \forall u \in D(Q),$$

i.e., Q_E is an extension of Q which is called the *energetic extension*. The term *energetic* will become clear later when we discuss examples.

We also assume that

(b)   the identity map from H_E into H is compact (i.e., H_E is compactly embedded into H).

Now we can state the following abstract spectral result.

Theorem 8.16. *Assume* **(a)** *and* **(b)** *above are fulfilled. Then there exist an increasing sequence* $(\lambda_n)_{n \geq 1}$ *in* $(0, \infty)$ *converging to* ∞, *and an orthonormal basis* $\{u_n\}_{n=1}^{\infty}$ *in* H *such that*

$$u_n \in D(Q), \quad Qu_n = \lambda_n u_n \quad \forall n \geq 1. \tag{8.6.26}$$

In addition, $\{\lambda_n^{-1/2} u_n\}_{n=1}^{\infty}$ *is an orthonormal basis in* H_E *(the energetic space defined above).*

Proof. We shall adapt the proof of Theorem 8.11 to the present abstract framework.

First of all, note that $Q : D(Q) \subset H \to H$ is bijective since its extension $Q_E : H_E \to H_E^*$ is. Denote $A = Q^{-1}$. Obviously, $A \in L(H)$, $N(A) = \{0\}$, and A is self-adjoint. Operator A is also compact. Indeed, if for some $M > 0$ we take $f \in H$ such that $\|f\| \leq M$, then we have for $u = Af$ (equivalently $Qu = f$),

$$\|u\|_E^2 = (Qu, u) \leq \|f\| \cdot \|u\| \leq M\|u\|. \tag{8.6.27}$$

Combining (8.6.27) with (8.6.25) yields

$$\|Af\|_E = \|u\|_E \leq \frac{M}{\sqrt{c}},$$

i.e., A sends bounded sets in H to bounded sets in H_E, hence A is compact (cf. **(b)**). According to the Hilbert–Schmidt Theorem there exists a sequence of eigenpairs for $A = Q^{-1}$, $((\mu_n, u_n))_{n \geq 1}$ with the known properties, with $\mu_n > 0$, $n \geq 1$, since Q is strongly positive. Thus, the first part of the theorem follows with $((\lambda_n, u_n))_{n \geq 1}$, where $\lambda_n = 1/\mu_n$, $n = 1, 2, \ldots$

In order to prove the second part, denote $w_n = \lambda_n^{-1/2} u_n$, $n \geq 1$. It follows from (8.6.26) that

$$(w_n, w_k)_E = (Qu_n, u_k) = \delta_{nk} \quad \forall n, k \geq 1,$$

i.e., the system $\{w_n\}_{n=1}^{\infty}$ is orthonormal in H_E. Now, since $\{u_n\}_{n=1}^{\infty}$ is complete in H, any $u \in H$ can be expressed as (cf. Theorem 6.21)

$$u = \sum_{n=1}^{\infty} (u, u_n) u_n.$$

So, by virtue of (8.6.26),

$$u = \sum_{n=1}^{\infty} (u, w_n)_E w_n,$$

and so we can conclude that $\{w_n\}_{n=1}^{\infty}$ is complete in H_E. $\qquad\square$

Remark 8.17. In addition to the conclusions of Theorem 8.16 one can show that $\{\lambda_n^{1/2} u_n\}_{n=1}^{\infty}$ is an orthonormal basis in H_E^*.

For more details on energetic spaces and extensions we refer the reader to [52, Chapter 5]. See also [22, Chapter 1, p. 18].

Remark 8.18. One can reobtain from Theorem 8.16 the previous statements related to $Q = -\Delta$ with Dirichlet, Neumann or Robin boundary condition.

In the Dirichlet case we have $H = L^2(\Omega)$ with its usual scalar product and induced norm, $D(Q) = H_0^1(\Omega) \cap H^2(\Omega)$, and $H_E = H_0^1(\Omega)$ with the energetic scalar product $(u, v)_E = \int_\Omega \nabla u \cdot \nabla v \, dx$ and $\|u\|_E^2 = (u, u)_E$. Note that H_E is equal to W_D defined above.

In the Neumann case, $H = V_1 := \{v \in L^2(\Omega); \int_\Omega v \, dx = 0\}$ with the scalar product and norm inherited from $L^2(\Omega)$, $D(Q) = V_1 \cap H^2(\Omega)$, and $H_E = V_1 \cap H^1(\Omega)$ (denoted above by W_N) with $(u, v)_E = \int_\Omega \nabla u \cdot \nabla v \, dx$, $\|u\|_E^2 = (u, u)_E$. Of course, in this case we have an additional eigenvalue $\lambda_0 = 0$ as specified before.

Finally, in the case of the Robin boundary condition, $H = L^2(\Omega)$ with its usual scalar product and norm, $D(Q) = H^2(\Omega)$, and $H_E = H^1(\Omega)$ (denoted above by W_R) with $(u, v)_E = \int_\Omega \nabla u \cdot \nabla v \, dx + \alpha \int_{\partial\Omega} uv \, ds$ and $\|u\|_E^2 = (u, u)_E$.

There are also many other specific examples covered by Theorem 8.16, in particular the case $Q = -\Delta$ with different conditions on parts of the boundary of Ω.

Remark 8.19. In order to develop the above theory on energetic extensions we can begin with an operator Q which satisfies all the assumptions in **(a)**, with one exception: Q is only symmetric, not self-adjoint. Everything works similarly and H_E and Q_E can be constructed by using the same arguments. Now define an operator $\hat{Q} : D(\hat{Q}) \subset H \to H$ as follows:

$$D(\hat{Q}) = \{v \in H_E; \, Q_E v \in H\}, \quad \hat{Q} v = Q_E v \quad \forall v \in D(\hat{Q}).$$

Obviously, $\hat{Q}$ is an extension of Q so $D(\hat{Q})$ is dense in H. It is also easily seen that $\hat{Q}$ is strongly positive. As Q_E is bijective so is $\hat{Q}$ since it is a restriction of Q_E. Note also that $\hat{Q}^{-1} \in L(H)$ and is symmetric, hence self-adjoint. Thus $\hat{Q}$ is self-adjoint as well. Operator $\hat{Q}$ is called the *Friedrichs extension* of Q. It is easily seen that the energetic space and the energetic extension defined by $\hat{Q}$ are exactly H_E and Q_E. Summarizing, we see that $\hat{Q}$ satisfies all the conditions in **(a)** and plays the role of the former Q. So assuming in **(a)** that Q is a self-adjoint operator (not a symmetric one) does not restrict the generality. In fact, in this case the Friedrichs extension of Q is Q itself.

For example, if we choose $H = L^2(\Omega)$ (where $\Omega \subset \mathbb{R}^N$ is an open bounded set with smooth boundary) and $D(Q) = C_0^\infty(\Omega)$, $Qu = -\Delta u$, then Q is symmetric in H (not self-adjoint), the corresponding energetic space is $H_E = H_0^1(\Omega)$, and $Q_E : H_E \to H_E^*$ is given by

$$Q_E(u)(v) = \int_\Omega \nabla u \cdot \nabla v \, dx \quad \forall u, v \in H_0^1(\Omega),$$

i.e., the same energetic extension we had before (see Remark 8.18). Obviously, the corresponding Friedrichs extension of Q is given by

$$D(\hat{Q}) = H_0^1(\Omega) \cap H^2(\Omega), \quad \hat{Q}u = -\Delta u \quad \forall u \in D(\hat{Q}).$$

8.7 Exercises

1. Let X denote the real linear space of all polynomials with real coefficients of degree ≤ 3. Define $A : X \to X$ by

$$(Ap)(x) = xp'(x), \quad x \in \mathbb{R}, \, p \in X,$$

where p' denotes the derivative of p.

   (a) Determine $N(A)$ and $R(A)$;

   (b) Find all the eigenpairs of A.

2. Let $X = C[0,1]$ be the usual real Banach space equipped with the sup-norm. Define on X the operator A by

$$(Au)(t) = (at + b)u(t), \quad t \in [0,1], \, u \in X,$$

where a, b are real constants.

   (i)   Show that $A \in L(X)$;

   (ii)   Find the eigenvalues and eigenvectors of A.

3. Let X be a Banach space over $\mathbb{K}$. Let A, $B \in L(X)$ and $\lambda \in \mathbb{K}$, $\lambda \neq 0$. Prove that λ is an eigenvalue of $AB := A \circ B$ if and only if λ is an eigenvalue of $BA := B \circ A$.

4. Let X denote the real Banach space $C[0,1]$ with the usual sup-norm. Let $k = k(t,s) \in C[0,1] \times C[0,1]$, with $\partial k / \partial t \in C[0,1] \times C[0,1]$, $k(t,t) \neq 0 \ \forall t \in [0,1]$. Define on X the operator A by

$$(Au)(t) = \int_0^t k(t,s)u(s)\,ds, \quad t \in [0,1].$$

Show that

(a) $A \in L(X)$;

(b) A has no eigenvalue.

Solve the same exercise for $X = L^2(0,1)$ with the usual norm.

5. Let $H = l^2$ be the usual Hilbert space of sequences $x = (x_1, x_2, \dots)$ in $\mathbb{C}$ satisfying $\sum_{n=1}^\infty |x_n|^2 < \infty$ with the inner product

$$\langle x, y \rangle = \sum_{n=1}^\infty x_i \bar{y}_i, \quad x = (x_1, x_2, \dots), \ y = (y_1, y_2, \dots) \in H,$$

and the corresponding Hilbertian norm. Define the multiplication operator A by

$$Ax = (\lambda_1 x_1, \lambda_2 x_2, \dots) \quad \forall x = (x_1, x_2, \dots) \in H,$$

where $(\lambda_n)_{n \in \mathbb{N}}$ is a given sequence in $\mathbb{C}$ with $\sup_{n \in \mathbb{N}} |\lambda_n| < \infty$.

(a) Show that $A \in L(H)$ and determine $\|A\|$;

(b) Show that A is symmetric (hence self-adjoint) $\Longleftrightarrow \lambda_n \in \mathbb{R}$ for all $n \in \mathbb{N}$;

(c) Find all the eigenvalues of A.

6. Let $H = L^2(0,1)$ be the real Hilbert space equipped with the usual scalar product and the induced norm, denoted $\|\cdot\|$. Define $A : H \to H$ by

$$(Au)(t) = t \int_t^1 u(s)\,ds + \int_0^t s u(s)\,ds, \quad 0 \le t \le 1, \ u \in H.$$

(a) Check that $A \in L(H)$;

(b) Prove that A is a compact operator;

(c) Prove that A is symmetric (hence self-adjoint);

    (d) Find all the eigenvalues and eigenvectors (eigenfunctions) of A and use this information to determine an orthonormal basis of H.

7. Let $(H, (\cdot, \cdot), \| \cdot \|)$ be a Hilbert space. Show that $x \in H \setminus \{0\}$ is an eigenvector of $A \in L(H) \iff |(Ax, x)| = \|Ax\| \cdot \|x\|$.

8. Let $(H, (\cdot, \cdot), \| \cdot \|)$ be a Hilbert space and let $u, v \in H \setminus \{0\}$ be two orthogonal vectors (i.e., $(u, v) = 0$). Define $A : H \to H$ by

$$Ax = (x, v)u + (x, u)v, \quad x \in H.$$

Obviously, $A \in L(H)$.

    (a) Calculate $\|A\|$;

    (b) Show that A is symmetric (hence self-adjoint);

    (c) Using (a) calculate $\|A\|$, where $A : L^2(-\pi, \pi) \to L^2(-\pi, \pi)$ is the linear operator defined by

$$(Af)(t) = \sin t \int_{-\pi}^{\pi} f(s) \cos s \, ds +$$

$$\cos t \int_{-\pi}^{\pi} f(s) \sin s \, ds, \ t \in [-\pi, \pi],$$

    for all $f \in L^2(-\pi, \pi)$;

    (d) Find all the eigenpairs of A.

9. Let $(H, (\cdot, \cdot), \| \cdot \|)$ be a Hilbert space and let $\{e_1, e_2, \ldots, e_m\} \subset H$ be an orthonormal system, where m is a given natural number. Define $A : H \to H$ by

$$Ax = \sum_{i=1}^{m} c_i(x, e_i)e_i, \quad x \in H,$$

where $c_i \in \mathbb{K} \setminus \{0\}$, $i = 1, \ldots, m$.

    (a) Show that $A \in L(H)$ and determine $\|A\|$, $R(A)$ and $N(A)$;

    (b) Show that A is symmetric $\iff c_i \in \mathbb{R} \ \forall i \in \{1, \ldots, m\}$;

    (c) Determine all the eigenvalues of A.

10. Let $H = L^2(0, 1)$ be the real Hilbert space equipped with the usual scalar product and norm. Define $A : H \to H$ by

$$(Au)(t) = \frac{t}{1+t} \int_0^1 \frac{s}{1+s} u(s)\, ds, \quad t \in [0, 1], \quad u \in H.$$

   (a) Show that $A \in L(H)$ and A is symmetric (hence self-adjoint);

   (b) Determine $R(A)$ and $N(A)$;

   (c) Determine all the eigenpairs of A.

11. Let $H = L^2(0, 1)$ be the real Hilbert space equipped with the usual scalar product and norm. For $u \in H$ consider the problem

$$\begin{cases} v''(t) = u(t) & \text{a.e. in } (0, 1), \\ v'(0) = 0, \ v(1) = 0. \end{cases}$$

Define $A : H \to H$ by $Au = v$, $u \in H$, where v is the solution of the above problem corresponding to u.

   (a) Show that $A \in L(H)$ and $N(A) = \{0\}$;

   (b) Prove that A is symmetric and compact;

   (c) Find all the eigenpairs of A and use this information to determine an orthonormal basis of H.

12. Solve the Dirichlet eigenvalue problem

$$\begin{cases} -\Delta u = \lambda u & \text{in } \Omega \subset \mathbb{R}^2, \\ u = 0 & \text{on } \partial\Omega, \end{cases}$$

where Ω is the rectangle $(0, a) \times (0, b) \subset \mathbb{R}^2$, $a, b \in (0, \infty)$.

13. Consider in $\Omega = (0, a) \times (0, b) \subset \mathbb{R}^2$, $a, b \in (0, \infty)$, the eigenvalue problem for $-\Delta$ with Neumann conditions on all sides of the rectangle Ω or combinations of Dirichlet and Neumann conditions on different sides of Ω. Solve all these eigenvalue problems.

Chapter 9

Semigroups of Linear Operators

Let A be an $n \times n$ matrix with entries $a_{ij} \in \mathbb{C}$ for all $i, j = 1, 2, \ldots, n$. Consider the Cauchy problem

$$u'(t) = Au(t), \ t \geq 0, \tag{E}$$

$$u(0) = x, \tag{IC}$$

where x is a given (column) vector in $\mathbb{C}^n$. It is well known that problem (E), (IC) has a unique solution given by

$$u(t) = e^{tA}x, \ t \geq 0, \tag{9.0.1}$$

where e^{tA} represents the fundamental matrix of the linear differential system (E) which equals I (the $n \times n$ identity matrix) for $t = 0$. We have

$$e^{tA} = \sum_{k=0}^{\infty} \frac{t^k}{k!} A^k, \tag{9.0.2}$$

which is valid for all $t \in \mathbb{R}$. Here A and e^{tA} can be interpreted as linear operators A, $e^{tA} \in L(X)$, where $X = \mathbb{C}^n$, equipped with one of its equivalent norms, and $L(X)$ denotes, as usual, the space of bounded linear operators from X into itself. As we will see later, the family of matrices (operators) $\{T(t) = e^{tA}; \ t \geq 0\}$ is a uniformly continuous semigroup on $X = \mathbb{C}^n$. What's more, the family $\{T(t); \ t \geq 0\}$ extends

© Springer Nature Switzerland AG 2019

G. Moroşanu, *Functional Analysis for the Applied Sciences*,

Universitext, https://doi.org/10.1007/978-3-030-27153-4_0

to a group of linear operators, $\{e^{tA}; t \in \mathbb{R}\}$. The representation of the solution $u(t)$ as

$$u(t) = T(t)x, \quad t \geq 0 \qquad (9.0.3)$$

allows the derivation of some properties of solutions from the properties of the family $\{T(t); t \geq 0\}$. This idea extends easily to the case when X is a general Banach space and A is a bounded (continuous) linear operator, $A \in L(X)$.

If A is not an element of $L(X)$, then the operator exponential e^{tA} no longer makes sense. This case is not trivial, rather it is much more interesting and very useful in applications. If $A : D(A) \subset X \to X$ satisfies certain conditions, then one can associate with A a so-called C_0-semigroup of linear operators $\{T(t); t \geq 0\} \subset L(X)$ (see Definition 9.1 below), so that the solution of the Cauchy problem (E), (IC) can again be represented by the above formula (9.0.3). Indeed, there is a central result in the linear semigroup theory, known as the Hille–Yosida theorem,[1] which establishes the necessary and sufficient conditions for a linear operator A to "generate" a C_0-semigroup of linear operators $\{T(t); t \geq 0\} \subset L(X)$. In this way, one can solve linear partial differential equations of the form (E), where A represents unbounded linear differential operators with respect to the space variables, defined on convenient function spaces.

The linear semigroup theory received considerable attention in the 1930s as a new approach in the study of parabolic and hyperbolic linear partial differential equations. This theory has since developed as an independent theory with applications in some other fields, such as ergodic theory, the theory of Markov processes, etc.

In this chapter we present some of the most important results of the linear semigroup theory and provide some related applications.

9.1 Definitions

Throughout this chapter X will be a Banach space over $\mathbb{K}$ with norm $\| \cdot \|$, where $\mathbb{K}$ is either $\mathbb{R}$ or $\mathbb{C}$. Denote as usual by $L(X)$ the space of all bounded (continuous) linear operators $T : X \to X$, which is a Banach space with respect to the operator norm

$$\|T\| = \sup \{\|Tx\| : x \in X, \|x\| \leq 1\}.$$

---

[1] Carl Einar Hille, American mathematician, 1894–1980; Kosaku Yosida, Japanese mathematician, 1909–1990.

Definition 9.1. *A one-parameter family* $\{T(t);\, t \geq 0\} \subset L(X)$ *is said to be a* **semigroup** *if*

(i) $T(0) = I$ *(the identity operator on X);*

(ii) $T(t + s) = T(t)T(s)$ *for all $t, s \geq 0$ (the semigroup property).*

If, in addition,

(iii) $\lim_{t \to 0+} \|T(t)x - x\| = 0$ *for all $x \in X$,*

then $\{T(t);\, t \geq 0\}$ *is called a* C_0*-semigroup (or a* **semigroup of class** C_0, *or a* **strongly continuous semigroup***).*

Condition (iii) says that the function $t \mapsto T(t)x$ is continuous at $t = 0$, that is why $\{T(t);\, t \geq 0\}$ is called a C_0-semigroup.

Definition 9.2. *A family* $\{T(t);\, t \geq 0\} \subset L(X)$ *is said to be a* **uniformly continuous semigroup** *if it satisfies conditions* (i) *and* (ii) *above, and*

(iii)' $\lim_{t \to 0+} \|T(t) - I\| = 0.$

Remark 9.3. Obviously, condition (iii)' is stronger than (iii). Indeed, for any $x \in X$, we have

$$\|T(t)x - x\| \leq \|T(t) - I\| \cdot \|x\|,$$

which proves the assertion.

Definition 9.4. *Let* $\{T(t);\, t \geq 0\} \subset L(X)$ *be a* C_0*-semigroup. Denote*

$$Ax := \lim_{h \to 0+} \frac{1}{h}[T(h)x - x], \qquad (9.1.4)$$

for all $x \in X$ *for which the above limit exists. If* $D(A)$ *is the set of all such x's, then we have a linear operator* $A : D(A) \subset X \to X$, *which is called the* **infinitesimal generator** *of the semigroup* $\{T(t);\, t \geq 0\}$.

Now let us state a first result on semigroups of linear operators:

Theorem 9.5. *For any operator* $A \in L(X)$ *the family* $\{T(t) = e^{tA};\, t \geq 0\}$ *is a uniformly continuous semigroup whose infinitesimal generator is A.*

Proof. Recall that

$$e^{tA} = \sum_{k=0}^{\infty} \frac{t^k}{k!} A^k,$$

meaning that for any $t \geq 0$ this series is convergent in $L(X)$ and its sum is e^{tA}. It is easily seen that the family $\{T(t) = e^{tA}; t \geq 0\}$ satisfies (i) and (ii). Condition $(iii)'$ is also satisfied since

$$\|T(t) - I\| \leq \| \sum_{k=1}^{\infty} \frac{t^k}{k!} A \| \leq t\|A\| \cdot e^{t\|A\|}$$

for all $t \geq 0$. Note also that for all $h > 0$

$$\|h^{-1}[T(h) - I] - A\| \leq h\|A\|^2 e^{h\|A\|},$$

which shows that A is the infinitesimal generator of $\{T(t) = e^{tA}; t \geq 0\}$. □

Remark 9.6. We will see later that, in fact, every uniformly continuous semigroup is a family of operator exponentials $\{e^{tA}; t \geq 0\}$ with $A \in L(X)$. Note that A can be obtained from the right derivative of $T(t) = e^{tA}$ calculated at $t = 0$. This explains the above definition of the generator of a C_0-semigroup $\{T(t); t \geq 0\}$. In this case, we can expect only the existence of the right derivative at $t = 0$ of $T(t)x$ for some points $x \in X$.

Examples of C_0-semigroups (that do not belong to the class of uniformly continuous semigroups) will be provided later.

9.2 Some Properties of C_0-Semigroups

We start this section with a basic result in the linear semigroup theory:

Theorem 9.7. *If $\{T(t); t \geq 0\} \subset L(X)$ is a C_0-semigroup, then the following hold:*

(a) there exist constants $M \geq 1$ and $\omega \in \mathbb{R}$ such that

$$\|T(t)\| \leq Me^{\omega t} \quad \forall t \geq 0; \tag{9.2.5}$$

(b) the function $t \mapsto T(t)x$ is continuous on $[0, \infty)$ for all $x \in X$.

Proof. Assertion (a): Let us first prove that there exists a constant $\delta > 0$ such that $\|T(t)\|$ is bounded on $[0, \delta]$, i.e.,

$$\sup\{\|T(t)\| : 0 \le t \le \delta\} =: C < \infty. \qquad (9.2.6)$$

Assume, by way of contradiction, that this is not the case, i.e., there exists a sequence of real numbers $t_k \searrow 0$ such that $\|T(t_k)\| \to \infty$. On the other hand, condition (iii) of Definition 9.1 implies that for each $x \in X$ there exists a natural number $N = N(x)$ such that

$$\|T(t_k)x\| \le \|T(t_k)x - x\| + \|x\| \le 1 + \|x\|, \ \forall k > N. \qquad (9.2.7)$$

By the Uniform Boundedness Principle, we derive from (9.2.7) that $\|T(t_k)\|$ is bounded, which contradicts the assumption above. Thus (9.2.6) holds true for some $\delta > 0$. Since $\|T(0)\| = \|I\| = 1$, we have $C \ge 1$.

Now, for all $t \ge 0$ we have the decomposition (division with remainder)

$$t = n\delta + r, \ \ n \in \mathbb{N}, \ 0 \le r < \delta.$$

So, by using condition (ii) of Definition 9.1, we can derive the estimate

$$\|T(t)\| \le \|T(\delta)\|^n \cdot \|T(r)\| \le C^{n+1}.$$

Therefore,
$$\|T(t)\| \le C \cdot C^{t/\delta}, \ t \ge 0,$$

which shows that (9.2.5) holds true with $M = C$ and $\omega = (\ln C)/\delta$.

Assertion (b): Let $t_0 > 0$ and $x \in X$ be arbitrary but fixed. For any $h > 0$ we have (cf. condition (ii) from Definition 9.1)

$$
\begin{aligned}
\|T(t_0 + h)x - T(t_0)x\| &= \|T(t_0)[T(h)x - x]\| \\
&\le \|T(t_0)\| \cdot \|T(h)x - x\|,
\end{aligned}
$$

which shows that the function $t \mapsto T(t)x$ is continuous from the right at $t = t_0$ (cf. condition (iii) of Definition 9.1). Now, for $0 < h < t_0$, we can write (cf. (ii) and (9.2.5))

$$
\begin{aligned}
\|T(t_0 - h)x - T(t_0)x\| &= \|T(t_0 - h)[x - T(h)x]\| \\
&\le M e^{\omega(t_0 - h)} \|x - T(h)x\|,
\end{aligned}
$$

which implies that $t \mapsto T(t)x$ is continuous from the left at $t = t_0$. $\square$

Remark 9.8. In fact, if $\{T(t); t \geq 0\} \subset L(X)$ is a C_0-semigroup, one can easily derive the following property that is stronger than (b) above: the map $(t, x) \mapsto T(t)x$ is continuous from $[0, \infty) \times X$ to X (see Exercise 9.4).

Remark 9.9. The constant ω in (9.2.5) determined in the proof above is nonnegative, but this is not the best constant. Indeed, sometimes ω can be negative (e.g., this is the case if $T(t) = e^{tA}$, where A is a real square matrix whose eigenvalues have negative real parts).

Theorem 9.10. *Let* $\{T(t) : t \geq 0\} \subset L(X)$ *be a* C_0-*semigroup and let* A *be its infinitesimal generator. Then,*

(c) A *is densely defined:* $\overline{D(A)} = X$;

(d) A *is a closed operator;*

(e) *for all* $t \geq 0$, $x \in D(A)$, *we have* $T(t)x \in D(A)$ *and*

$$\frac{d}{dt}T(t)x = AT(t)x = T(t)Ax. \qquad (9.2.8)$$

Proof of (c): Obviously,

$$x = \lim_{t \to 0^+} \frac{1}{t} \int_0^t T(s)x\,ds, \ \forall x \in X.$$

Since $D(A)$ is a linear subspace of X, to prove (c) it suffices to show that

$$\int_0^t T(s)x\,ds \in D(A), \ \forall t > 0, \ x \in X. \qquad (9.2.9)$$

Indeed, for some given $t > 0$, $x \in X$, and for all $h > 0$, we have

$$h^{-1}[T(h) - I] \int_0^t T(s)x\,ds = h^{-1} \int_0^t [T(s+h)x - T(s)x]\,ds$$

$$= h^{-1}\left(\int_h^{t+h} T(s)x\,ds \right.$$

$$\left. - \int_0^t T(s)x\,ds \right)$$

$$= h^{-1} \int_t^{t+h} T(s)x\,ds$$

$$- h^{-1} \int_0^h T(s)x\,ds.$$

Therefore, there exists

$$\lim_{h\to 0^+} h^{-1}[T(h) - I] \int_0^t T(s)x\, ds = T(t)x - x, \qquad (9.2.10)$$

which implies (9.2.9).

Proof of (d): Let (x_n) be a sequence in $D(A)$ such that $x_n \to x$ and $Ax_n \to y$. Using (9.2.10), we can write

$$
\begin{aligned}
T(t)x_n - x_n &= \lim_{h\to 0^+} \int_0^t T(s)h^{-1}[T(h)x_n - x_n]\, ds \\
&= \int_0^t T(s)Ax_n\, ds \ \ \forall t > 0.
\end{aligned}
$$

It follows that

$$T(t)x - x = \int_0^t T(s)y\, ds \ \ \forall t > 0,$$

so

$$\lim_{t\to 0^+} t^{-1}[T(t)x - x] = y.$$

It follows that $x \in D(A)$ and $y = Ax$.

Proof of (e): Let $t \geq 0$ and $x \in D(A)$. We have

$$
\begin{aligned}
T(t)Ax &= \lim_{h\to 0^+} T(t)\{h^{-1}[T(h)x - x]\} \\
&= \lim_{h\to 0^+} h^{-1}[T(h)T(t)x - T(t)x],
\end{aligned}
$$

which shows that $T(t)x \in D(A)$ and

$$T(t)Ax = AT(t)x. \qquad (9.2.11)$$

On the other hand,

$$
\begin{aligned}
\lim_{h\to 0^+} h^{-1}[T(t+h)x - T(t)x] &= \lim_{h\to 0^+} T(t)\{h^{-1}[T(h)x - x]\} \\
&= T(t)Ax. \qquad (9.2.12)
\end{aligned}
$$

From (9.2.11) and (9.2.12) we derive

$$\frac{d^+}{dt}T(t)x = AT(t)x = T(t)Ax. \qquad (9.2.13)$$

We have used $\frac{d^+}{dt}$ to denote the right derivative. To conclude, we need to show that the left derivative of $T(t)x$ exists and equals its right derivative at any $t > 0$. For $0 < h < t$, we have

$$\| - h^{-1}[T(t-h)x - T(t)x] - T(t)Ax\|$$
$$= \|T(t-h)\{h^{-1}[T(h)x - x] - T(h)Ax\}\|$$
$$\leq Me^{\omega(t-h)}\{\|h^{-1}[T(h)x - x] - Ax\| + \|Ax - T(h)Ax\|\}.$$

It follows that for all $t > 0$ and $x \in D(A)$,

$$\frac{d^-}{dt}T(t)x = T(t)Ax. \tag{9.2.14}$$

Obviously, (e) follows from (9.2.13) and (9.2.14). $\square$

Remark 9.11. In fact, if A is the generator of a C_0-semigroup $\{T(t); t \geq 0\} \subset L(X)$, then the subspace $Y := \cap_{n=1}^{\infty} D(A^n)$ is dense in X, where the operators $A^n : D(A^n) \to X$ are inductively defined as follows:

$$D(A^n) = \{x \in D(A^{n-1}); A^{n-1}x \in D(A)\},$$
$$A^n x = A(A^{n-1}x) \quad \forall x \in D(A^n),$$

for all $n \in \mathbb{N}$, $n \geq 2$. Now, for any $x \in X$ and $\phi \in C_0^{\infty}(\mathbb{R})$, with $\text{supp}\,\phi \subset (0, +\infty)$, define

$$x(\phi) = \int_0^{\infty} \phi(t)T(t)x\,dt.$$

For $h > 0$ we have

$$\frac{1}{h}(T(h) - I)x(\phi) = \frac{1}{h}\Big(\int_0^{\infty} \phi(t)T(t+h)x\,dt$$
$$- \int_0^{\infty} \phi(t)T(t)x\,dt\Big)$$
$$= \frac{1}{h}\Big(\int_h^{\infty} \phi(t-h)T(t)x\,dt$$
$$- \int_0^{\infty} \phi(t)T(t)x\,dt\Big)$$
$$= \int_0^{\infty} \frac{\phi(t-h) - \phi(t)}{h}T(t)x\,dt,$$

which converges to $-x(\phi')$ as $h \to 0^+$. Hence $x(\phi) \in D(A)$ and $Ax = -x(\phi')$. We infer by induction that $x(\phi) \in D(A^n)$ and $A^n x(\phi) =$

$(-1)^n x(\phi^{(n)})$ for all $n \in \mathbb{N}$, hence $x(\phi) \in Y$. Now, let us prove that any $x \in X$ can be approximated by $x(\phi)$ for suitable ϕ's (see [49, p. 44]). If $\omega \in C_0^\infty(\mathbb{R})$ is the usual test function with $\operatorname{supp}\omega = [-1, +1]$ and $\int_{-1}^{+1} \omega(t)\, dt = 1$, define the mollifier

$$\phi_\varepsilon(t) = \frac{1}{\varepsilon}\omega\left(\frac{t}{\varepsilon} - 2\right) \quad \forall t \in \mathbb{R},\ \varepsilon > 0.$$

Since

$$
\begin{aligned}
\|x(\phi_\varepsilon) - x\| &= \left\| \int_\varepsilon^{3\varepsilon} \phi_\varepsilon(t)[T(t)x - x]\, dt \right\| \\
&\leq \int_\varepsilon^{3\varepsilon} \phi_\varepsilon(t)\|T(t)x - x\|\, dt \\
&\leq \sup_{t \in [\varepsilon, 3\varepsilon]} \|T(t)x - x\| \int_\varepsilon^{3\varepsilon} \phi_\varepsilon(t)\, dt \\
&= \sup_{t \in [\varepsilon, 3\varepsilon]} \|T(t)x - x\|.
\end{aligned}
$$

Therefore

$$\lim_{\varepsilon \to 0^+} \|x(\phi_\varepsilon) - x\| = 0.$$

Theorem 9.12. *If two C_0-semigroups have the same infinitesimal generator, then they coincide.*

Proof. Let A be the common generator of two C_0-semigroups, say $\{T(t);\ t \geq 0\}$ and $\{S(t);\ t \geq 0\}$. For any $t > 0$ and $x \in D(A)$ we have (see Theorem 9.10, (e))

$$
\begin{aligned}
\frac{d}{ds}&[T(t-s)S(s)x] \\
&= -T(t-s)AS(s)x + T(t-s)AS(s)x = 0,\ \forall\, 0 \leq s < t.
\end{aligned}
$$

Hence, for all $t > 0$ and $x \in D(A)$, the function $s \mapsto T(t-s)S(s)x$ is constant on the interval $[0, t]$. In particular, $T(t)x = S(t)x$ on $D(A)$ for all $t \geq 0$. This concludes the proof since $\overline{D(A)} = X$. $\qquad\square$

Remark 9.13. Property (e) of Theorem 9.10 says that for every $x \in D(A)$ the function $u(t) = T(t)x$ is continuously differentiable on $[0, \infty)$ and satisfies the Cauchy problem

$$u'(t) = Au(t),\ t \geq 0;\quad u(0) = x. \qquad (CP)$$

This u, which is a C^1-solution of problem (CP) (hence a *classical solution* on every bounded interval $[0, r]$ in the sense of Definition 9.44 below) is unique. Indeed, if $\tilde{u}$ is also a C^1-solution of problem (CP), then for any $t > 0$ we have

$$\frac{d}{ds}T(t - s)\tilde{u}(s) = -T(t - s)A\tilde{u}(s) + T(t - s)\tilde{u}'(s) = 0 \quad \forall s \in (0, t),$$

hence $s \mapsto T(t - s)\tilde{u}(s)$ is a constant function on $[0, t]$. In particular, its values at $s = 0$ and $s = t$ coincide:

$$\tilde{u}(t) = T(t)\tilde{u}(0) = T(t)x,$$

which proves that the solution of (CP) is unique and is given by $u(t) = T(t)x$, $t \geq 0$.

Now, if $x \in X \backslash D(A)$, then the function $u(t) = T(t)x$ satisfies the initial condition $u(0) = x$, but is no longer differentiable (see Sect. 9.5 below), so it cannot satisfy the Cauchy problem above in a classical sense. However, u can be regarded as a generalized solution (or mild solution, as it will be called later, see Sect. 9.11) since the initial condition is still satisfied, $u(0) = x$, and there exists a sequence (u_n) of C^1-solutions of equation $(CP)_1$, such that $u_n \to u$ in $C([0, r]; X)$ for all $r > 0$. Indeed, one can choose a sequence (x_n) in $D(A)$, such that $x_n \to x$ (cf. Theorem 9.10, (c)), and obviously $u_n(t) = T(t)x_n$ are all C^1-solutions satisfying the required condition:

$$\|T(t)x_n - T(t)x\| \leq \|T(t)\| \cdot \|x_n - x\| \leq Me^{\omega r}\|x_n - x\|,$$

for all $t \in [0, r]$. Clearly, the definition of the generalized solution is independent of the choice of the sequence (u_n) (or $(x_n = u_n(0))$).

It is worth pointing out that in the discussion above A was assumed to be the infinitesimal generator of a C_0- semigroup. Now, given a linear operator A we want to know the conditions on A ensuring the existence of a C_0-semigroup whose generator is precisely A. This will allow us to solve Cauchy problems like (CP) above. From Theorem 9.10 we know that such an A has to necessarily be densely defined and closed. The complete answer will be provided later.

9.3 Uniformly Continuous Semigroups

Uniformly continuous semigroups have been defined before. We have also seen that for any $A \in L(X)$, the family $\{T(t) = e^{tA}; t \geq 0\}$ is a uniformly continuous semigroup whose generator is A. According to Theorem 9.12, this is the unique C_0-semigroup, hence the unique uniformly continuous semigroup, having A as its generator. The next result shows that, in fact, the class of uniformly continuous semigroups reduces to $\{\{e^{tA}; t \geq 0\}; A \in L(X)\}$.

Theorem 9.14. *Let $\{T(t); t \geq 0\} \subset L(X)$ be a uniformly continuous semigroup. If A is its infinitesimal generator, then $A \in L(X)$.*

Proof. Since

$$\lim_{t \to 0^+} \left\| I - \frac{1}{t} \int_0^t T(s) \, ds \right\| = 0,$$

there exists a $t_0 > 0$ such that

$$\|I - B\| < 1, \quad \text{where } B = \frac{1}{t_0} \int_0^{t_0} T(s) \, ds.$$

Therefore, B is invertible and $B^{-1} = \left(I - (I - B)\right)^{-1} \in L(X)$. Now, for all $h > 0$, we have

$$\frac{1}{h}[T(h) - I]B = \frac{1}{ht_0}\left(\int_0^{t_0} T(s+h) \, ds - \int_0^{t_0} T(s) \, ds\right)$$

$$= \frac{1}{t_0}\left(\frac{1}{h}\int_{t_0}^{t_0+h} T(s) \, ds - \frac{1}{h}\int_0^h T(s) \, ds\right).$$

Therefore, there exists

$$\lim_{h \to 0^+} \frac{1}{h}[T(h) - I]B = \frac{1}{t_0}[T(t_0) - I], \tag{9.3.15}$$

with respect to the topology of $L(X)$. Since the generator of $\{T(t); t \geq 0\}$ is A, it follows from (9.3.15) that

$$AB = \frac{1}{t_0}[T(t_0) - I]. \tag{9.3.16}$$

Since B is invertible and $B^{-1} \in L(X)$, we infer from (9.3.16)

$$A = \frac{1}{t_0}[T(t_0) - I]B^{-1} \in L(X). \qquad \square$$

In fact, every uniformly continuous semigroup $\{e^{tA}; t \geq 0\}$, $A \in L(X)$, can naturally be extended to the group $\{e^{tA}; t \in \mathbb{R}\}$ (see the next section).

Remark 9.15. Let $\{T(t); t \geq 0\} \subset L(X)$ be a C_0-semigroup whose infinitesimal generator $A : D(A) \subset X \to X$ is *bounded*, i.e., there exists a constant $c > 0$ such that $\|Ax\| \leq c\|x\|$ for all $x \in D(A)$. Then, $D(A) = X$, $A \in L(X)$ and so the semigroup is in fact uniformly continuous: $T(t) = e^{tA}$, $t \geq 0$. Indeed, since $\overline{D(A)} = X$, A has an extension $\tilde{A} \in L(X)$. Denote by $\{\tilde{T}(t) = e^{t\tilde{A}}; t \geq 0\}$ the (uniformly continuous) semigroup with generator $\tilde{A}$. For an arbitrary $t > 0$ and $x \in D(A)$, we have

$$\frac{d}{ds}[\tilde{T}(t-s)T(s)x]$$
$$= -\tilde{T}(t-s)\tilde{A}T(s)x + \tilde{T}(t-s)AT(s)x = 0 \quad \forall s \in (0, t),$$

since $T(s)x \in D(A) \subset D(\tilde{A}) = X$ and $\frac{d}{ds}T(s)x = AT(s)x$ for all $s \in (0, t)$. It follows that the function $s \mapsto \tilde{T}(t-s)T(s)x$ is constant on $[0, t]$, and hence $\tilde{T}(t)x = T(t)x$ for all $x \in D(A)$ which shows that $\tilde{T}(t)x = T(t)x$ for all $x \in X$. Therefore, A coincides with $\tilde{A}$ and the assertion follows.

9.4   Groups of Linear Operators. Definitions and Link to Operator Semigroups

Definition 9.16. *A family* $\{G(t); t \in \mathbb{R}\} \subset L(X)$ *is called a* **group** *if*

   (j) $G(0) = I$ *(the identity operator on X);*

   (jj) $G(t + s) = G(t)G(s)$ *for all* $t, s \in \mathbb{R}$ *(the group property).*

If, in addition,

   (jjj) $\lim_{t \to 0} \|G(t)x - x\| = 0$ *for all* $x \in X$,

then $\{G(t); t \in \mathbb{R}\}$ *is called a* C_0**-group** *(or a* **group of class** C_0*).*

The **infinitesimal generator** A of a group $\{G(t); t \in \mathbb{R}\}$ is defined by

$$Ax = \lim_{h \to 0} \frac{1}{h}[G(h)x - x] \quad \forall x \in D(A),$$

where $D(A)$ is the set of all $x \in X$ for which the limit above exists. If $\{G(t); t \in \mathbb{R}\}$ satisfies conditions (j), (jj) and, in addition, (jjj)' $\lim_{t \to 0} \|G(t) - I\| = 0$,

(which is stronger than (jjj)), then $\{G(t); t \in \mathbb{R}\}$ is called a **uniformly continuous group**.

Remark 9.17. If $\{G(t); t \in \mathbb{R}\}$ is a C_0-group, then the families $\{G(t); t \geq 0\}$ and $\{G(-t); t \geq 0\}$ are both C_0-semigroups, with generators A and $-A$, respectively (prove it!). Conversely, if $\{T_+(t); t \geq 0\}$, $\{T_-(t); t \geq 0\}$ are C_0-semigroups with generators A and $-A$, respectively, then one can define a C_0-group

$$G(t) = \begin{cases} T_+(t) & \text{if } t \geq 0, \\ T_-(-t) & \text{if } t < 0, \end{cases}$$

having A as its generator. The proof of this assertion relies on the identity

$$T_+(t)T_-(t) = T_-(t)T_+(t) = I \quad \forall t \geq 0. \qquad (9.4.17)$$

Indeed, for any $x \in D(A) = D(-A)$ and $t \geq 0$, we have (cf. Theorem 9.10, (e))

$$\frac{d}{dt} T_+(t)T_-(t)x = T_+(t)AT_-(t)x - T_+(t)AT_-(t)x = 0,$$

hence $t \mapsto T_+(t)AT_-(t)x$ is a constant function. Since it takes the value x for $t = 0$, it follows that

$$T_+(t)T_-(t)x = x \quad \forall t \geq 0, \ x \in D(A). \qquad (9.4.18)$$

We know that $\overline{D(A)} = X$, therefore (9.4.18) holds for all $x \in X$, i.e.,

$$T_+(t)T_-(t) = I \quad \forall t \geq 0.$$

Similarly,

$$T_-(t)T_+(t) = I \quad \forall t \geq 0,$$

so (9.4.17) holds true. Identity (9.4.17) shows that $T_+(t)$ and $T_-(t)$ are invertible for all $t \geq 0$, being inverse to each other. Thus $\{G(t); t \in \mathbb{R}\}$ satisfies the group property (jj). Since (j) and (jjj) are trivially satisfied, we conclude that $\{G(t); t \in \mathbb{R}\}$ constructed above is indeed a C_0-group, and its generator is A, as claimed.

Note that all the members $G(t)$ of any group are necessarily invertible operators, since $G(t)G(-t) = I = G(-t)G(t)$. The next result shows that invertibility allows one to extend any C_0-semigroup to a C_0-group.

It is worth pointing out that if $\{T(t); t \geq 0\} \subset L(X)$ is a semigroup and $T(t_0)$ is a bijection from X to itself (hence $T(t_0)$ is invertible) for some $t_0 > 0$, then so is $T(t)$ for all $t \geq 0$. Indeed, for $t \in (0, t_0)$, we have

$$T(t_0) = T(t)T(t_0 - t) = T(t_0 - t)T(t),$$

which shows that $T(t)$ is bijective. For $t > t_0$ we write t as $t = nt_0 + s$, where $n \in \mathbb{N}$ and $0 \leq s < t_0$ (division with remainder) and so $T(t) = T(t_0)^n T(s)$, which clearly shows that $T(t)$ is also bijective in this case.

Theorem 9.18. *Let $\{T(t); t \geq 0\}$ be a C_0-semigroup and let A denote its infinitesimal generator. If $T(t)$ is a bijection from X to itself for all $t > 0$ (equivalently, $T(t_0)$ is a bijection for some $t_0 > 0$), then $\{T(t)^{-1}; t \geq 0\}$ is a C_0-semigroup with the generator $-A$, so $\{G(t); t \in \mathbb{R}\}$ defined by*

$$G(t) := \begin{cases} T(t) & \text{if } t \geq 0, \\ T(-t)^{-1} & \text{if } t < 0, \end{cases}$$

is a C_0-group whose generator is A.

Proof. Denote $S(t) = T(t)^{-1}$, $t \geq 0$. Obviously, $S(0) = I$ and

$$S(t + s) = [T(s)T(t)]^{-1} = T(t)^{-1}T(s)^{-1} = S(t)S(s),$$

for all t, $s \geq 0$. Thus, the family $\{S(t) = T(t)^{-1}; t \geq 0\}$ is a semigroup, and $\{G(t); t \in \mathbb{R}\}$ defined in the statement above is a group. Now, let us prove that the semigroup $\{S(t) = G(-t); t \geq 0\}$ satisfies condition (iii) of Definition 9.1. Let $x \in X$ and $s > 1$. Denote $y := T(s)^{-1}x$. For $0 < t < 1$, we have

$$
\begin{aligned}
\|S(t)x - x\| &= \|G(-t)x - x\| \\
&= \|G(-t)G(s)y - T(s)y\| \\
&= \|T(s - t)y - T(s)y\| \to 0 \quad \text{as } t \to 0^+,
\end{aligned}
$$

since $t \mapsto T(t)y$ is continuous on $[0, \infty)$. Therefore $\{S(t); t \geq 0\}$ satisfies condition (iii) as claimed, i.e., it is a C_0-semigroup. Let B be the infinitesimal generator of $\{S(t) = T(t)^{-1}; t \geq 0\}$. For $x \in D(A)$ we have

$$\lim_{h \to 0^+} \left\{ \frac{1}{h}[x - T(h)x] + Ax \right\} = 0.$$

This implies that

$$\lim_{h \to 0^+} S(h)\{\frac{1}{h}[x - T(h)x] + Ax\} = 0,$$

since $\|S(h)\| \le M_1 e^{\omega_1 h}$ for some $M_1 \ge 0$ and $\omega_1 \in \mathbb{R}$ (cf. Theorem 9.7, (a)). Therefore,

$$\lim_{h \to 0^+} h^{-1}[T(h)^{-1}x - x] = -Ax,$$

i.e., $D(A) \subset D(B)$ and $Bx = -Ax \ \forall x \in D(A)$. Since $T(t) = \left(T(t)^{-1}\right)^{-1}$, $t \ge 0$, we also have $D(B) \subset D(A)$. Hence, $D(A) = D(B)$ and $Bx = -Ax \ \forall x \in D(A)$, i.e., $B = -A$. $\qquad \square$

Remark 9.19. Let $\{G(t); t \in \mathbb{R}\} \subset L(X)$ be a group. If for all $x \in X$ the function $t \mapsto G(t)x$ is continuous from the right (or from the left) at some point $t = t_0 \in \mathbb{R}$, then there exist constants $M \ge 1$ and $\omega \in \mathbb{R}$ such that

$$\|G(t)\| \le M e^{\omega |t|} \quad \forall t \in \mathbb{R}. \tag{9.4.19}$$

This follows by the Uniform Boundedness Principle (see the proof of Theorem 9.7). Moreover, using this estimate and the invertibility of every $G(t)$, one can easily see that $t \mapsto G(t)x$ is continuous on $\mathbb{R}$; even more, the function $(t, x) \mapsto G(t)x$ is continuous from $\mathbb{R} \times X$ to X.

Remark 9.20. If $A \in L(X)$, then $\{G(t) = e^{tA}; t \in \mathbb{R}\}$ is a uniformly continuous group. In fact, it follows from the discussion above that the class of uniformly continuous groups is precisely $\{\{e^{tA}; t \in \mathbb{R}\}; A \in L(X)\}$.

9.5 Translation Semigroups

In this section we present the first examples of C_0-semigroups which are not uniformly continuous ones.

Let X be the space of all functions $f : [0, \infty) \to \mathbb{R}$ which are uniformly continuous and bounded. The space X is a real Banach space with respect to the norm

$$\|f\|_\infty = \sup_{t \ge 0} |f(t)|.$$

For each $t \ge 0$ define $T(t) : X \to X$ by

$$(T(t)f)(s) = f(t + s), \quad s \in [0, \infty), \ f \in X.$$

It is easily seen that the family $\{T(t); t \geq 0\}$ is a C_0-semigroup. Its infinitesimal generator is defined by

$$D(A) = \{f \in X; f \text{ is differentiable on } [0, \infty) \text{ and } f' \in X\}, \quad (9.5.20)$$

$$Af = f' \quad \forall f \in D(A). \quad\quad\quad\quad\quad\quad (9.5.21)$$

Indeed, if $f \in X$, f is differentiable on $[0, \infty)$, and $f' \in X$, then for all $h > 0$ and $s \geq 0$

$$\left(h^{-1}[T(h)f - f]\right)(s) = h^{-1}[f(s+h) - f(s)] = f'(\theta),$$

for some $\theta \in (s, s+h)$, so

$$\left(h^{-1}[T(h)f - f]\right)(s) - f'(s) = f'(\theta) - f'(s) \to 0,$$

as $h \to 0^+$, uniformly in s (since f' is uniformly continuous). Therefore, $f \in D(A)$ and $Af = f'$.

To conclude the proof, we need to show that (9.5.20) holds true, i.e., the converse inclusion relation is valid. To this end, let $f \in D(A)$, which means there exists

$$\lim_{h \to 0^+} h^{-1}[T(h)f - f] = \lim_{h \to 0^+} h^{-1}[f(\cdot + h) - f(\cdot)] = f'_+ \in X,$$

where f'_+ denotes the right derivative of f. It remains to prove that f is differentiable on $[0, \infty)$ so that $f'_+ = f'$. For an arbitrary $\varepsilon > 0$ define

$$g(t) = f(t) - f(0) - \int_0^t f'_+(s)\, ds - \varepsilon t.$$

We have $g(0) = 0$ and

$$g'_+(t) = -\varepsilon < 0 \quad \forall t \geq 0,$$

which implies $g(t) \leq 0$, which in turn means

$$f(t) \leq f(0) + \int_0^t f'_+(s)\, ds,$$

for all $t \geq 0$ (since ε was arbitrarily chosen). Similarly, replacing $-\varepsilon$ by $+\varepsilon$, we obtain the converse inequality, so

$$f(t) = f(0) + \int_0^t f'_+(s)\, ds \quad \forall t \geq 0,$$

which shows that f is indeed differentiable on $[0, \infty)$ and so $f' = f'_+ \in X$, as claimed.

The semigroup defined above is called a *translation semigroup*. Obviously,

$$\|T(t)f\|_\infty \leq \|f\|_\infty \quad \forall t \geq 0,$$

which shows that $\|T(t)\| \leq 1$ for all $t \geq 0$, i.e., the estimate in Theorem 9.7 holds with $M = 1$ and $\omega = 0$.

It is worth pointing out that A is not a member of $L(X)$ in this case, so $\{T(t); t \geq 0\}$ is **not** a uniformly continuous semigroup (see Theorem 9.14). This confirms the fact that the unit sphere of X is not equicontinuous (equivalently, condition $(iii)'$ is not valid).

Remark 9.21. If $f \in D(A)$ (see (9.5.20)), then

$$u(t) = u(t, \cdot) = \big(T(t)f\big)(\cdot) = f(t + \cdot)$$

satisfies the Cauchy problem in X

$$\begin{cases} u'(t) = Au(t) & \forall t \geq 0, \\ u(0) = f, \end{cases}$$

i.e.,

$$\begin{cases} \frac{\partial u}{\partial t}(t, s) = \frac{\partial u}{\partial s}(t, s), & t, s \geq 0, \\ u(0, s) = f(s), & s \geq 0. \end{cases}$$

If $f \in X$ is not differentiable, then $u(t, s) = f(t + s)$ does not satisfy the above partial differential equation in a classical sense; it has to be interpreted as a generalized solution of the Cauchy problem above.

If X is replaced by the space of all functions $f : \mathbb{R} \to \mathbb{R}$ which are uniformly continuous and bounded, with the norm

$$\|f\|_\infty = \sup_{t \in \mathbb{R}} |f(t)|,$$

then one can define similarly a semigroup of translations, $T(t) : X \to X$, $t \geq 0$,

$$\big(T(t)f\big)(s) = f(t + s) \quad \forall s \in \mathbb{R}, \ f \in X.$$

In this case, the family $\{T(t); t \geq 0\}$ is again a C_0-semigroup, with $\|T(t)\| = 1$ for all $t \geq 0$, and its infinitesimal generator A is given by

$$D(A) = \{f \in X; \ f \text{ is differentiable on } \mathbb{R} \text{ and } f' \in X\},$$

$$Af = f' \ \ \forall f \in D(A).$$

It is worth mentioning that this C_0-semigroup can be extended to a C_0-group $\{G(t); \ t \in \mathbb{R}\}$ defined by

$$(G(t)f)(s) = f(t+s) \ \ \forall t, s \in \mathbb{R}, \ f \in X.$$

This is **not** a uniformly continuous group, since its infinitesimal generator does not belong to $L(X)$.

9.6   The Hille–Yosida Generation Theorem

Let X be a Banach space and let $A : D(A) \subset X \to X$ be a linear closed operator, not necessarily bounded. The set

$$\rho(A)$$
$$= \{\lambda \in \mathbb{K}; \ \lambda I - A \text{ is a bijective operator from } D(A) \text{ to } X\} \tag{9.6.22}$$

is called the *resolvent set of* A. If $\rho(A)$ is nonempty, then, for $\lambda \in \rho(A)$, denote

$$R(\lambda, A) = (\lambda I - A)^{-1}, \tag{9.6.23}$$

which is called the *resolvent of* A. Since A is a closed operator, so is $R(\lambda, A)$ for all $\lambda \in \rho(A)$. If we also take into account the fact that $D(R(\lambda, A)) = X$, we infer that $R(\lambda, A) \in L(X)$ for all $\lambda \in \rho(A)$ (cf. Theorem 4.10 (Closed Graph Theorem)).

Now, let us state a central result in the theory of semigroups of linear operators, which belongs to E. Hille and K. Yosida.

Theorem 9.22. *A linear operator* $A : D(A) \subset X \to X$ *is the infinitesimal generator of a* C_0-*semigroup of contractions* $\{T(t); \ t \geq 0\}$ *(i.e.,* $\|T(t)\| \leq 1 \ \forall t \geq 0$*) if and only if*

(k) $\overline{D(A)} = X$ *and* A *is closed;*

(kk) $(0, \infty) \subset \rho(A)$ *and* $\|R(\lambda, A)\| \leq \frac{1}{\lambda} \ \ \forall \lambda > 0$.

Proof. Necessity: If A is the generator of a C_0-semigroup, then the two conditions of (k) are fulfilled (cf. Theorem 9.10). It remains to prove (kk), under the assumption that $\{T(t); t \geq 0\}$ is a C_0-semigroup of contractions. To this purpose, define

$$R_\lambda x = \int_0^\infty e^{-\lambda t} T(t) x \, dt \quad \forall \lambda > 0, \, x \in X. \tag{9.6.24}$$

Note that R_λ is well defined, since

$$\|e^{-\lambda t} T(t) x\| \leq e^{-\lambda t} \|x\| \quad \forall t \geq 0.$$

Furthermore, $R_\lambda \in L(X)$ and

$$\begin{aligned}
\|R_\lambda x\| &\leq \int_0^\infty e^{-\lambda t} \|T(t)\| \cdot \|x\| \, dt \\
&\leq \left(\int_0^\infty e^{-\lambda t} dt \right) \|x\| \\
&= \frac{1}{\lambda} \|x\| \quad \forall x \in X, \, \lambda > 0,
\end{aligned}$$

which implies that

$$\|R_\lambda\| \leq \frac{1}{\lambda} \quad \forall \lambda > 0. \tag{9.6.25}$$

Let us prove that for all $\lambda > 0$ and $x \in X$, $R_\lambda x \in D(A)$. For all $h > 0$ we have

$$\begin{aligned}
h^{-1}[T(h) - I] R_\lambda x &= h^{-1} \Big(\int_0^\infty e^{-\lambda t} T(t+h) x \, dt \\
&\qquad - \int_0^\infty e^{-\lambda t} T(t) x \, dt \Big) \\
&= \frac{e^{\lambda h} - 1}{h} R_\lambda x - \frac{e^{\lambda h}}{h} \int_0^h e^{-\lambda \tau} T(\tau) x \, d\tau.
\end{aligned}$$

Observe that the right-hand side of the last equality converges to $\lambda R_\lambda x - x$ as $\lambda \to 0^+$. Therefore, $R_\lambda x \in D(A)$ and

$$A R_\lambda x = \lambda R_\lambda x - x,$$

i.e.,

$$(\lambda I - A) R_\lambda = I \quad \forall \lambda > 0. \tag{9.6.26}$$

On the other hand, for all $x \in D(A)$ and $t \geq 0$, $T(t)x \in D(A)$ (cf. Theorem 9.10, (e)) and

$$
\begin{aligned}
AR_\lambda x &= \lim_{h \to 0^+} \int_0^\infty e^{-\lambda t} h^{-1}[T(t+h)x - T(t)x]\, dt \\
&= \int_0^\infty e^{-\lambda t} T(t) Ax\, dt \\
&= R_\lambda Ax,
\end{aligned}
$$

hence (see also (9.6.26))

$$
R_\lambda(\lambda I - A) = I_{D(A)} \quad \forall \lambda > 0, \tag{9.6.27}
$$

where $I_{D(A)}$ is the identity operator on $D(A)$. From (9.6.26) and (9.6.27) we infer that $\lambda I - A$ is a bijective operator from $D(A)$ to X and

$$
R_\lambda = (\lambda I - A)^{-1} \quad \forall \lambda > 0.
$$

Therefore, $(0, \infty) \subset \rho(A)$ and

$$
R_\lambda = R(\lambda, A) \quad \forall \lambda > 0,
$$

so (9.6.25) implies that

$$
\|R(\lambda, A)\| \leq \frac{1}{\lambda} \quad \forall \lambda > 0.
$$

Thus the proof of necessity is complete.

Sufficiency: Assume that both (k) and (kk) hold. For the convenience of the reader, the proof will be divided into several steps.

Step 1: $\lim_{\lambda \to \infty} \lambda R(\lambda, A)x = x \quad \forall x \in X$.
If $x \in D(A)$, then, according to (kk), we have

$$
\|\lambda R(\lambda, A)x - x\| = \|R(\lambda, A)Ax\| \leq \frac{1}{\lambda}\|Ax\|,
$$

which shows that

$$
\lim_{\lambda \to \infty} \lambda R(\lambda, A)x = x \quad \forall x \in D(A). \tag{9.6.28}
$$

Now, if $x \in X$, according to (k), there exists a sequence (x_n) in $D(A)$ such that $x_n \to x$. Since

$$
\begin{aligned}
\|\lambda R(\lambda, A)x - x\| &\leq \|\lambda R(\lambda, A)(x - x_n)\| + \|\lambda R(\lambda, A)x_n - x_n\| \\
&\quad + \|x_n - x\| \\
&\leq \|\lambda R(\lambda, A)x_n - x_n\| + 2\|x_n - x\|,
\end{aligned}
$$

we have (see (9.6.28))

$$\limsup_{\lambda\to\infty}\|\lambda R(\lambda,A)x - x\| \leq 2\|x_n - x\|,$$

which concludes Step 1.

Step 2: Define $A_\lambda := \lambda AR(\lambda,A)$, $\lambda > 0$ (the *Yosida approximation* of A); then, for all $\lambda > 0$, $A_\lambda \in L(X)$, and

$$\lim_{\lambda\to\infty} A_\lambda x = Ax \quad \forall x \in D(A). \tag{9.6.29}$$

Indeed, since

$$A_\lambda x = \lambda^2 R(\lambda,A)x - \lambda x \quad \forall x \in X,\ \lambda > 0,$$

we have $A_\lambda \in L(X)$ and, if $x \in D(A)$,

$$A_\lambda x = \lambda R(\lambda,A)Ax \quad \forall \lambda > 0.$$

According to Step 1, this implies (9.6.29), thus the proof of Step 2 is complete.

Step 3: For all $t \geq 0$, $x \in X$, and $\lambda,\nu > 0$, we have

$$\|e^{tA_\lambda}x - e^{tA_\nu}x\| \leq t\|A_\lambda x - A_\nu x\|. \tag{9.6.30}$$

First of all, note that for all $t \geq 0$ and $\lambda > 0$,

$$\begin{aligned}
\|e^{tA_\lambda}\| &= e^{-\lambda t}\|e^{t\lambda^2 R(\lambda,A)}\| \\
&\leq e^{-\lambda t}\cdot e^{t\lambda^2\|R(\lambda,A)\|} \\
&\leq 1.
\end{aligned}$$

It is also easily seen that e^{tA_λ}, e^{tA_ν}, A_λ, A_ν commute with each other. Using this information, we infer that

$$\begin{aligned}
\|e^{tA_\lambda}x - e^{tA_\nu}x\| &\leq \left\|\int_0^1 \frac{d}{ds}\left(e^{tsA_\lambda}e^{t(1-s)A_\nu}x\right)ds\right\| \\
&\leq \int_0^1 t\|e^{tsA_\lambda}e^{t(1-s)A_\nu}(A_\lambda x - A_\nu x)\|\,ds \\
&\leq t\|A_\lambda x - A_\nu x\|, \tag{9.6.31}
\end{aligned}$$

as claimed.

Step 4: The limit $\lim_{\lambda\to\infty} e^{tA_\lambda}x =: T(t)x$, $t \geq 0$, $x \in X$ exists, and $\{T(t);\ t \geq 0\}$ is a C_0-semigroup of contractions having A as its generator.

First of all, according to Steps 2 and 3, the above limit exists for each $x \in D(A)$, uniformly on compact subintervals of $[0, \infty)$, thus $t \mapsto T(t)x$ is a continuous function on $[0, \infty)$. It is also easily seen that

$$T(0)x = x, \ \|T(t)x\| \le \|x\| \ \forall x \in D(A), \ t, s \ge 0. \tag{9.6.32}$$

Obviously, $T(t)$ extends to $\overline{D(A)} = X$ as a bounded (continuous) operator and (9.6.32) are satisfied for all $x \in X$. Moreover, $t \mapsto T(t)x$ is continuous on $[0, \infty)$ for all $x \in X$. Indeed, if (x_n) is a sequence in $D(A)$ converging to x, then

$$\|T(t)x_n - T(t)x_m\| = \|T(t)(x_n - x_m)\| \le \|x_n - x_m\|,$$

hence $T(t)x_n \to T(t)x$ uniformly and so the function $t \mapsto T(t)x$ is indeed continuous on $[0, \infty)$. On the other hand,

$$\|T(t)x - e^{tA_\lambda}x\| \le \|T(t)x - T(t)x_n\| + \|T(t)x_n - e^{tA_\lambda}x_n\|$$
$$+ \|e^{tA_\lambda}(x_n - x)\|$$
$$\le 2\|x - x_n\| + \|T(t)x_n - e^{tA_\lambda}x_n\|,$$

which implies that

$$T(t)x = \lim_{\lambda \to \infty} e^{tA_\lambda}x \ \forall x \in X,$$

uniformly on compact subintervals of $[0, \infty)$. Since

$$e^{(t+s)A_\lambda}x = e^{tA_\lambda}e^{sA_\lambda}x \ \forall \lambda > 0, \ t, s \ge 0, \ x \in X,$$

we have

$$T(t + s)x = T(t)T(s)x \ \forall t, s \ge 0, \ x \in X.$$

Thus we have already proved that $\{T(t); t \ge 0\}$ is a C_0-semigroup of contractions, and all we have to prove next is that its generator, say B, coincides with the given operator A. If $x \in D(A)$, we have

$$\begin{aligned} T(t)x - x &= \lim_{\lambda \to \infty} \left(e^{tA_\lambda}x - x \right) \\ &= \lim_{\lambda \to \infty} \int_0^t e^{sA_\lambda}A_\lambda x \, ds \\ &= \int_0^t T(s)Ax \, ds, \end{aligned} \tag{9.6.33}$$

since $e^{sA_\lambda}A_\lambda x \to T(s)Ax$ uniformly on bounded subintervals of $[0, \infty)$, as $\lambda \to \infty$. From (9.6.33) we easily see that $D(A) \subset D(B)$ and $Bx = Ax$ for all $x \in D(A)$. Now, by assumption $1 \in \rho(A)$. On the other hand, according to the forward implication, we also have $1 \in \rho(B)$ (since B is the generator of a C_0-semigroup of contractions). So both A and B are bijections from $D(A)$ and respectively $D(B)$ to X. Since $Ax = Bx$ for all $x \in D(A)$, it follows that $D(A) = D(B)$ and $A = B$. $\square$

9.7 The Lumer–Phillips Theorem

In this section we discuss another result which also provides necessary and sufficient conditions for a linear operator to generate a C_0-semigroup of contractions. This result belongs to Lumer and Phillips[2] and is useful in applications. Before stating this result we need the following definition.

Definition 9.23. *A linear operator* $A : D(A) \subset X \to X$ *is said to be dissipative if*

$$\lambda\|x\| \le \|\lambda x - Ax\| \quad \forall \lambda > 0, \, x \in D(A). \tag{9.7.34}$$

If in addition $R(\lambda I - A) = X$ *for all* $\lambda > 0$, *then* A *is called m-dissipative.*

Remark 9.24. If A is a dissipative linear operator, then it is m-dissipative if and only if there exists a $\lambda_0 > 0$ such that $R(\lambda_0 I - A) = X$. Indeed, by the dissipativity condition (9.7.34) it follows that $\lambda_0 I - A$ is a bijection between $D(A)$ and X, $(\lambda_0 I - A)^{-1} \in L(X)$ and $\|(\lambda_0 I - A)^{-1}\| \le 1/\lambda_0$. Using this information and Banach's Fixed Point Theorem it follows easily that $R(\lambda I - A) = X$ for all $\lambda \in (0, 2\lambda_0)$. Obviously, this interval can be extended indefinitely to the right and so $R(\lambda I - A) = X$ for all $\lambda > 0$.

Theorem 9.25 (Lumer–Phillips). *A linear operator* $A : D(A) \subset X \to X$ *is the infinitesimal generator of a* C_0-*semigroup of contractions if and only if the following conditions hold: (a)* $\overline{D(A)} = X$, *and (b)* A *is m-dissipative.*

---

[2]Günter Lumer, German-born mathematician, 1929–2005; Ralph S. Phillips, American mathematician, 1913–1998.

Proof. Sufficiency: Assume that both (a) and (b) hold. From (b) it follows that for every $\lambda > 0$ we have $\lambda \in \rho(A)$, $R(\lambda, A) \in L(X)$, and $\|R(\lambda, A)\| \le 1/\lambda$ (see the remark above). Also, A is a closed operator since $(\lambda I - A)^{-1} \in L(X)$ for all $\lambda > 0$. It follows by the Hille–Yosida Theorem that A generates a C_0-semigroup of contractions.

Necessity: Assume that A is the generator of a C_0-semigroup of contractions $\{T(t); t \ge 0\}$. According to the Hille–Yosida Theorem, it suffices to show that A is dissipative. Let $x \in D(A)$ and $x^* \in J(x)$, where J is the duality mapping of X. We have

$$
\begin{aligned}
\operatorname{Re} x^*(Ax) &= \lim_{h \to 0^+} \operatorname{Re} x^*(h^{-1}[T(h)x - x]) \\
&= \lim_{h \to 0^+} h^{-1}\big(\operatorname{Re} x^*(T(h)x) - \|x\|^2\big) \\
&\le 0,
\end{aligned}
$$

since

$$
\operatorname{Re} x^*(T(h)x) \le \|x^*\| \cdot \|T(h)\| \cdot \|x\| \le \|x\|^2,
$$

where Re denotes the real part. Therefore,

$$
\begin{aligned}
\operatorname{Re} x^*(\lambda x - Ax) &= \lambda\|x\|^2 - \operatorname{Re} x^*(Ax) \\
&\ge \lambda\|x\|^2 \quad \forall \lambda > 0,
\end{aligned}
$$

which obviously implies (9.7.34).   □

Remark 9.26. A linear operator $A : D(A) \subset X \to X$ is dissipative if and only if

$$
\forall x \in D(A) \ \exists x^* \in J(x) \ \text{ such that } \operatorname{Re} x^*(Ax) \le 0. \qquad (9.7.35)
$$

From the proof above we see that (9.7.35) implies (9.7.34). For the proof of the converse implication, see [13, p. 81] or [39, p. 14]. If X is a Hilbert space, then this implication follows easily. If X is a real Hilbert space, then (9.7.35) means that A is negative semidefinite: $(Ax, x) \le 0 \ \forall x \in D(A)$ (equivalently, $-A$ is positive semidefinite or monotone).

Note that if X is assumed to be reflexive then condition (a) in the Lumer–Phillips Theorem becomes superfluous, so we have

Theorem 9.27. *Assume X is a reflexive Banach space. Then a linear operator $A : D(A) \subset X \to X$ is the infinitesimal generator of a C_0-semigroup of contractions if and only if A is m-dissipative.*

Proof. Bearing in mind the Lumer–Phillips Theorem, we need to prove that if X is reflexive and A is m-dissipative (equivalently, A satisfies (9.7.35) and $R(\lambda_0 I - A) = X$ for some $\lambda_0 > 0$), then $\overline{D(A)} = X$. Obviously, $(0, \infty) \subset \rho(A)$, $R(\lambda, A) \in L(X)$, and $\|R(\lambda, A)\| \leq 1/\lambda$ for all $\lambda > 0$. Now, for $x \in D(A)$ and $\lambda > 0$ denote $x_\lambda := \lambda R(\lambda, A)x$. As in the proof of the Hille–Yosida Theorem, we can see that

$$\|x_\lambda - x\| \leq \frac{1}{\lambda}\|Ax\|,$$

hence x_λ converges to x as $\lambda \to +\infty$. (Note that this property cannot be extended, for the time being, to all $x \in X$, as in the proof of the Hille–Yosida Theorem, since $\overline{D(A)} = X$ is now a target, not a hypothesis). It is also easily seen that for $x \in D(A)$ and $\lambda > 0$

$$\|A_\lambda x\| = \lambda\|R(\lambda, A)Ax\| \leq \|Ax\|.$$

Now, by the reflexivity assumption on X we derive the existence of a sequence $\lambda_n \to \infty$ such that (Ax_{λ_n}) converges weakly. Moreover, since A is m-dissipative, its graph is closed in $X \times X$, hence weakly closed, so we have

$$\lim_{n \to \infty} x^*(Ax_{\lambda_n}) = x^*(Ax) \quad \forall x^* \in X^*, \tag{9.7.36}$$

where X^* denotes the dual of X. Now, let $x^* \in X^*$ such that $x^*(x) = 0$ for all $x \in D(A)$. Since $Ax_\lambda = \lambda R(\lambda, A)Ax \in D(A)$ for all $\lambda > 0$, we derive from (9.7.36) that

$$x^*(Ax) = 0 \quad \forall x \in D(A). \tag{9.7.37}$$

Taking into account (9.7.37) and $R(\lambda_0 I - A) = X$, we infer that $x^* = 0$. Therefore $\overline{D(A)} = X$ as claimed. $\qquad\square$

Remark 9.28. The reflexivity of X is an essential assumption in Theorem 9.27, as the following counterexample shows: $X = C[0, 1]$ equipped with the usual sup-norm (which is a non-reflexive Banach space), $A : D(A) \subset X \to X$, $D(A) = \{u \in C^1[0, 1]; u(0) = 0\}$, $Au = -u'$. It is easily seen that A is m-dissipative, but not densely defined ($\overline{D(A)} =$

$\{u \in C[0,1]; u(0) = 0\}$). Hence, according to the Lumer–Phillips Theorem, A cannot be the generator of a C_0-semigroup in X. This counterexample clearly shows that Theorem 9.27 fails to hold in non-reflexive Banach spaces.

We close this section with the following result which is valid in a general Banach space X.

Theorem 9.29. *If $A : D(A) \subset X \to X$ is a closed linear operator such that $\overline{D(A)} = X$ and both A and A^* are dissipative (where A^* denotes the adjoint of A), then A is m-dissipative (hence, according to the Lumer–Phillips Theorem, A is the generator of a C_0-semigroup of contractions).*

Proof. Let $x^* \in X^*$ be such that $x^*(x - Ax) = 0$ for all $x \in D(A)$. It follows that $x^* \in D(A^*)$ and $x^* - A^* x^* = 0$. Since A^* is assumed to be dissipative, we infer that $x^* = 0$, so $\overline{R(I - A)} = X$. In fact, $R(I - A)$ is a closed subspace of X (since A is dissipative and closed), hence $R(I - A) = X$. $\qquad\qquad\square$

9.8   The Feller–Miyadera–Phillips Theorem

The Hille–Yosida theorem has the following significant generalization that belongs to Feller, Miyadera, and Phillips.[3]

Theorem 9.30. *A linear operator $A : D(A) \subset X \to X$ is the infinitesimal generator of a C_0-semigroup $\{T(t); t \geq 0\}$ satisfying $\|T(t)\| \leq M e^{\omega t}$, $t \geq 0$, with $M \geq 1$, $\omega \in \mathbb{R}$, if and only if*

(k) $\overline{D(A)} = X$ and A is closed;

(kk)' $(\omega, \infty) \subset \rho(A)$ and $\|R(\lambda, A)^n\| \leq \frac{M}{(\lambda - \omega)^n}$ $\forall \lambda > \omega, n = 1, 2, \ldots$

Proof. Necessity: This is similar to the necessity part of the proof of the Hille–Yosida Theorem. Here R_λ is well defined for $\lambda > \omega$ and one can similarly prove that $(\omega, \infty) \subset \rho(A)$ and $R(\lambda, A) = R_\lambda$, $\|R(\lambda, A)\| \leq \frac{M}{\lambda - \omega}$ for all $\lambda > \omega$. Then, for all $\lambda > \omega$ and $x \in X$, we have

---

[3]William S. Feller, Croatian-American mathematician, 1906–1970; Isao Miyadera, Japanese mathematician, born 1926.

$$
\begin{aligned}
R(\lambda, A)^2 x &= \int_0^\infty e^{-\lambda t} T(t) R_\lambda x \, dt \\
&= \int_0^\infty e^{-\lambda t} T(t) \left(\int_0^\infty e^{-\lambda s} T(s) x \, ds \right) dt \\
&= \int_0^\infty \left(\int_0^\infty e^{-\lambda(t+s)} T(t+s) x \, ds \right) dt \\
&= \int_0^\infty \left(\int_t^\infty e^{-\lambda r} T(r) x \, dr \right) dt \\
&= \int_0^\infty t e^{-\lambda t} T(t) x \, dt.
\end{aligned}
$$

It follows by induction that for all $\lambda > \omega$, $x \in X$ and $n = 1, 2, \ldots$

$$
R(\lambda, A)^n x = \frac{1}{(n-1)!} \int_0^\infty t^{n-1} e^{-\lambda t} T(t) x \, dt. \tag{9.8.38}
$$

We derive from (9.8.38) and the exponential estimate satisfied by the semigroup that for all $\lambda > \omega$, $x \in X$ and $n = 1, 2, \ldots$

$$
\begin{aligned}
\| R(\lambda, A)^n x \| &\leq \frac{M}{(n-1)!} \int_0^\infty t^{n-1} e^{(\omega - \lambda)t} \, dt \cdot \|x\| \\
&= \frac{M}{(\lambda - \omega)^n} \cdot \|x\|,
\end{aligned}
$$

which completes the proof of necessity.

Sufficiency: To simplify the proof, we note that, in general, if $\{T(t); t \geq 0\}$ is a C_0-semigroup satisfying $\|T(t)\| \leq M e^{\omega t}$, $t \geq 0$, for some $M \geq 1$ and $\omega \in \mathbb{R}$, with generator A, then the family $\{S(t) = e^{-\omega t} T(t); t \geq 0\}$ is also a C_0-semigroup with the generator $A - \omega I$. Thus, one can assume in the following that $\omega = 0$ (i.e., $(0, \infty) \subset \rho(A)$ and $\|\lambda^n R(\lambda, A)^n\| \leq M$ for all $\lambda > 0$ and $n = 1, 2, \ldots$). The idea that can be used to complete the proof is to define a new norm on X, say $\| \cdot \|_*$, equivalent to the original, such that the corresponding operator norm of $R(\lambda, A)$ be less than or equal to $1/\lambda$ for all $\lambda > 0$. Then the conclusion will follow from the Hille–Yosida theorem. First, define for $\nu > 0$ the following norm on X

$$
\|x\|_\nu = \sup\{\|\nu^n R(\nu, A)^n x\|; \, n \in \mathbb{N} \cup \{0\}\}.
$$

Obviously, the new norm is equivalent to the original one because

$$
\|x\| \leq \|x\|_\nu \leq M \|x\| \quad \forall x \in X, \tag{9.8.39}
$$

and the operator norm of $R(\nu, A)$ with respect to the new norm satisfies

$$\|R(\nu, A)\|_\nu \leq \frac{1}{\nu} \quad \forall \nu > 0. \tag{9.8.40}$$

In addition,

$$\|R(\lambda, A)\|_\nu \leq \frac{1}{\lambda} \quad \text{for all } 0 < \lambda \leq \nu. \tag{9.8.41}$$

This follows easily from (9.8.40) and the so-called *resolvent identity*:

$$R(\lambda, A) - R(\nu, A) = (\nu - \lambda)R(\nu, A)R(\lambda, A). \tag{9.8.42}$$

Now, define

$$\|x\|_* = \sup\{\|x\|_\nu; \, \nu > 0\},$$

and observe that (see (9.8.39) and (9.8.41))

$$\|x\| \leq \|x\|_* \leq M\|x\| \quad \text{and} \quad \|R(\lambda, A)\|_* \leq \frac{1}{\lambda} \quad \forall \lambda > 0.$$

So, according to the Hille–Yosida Theorem, A generates a C_0-semigroup $\{T(t); \, t \geq 0\} \subset L(X)$ satisfying

$$\|T(t)\|_* \leq 1 \quad \forall t \geq 0,$$

hence

$$\|T(t)\| \leq M \quad \forall t \geq 0. \qquad \square$$

Remark 9.31. Obviously, if $\|\cdot\|_\nu$ is the norm defined in the proof of Theorem 9.30, then

$$\|\lambda^n R(\lambda, A)^n x\| \leq \|\lambda^n R(\lambda, A)^n x\|_\nu$$
$$\leq \|x\|_\nu \quad \forall 0 < \lambda \leq \nu, \, x \in X, \, n = 0, 1, 2, \ldots$$

which implies

$$\|x\|_\lambda \leq \|x\|_\nu \quad \forall x \in X, \, 0 < \lambda \leq \nu.$$

Therefore, the norm $\|\cdot\|_*$ can be obtained as a limit

$$\|x\|_* = \lim_{\lambda \to \infty} \|x\|_\lambda.$$

Taking into account the above discussion on groups and their relationship with semigroups, one can easily derive the following extension to groups of the Feller–Miyadera–Phillips generation theorem.

Theorem 9.32. *A linear operator $A : D(A) \subset X \to X$ is the infinitesimal generator of a C_0-group $\{G(t); t \in \mathbb{R}\}$ satisfying $\|G(t)\| \leq Me^{\omega|t|}$, $t \in \mathbb{R}$, with $M \geq 1$, $\omega \in \mathbb{R}$, if and only if*

(k) $\overline{D(A)} = X$ and A is closed;

(kk)'' for every $\lambda \in \mathbb{R}$ with $|\lambda| > \omega$ one has $\lambda \in \rho(A)$ and $\|R(\lambda, A)^n\| \leq \frac{M}{(|\lambda|-\omega)^n}$ $\forall n = 1, 2, \ldots$

Remark 9.33. Obviously, if $M = 1$ in the above theorem, then the inequality $\|R(\lambda, A)^n\| \leq \frac{1}{(|\lambda|-\omega)^n}$ $\forall n = 1, 2, \ldots$ is equivalent to $\|R(\lambda, A)\| \leq \frac{1}{|\lambda|-\omega}$. If $M = 1$ and $\omega = 0$, then $\|G(t)\| = 1$ for all $t \in \mathbb{R}$, or equivalently $\|G(t)x\| = \|x\|$ for all $t \in \mathbb{R}$ (i.e., all $G(t)$'s are isometries). Summarizing, we have the following result.

Corollary 9.34. *A linear operator $A : D(A) \subset X \to X$ is the infinitesimal generator of a C_0-group of isometries $\{G(t); t \in \mathbb{R}\}$ if and only if*

(k) $\overline{D(A)} = X$ and A is closed;

(kk)* for every $\lambda \in \mathbb{R} \setminus \{0\}$ one has $\lambda \in \rho(A)$ and $\|R(\lambda, A)\| \leq \frac{1}{|\lambda|}$.

9.9 A Perturbation Result

It is intuitive that perturbing the generator A of a C_0-semigroup with any operator $B \in L(X)$ yields a generator. Indeed, the following result holds.

Theorem 9.35. *Let $A : D(A) \subset X \to X$ be the generator of a C_0-semigroup $\{T(t); t \geq 0\} \subset L(X)$ satisfying $\|T(t)\| \leq Me^{\omega t}$ for all $t \geq 0$, with $M \geq 1$, $\omega \in \mathbb{R}$, and let $B \in L(X)$. Then the operator $C = A + B$ with $D(C) = D(A)$ is the generator of a C_0-semigroup $\{S(t); t \geq 0\} \subset L(X)$ satisfying $\|S(t)\| \leq Me^{(\omega+M\|B\|)t}$ for all $t \geq 0$.*

Proof. As in the proof of the Feller–Miyadera–Phillips Theorem (Theorem 9.30), one can assume that $\omega = 0$. Next, we also assume that $M = 1$. Then $(0, \infty) \subset \rho(A)$ and for all $\lambda > 0$ we can write

$$\lambda I - C = (I - BR(\lambda, A))(\lambda I - A). \qquad (9.9.43)$$

For all $\lambda > \|B\|$ we have $\|BR(\lambda, A)\| \leq \|B\| \cdot \|R(\lambda, A)\| < 1$, so $I - BR(\lambda, A)$ is invertible in $L(X)$. Thus, taking into account (9.9.43), we can see that $(\|B\|, \infty) \subset \rho(C)$ and for all $\lambda > \|B\|$

$$
\begin{aligned}
R(\lambda, C) &= R(\lambda, A)\big(I - BR(\lambda, A)\big)^{-1} \\
&= R(\lambda, A) \sum_{n=0}^{\infty} \big(BR(\lambda, A)\big)^n,
\end{aligned}
$$

which shows that

$$
\|R(\lambda, C)\| \leq \frac{1}{\lambda - \|B\|} \quad \forall \lambda > \|B\|.
$$

This is enough to conclude that C generates a C_0-semigroup $\{S(t); t \geq 0\}$ satisfying $\|S(t)\| \leq e^{\|B\|t}$ for all $t \geq 0$.

Now, let us consider the general case $M \geq 1$ (and $\omega = 0$). Define the norm

$$
\|x\|_* = \sup_{t \geq 0} \|T(t)\|,
$$

which is equivalent to the original norm of X:

$$
\|x\| \leq \|x\|_* \leq M\|x\| \quad \forall x \in X.
$$

Obviously, $\|T(t)\|_* \leq 1$ for all $t \geq 0$ and

$$
\|Bx\|_* \leq M\|B\| \cdot \|x\| \leq M\|B\| \cdot \|x\|_* \quad \forall x \in X.
$$

By the above proof for the case $M = 1$, $C = A + B$ generates a C_0-semigroup $\{S(t); t \geq 0\}$ satisfying

$$
\|S(t)\|_* \leq e^{\|B\|_* t} \quad t \geq 0.
$$

Therefore,

$$
\begin{aligned}
\|S(t)x\| &\leq \|S(t)x\|_* \\
&\leq e^{\|B\|_* t}\|x\|_* \\
&\leq M e^{M\|B\|t}\|x\| \quad \forall t \geq 0,
\end{aligned}
$$

which concludes the proof. $\square$

9.10   Approximation of Semigroups

An example of approximation has already been encountered in the proof of Theorem 9.22. Specifically, we saw that if $\{T(t);\, t \geq 0\} \subset L(X)$ is a C_0-semigroup of contractions with generator A, then $S(t)x$ can be approximated (uniformly for t in compact intervals) by $e^{tA_\lambda}x$ as $\lambda \to \infty$, where A_λ denotes the Yosida approximation of A. Definitely, this approximation result extends to any C_0-semigroup.

In what follows, we present another approximation result, known as the Trotter Theorem,[4] which is relevant for applications. As in [39], for $M \geq 1$ and $\omega \in \mathbb{R}$ denote by $G(M,\omega)$ the class of operators which generate C_0-semigroups $\{T(t);\, t \geq 0\}$ satisfying $\|T(t)\| \leq Me^{\omega t}$, $\forall t \geq 0$. The Trotter Theorem [48] says that the convergence of a sequence $A_n \in G(M,\omega)$ to $A \in G(M,\omega)$ in some sense (see below) is equivalent to the convergence of the corresponding semigroups.

Theorem 9.36. *If A, $A_n \in G(M,\omega)$ and $\{T(t);\, t \geq 0\}$, $\{T_n(t);\, t \geq 0\}$ are the C_0-semigroups generated by A, A_n $(n = 1, 2, \dots)$, then the following conditions are equivalent:*

(a) *for some $\lambda > \omega$ and for all $x \in X$, $R(\lambda, A_n)x \to R(\lambda, A)x$ as $n \to \infty$;*

(b) *for all $x \in X$ and $t \geq 0$, $T_n(t)x \to T(t)x$ as $n \to \infty$, uniformly for t in compact subintervals of $[0, \infty)$.*

Proof. We first prove that (a) implies (b). For a given $t > 0$, every $s \in (0, t)$, and every $x \in X$, we have

$$\frac{d}{ds}[T_n(t - s)R(\lambda, A_n)T(s)R(\lambda, A)x]$$

$$= -T_n(t - s)A_n R(\lambda, A_n)T(s)R(\lambda, A)x$$

$$\quad + T_n(t - s)R(\lambda, A_n)AT(s)R(\lambda, A)x$$

$$= T_n(t - s)[-A_n R(\lambda, A_n)R(\lambda, A) + R(\lambda, A_n)AR(\lambda, A)]T(s)x$$

$$= T_n(t - s)[R(\lambda, A) - R(\lambda, A_n)]T(s)x.$$

Note that all the above operations are allowed. Integrating the above equality over $[0, t]$ yields

$$R(\lambda, A_n)[T_n(t) - T(t)]R(\lambda, A)x$$

$$= \int_0^t T_n(t - s)[R(\lambda, A) - R(\lambda, A_n)]T(s)x\, ds. \quad (9.10.44)$$

---

[4]Hale F. Trotter, Canadian mathematician, born 1931.

It follows from (9.10.44) that, for all t in an arbitrary compact interval $[0, t_1]$, one has

$$\|R(\lambda, A_n)[T_n(t) - T(t)]R(\lambda A)x\|$$

$$\leq \int_0^t \|T_n(t - s)\| \cdot \|[R(\lambda, A_n) - R(\lambda, A)]T(s)x\| \, ds$$

$$\leq \int_0^{t_1} \|T_n(t - s)\| \cdot \|[R(\lambda, A_n) - R(\lambda, A)]T(s)x\| \, ds. \qquad (9.10.45)$$

Note that the sequence of the integrands in (9.10.44) converges pointwise to zero in $[0, t_1]$ and it has in this interval the upper bound $2M^3 e^{\omega t_1} \|x\| (\lambda - \omega)^{-1}$. Thus, according to the Lebesgue Dominated Convergence Theorem, one gets from (9.10.45)

$$\lim_{n \to \infty} \|R(\lambda, A_n)[T_n(t) - T(t)]R(\lambda, A)x\| = 0,$$

uniformly for t in every compact subinterval of $[0, \infty)$. In fact, since the range of $R(\lambda, A) = D(A)$, we have

$$\lim_{n \to \infty} \|R(\lambda, A_n)[T_n(t) - T(t)]x\| = 0 \quad \forall x \in D(A), \qquad (9.10.46)$$

uniformly for t in every compact subinterval of $[0, \infty)$. Now, let us estimate

$$\|[T_n(t) - T(t)]R(\lambda, A)x\| \leq \|T_n(t)[R(\lambda, A) - R(\lambda, A_n)x]\|$$
$$+ \|R(\lambda, A_n)[T_n(t) - T(t)]x\|$$
$$+ \|[R(\lambda, A_n) - R(\lambda, A)]T(t)x\|.$$
$$\qquad (9.10.47)$$

The right-hand side of (9.10.47) has three terms, say $S_i = S_i(t, n, x)$, $i = 1, 2, 3$. Using our assumption (a) and the estimate $\|T_n(t)\| \leq M e^{\omega t}$, $t \geq 0$, we can see that, for each $x \in X$, $S_1(t, n, x)$ converges to zero as $n \to \infty$, uniformly for t in every compact subinterval of $[0, \infty)$. A similar conclusion holds for $S_2(t, n, x)$, $x \in D(A)$ (see (9.10.46)). Taking again assumption (a) into account, it follows that $S_3(t, n, x)$, $x \in X$, also converges to zero as $n \to \infty$, uniformly for t in every compact subinterval of $[0, \infty)$ (here we use the fact that $\{T(t)x; 0 \leq t \leq t_1\}$ is a compact set for each $t_1 > 0$). Summarizing, we derive from (9.10.47) that

$$\lim_{n \to \infty} \|[T_n(t) - T(t)]R(\lambda, A)x\| = 0, \quad x \in D(A).$$

Hence,

$$\lim_{n\to\infty} \|[T_n(t) - T(t)]z\| = 0, \quad \forall z \in D(A^2),$$

uniformly on every compact subinterval of $[0, \infty)$. Since $D(A^2)$ is dense in X (see Remark 9.11), this conclusion extends to all $x \in X$, so (b) holds.

Conversely, assuming now that (b) is satisfied, we have for any $\lambda > \omega$ and $x \in X$

$$\|R(\lambda, A_n)x - R(\lambda, A)x\| = \|\int_0^\infty e^{-\lambda t}[T_n(t)x - T(t)x]\, dt\|$$

$$\leq \int_0^\infty e^{-\lambda t}\|T_n(t)x - T(t)x\|\, dt.$$

$$(9.10.48)$$

Using again Lebesgue's Dominated Convergence Theorem for the right-hand side of the above inequality, we conclude that indeed (b) implies (a). $\square$

Remark 9.37. It is obvious from the proof above that condition (a) is equivalent to (a)': for all $x \in X$ and all $\lambda > \omega$, $R(\lambda, A_n)x \to R(\lambda, A)x$ as $n \to \infty$.   If one assumes that, for some $\lambda > \omega$, $R(\lambda, A_n)x$ converges as $n \to \infty$ to some $R_\lambda x$ for all $x \in X$, and if in addition the range of R_λ is assumed to be dense in X, then R_λ is the resolvent $R(\lambda, A)$ of an operator $A \in G(M, \omega)$. For the proof of this implication, see [24] and [39, p. 86]. This implication can be used to replace Theorem 9.36 by an improved version, in which the existence of $A \in G(M, \omega)$ is no longer assumed. The reformulation of the Trotter Theorem in view of the above information is left to the reader.

Remark 9.38. It is worth pointing out that the Trotter Theorem or suitable versions of it can be used successfully in the numerical analysis of various initial-boundary value problems.

We continue this section with a result known as the Chernoff product formula.[5]

Theorem 9.39. *Let $A \in G(M, \omega)$ for some $M \geq 1$ and $\omega \in \mathbb{R}$ and let $F : [0, \infty) \to L(X)$ be a function satisfying $F(0) = I$ and $\|F(t)^k\| \leq Me^{k\omega t}$ for all $t \geq 0$, $k \in \mathbb{N}$. Assume that*

$$Ax = \lim_{s\to 0^+} s^{-1}[F(s)x - x], \quad \forall x \in D(A). \tag{9.10.49}$$

---

[5]Paul R. Chernoff, American mathematician, born 1942.

Then,

$$T(t)x = \lim_{n \to \infty} F(t/n)^n x, \qquad (9.10.50)$$

for all $x \in X$, uniformly for t in compact subintervals of $[0, \infty)$, where $\{T(t); \, t \geq 0\}$ is the C_0-semigroup generated by A.

In order to prove this theorem we need the following lemma.

Lemma 9.40. *Let $Q \in L(X)$ such that $\|Q^j\| \leq M$ for some $M \geq 1$ and all $j \in \mathbb{N}$. Then we have*

$$\|e^{n(Q-I)}x - Q^n x\| \leq M\sqrt{n}\,\|Qx - x\|, \quad \forall n \in \mathbb{N}, \, x \in X.$$

Proof. Let $n \in \mathbb{N}$ be arbitrary but fixed. We have

$$
\begin{aligned}
e^{n(Q-I)} - Q^n &= e^{-n}\Big(e^{nQ} - e^n Q^n\Big) \\
&= e^{-n}\sum_{k=0}^{\infty}\frac{n^k}{k!}\big(Q^k - Q^n\big). \qquad (9.10.51)
\end{aligned}
$$

Note that for $k > n$ we have

$$Q^k - Q^n = \Big(\sum_{j=n}^{k-1} Q^j\Big)(Q - I),$$

and a similar identity holds for $k < n$. So we obtain by using $\|Q^j\| \leq M$

$$\|Q^k x - Q^n x\| \leq M|n - k| \cdot \|Qx - x\|. \qquad (9.10.52)$$

Now, using (9.10.51), (9.10.52), and the Bunyakovsky–Cauchy–Schwarz inequality, we derive

$$
\begin{aligned}
\|e^{n(Q-I)}x - Q^n x\| &\leq e^{-n}\sum_{k=0}^{\infty}\frac{n^k}{k!}M|n-k|\cdot\|Qx - x\| \\
&\leq Me^{-n}\|Qx - x\|\Big(\sum_{k=0}^{\infty}\frac{n^k}{k!}\Big)^{1/2} \\
&\quad \times \Big(\sum_{k=0}^{\infty}(n-k)^2\frac{n^k}{k!}\Big)^{1/2} \\
&= Me^{-n}\|Qx - x\|\big(e^n\big)^{1/2}\big(ne^n\big)^{1/2} \\
&= M\sqrt{n}\,\|Qx - x\|. \qquad \qquad \square
\end{aligned}
$$

Proof of Theorem 9.39. We consider first the case $\omega = 0$. Define for $s > 0$

$$A_s x = s^{-1}[F(s) - I]x, \quad x \in X.$$

Obviously, $A_s \in L(X)$ for all $s > 0$ and (cf. (9.10.49))

$$\lim_{s \to 0^+} A_s x = A x, \quad \forall x \in D(A). \tag{9.10.53}$$

Note also that for each $s > 0$,

$$\|e^{tA_s}\| \leq e^{-t/s} \sum_{k=0}^{\infty} \frac{t^k}{s^k k!} \|F(s)^k\| \leq M, \quad \forall t \geq 0, \tag{9.10.54}$$

i.e., $A_s \in G(M, 0)$. For $\lambda > \omega = 0$ and $y = (\lambda I - A)x$, $x \in D(A)$, we have

$$\begin{aligned}
R(\lambda, A_s)y &= R(\lambda, A_s)\big[(\lambda I - A_s)x - (\lambda I - A_s)x + (\lambda I - A)x\big] \\
&= x + R(\lambda, A_s)\big[A_s x - A x\big].
\end{aligned}$$

Therefore, according to (9.10.53), we have

$$R(\lambda, A_s)y \to R(\lambda, A)y, \quad \text{as } s \to 0^+, \quad \forall y \in X, \tag{9.10.55}$$

since $\|R(\lambda, A_s)\| \leq M/\lambda$. Now, using (9.10.53), (9.10.54), and (9.10.55), it follows by Theorem 9.36 (which also works with s instead of n),

$$\|T(t)x - e^{tA_s}x\| \to 0, \quad \text{as } s \to 0^+, \quad \forall x \in X,$$

uniformly for t in compact subintervals of $[0, \infty)$, and hence

$$\|T(t)x - e^{tA_{t/n}}x\| \to 0, \quad \text{as } n \to \infty, \quad \forall x \in X, \tag{9.10.56}$$

uniformly for t in compact subintervals of $[0, \infty)$. On the other hand, by Lemma 9.40, we have

$$\begin{aligned}
\|e^{tA_{t/n}}x - F(t/n)^n x\| &= \|e^{n[F(t/n)-I]}x - F(t/n)^n x\| \\
&\leq M\sqrt{n}\|F(t/n)x - x\| \\
&= \frac{Mt}{\sqrt{n}}\|A_{t/n}x\| \to 0, \quad \text{as } n \to \infty, \tag{9.10.57}
\end{aligned}$$

for all $x \in D(A)$, uniformly for t in compact subintervals of $[0, \infty)$. Combining (9.10.56) and (9.10.57), we derive (9.10.50) for all $x \in D(A)$. Since $D(A)$ is dense in X, (9.10.50) extends to the whole of X.

The case $\omega \neq 0$ can be reduced to the previous one. Indeed, the function $\tilde{F}$, defined by $\tilde{F}(t) = e^{-\omega t} F(t)$, satisfies $\tilde{F}(0) = I$, $\|\tilde{F}(t)^k\| \leq M$ for all $t \geq 0$ and $k \in \mathbb{N}$. Moreover, (9.10.49) is satisfied with $\tilde{F}$ instead of F, and $A - \omega I$ instead of A. So the conclusion of Theorem 9.39 follows easily. $\square$

Corollary 9.41. *For every* $A \in G(M, \omega)$, $M \geq 1$, $\omega \in \mathbb{R}$, *we have*

$$T(t)x = \lim_{n \to \infty} \left(I - \frac{t}{n} A \right)^{-n} x, \quad \forall x \in X, \tag{9.10.58}$$

uniformly for t *in compact subintervals of* $[0, \infty)$, *where* $\{T(t);\ t \geq 0\} \subset L(X)$ *is the* C_0*-semigroup generated by* A.

Proof. We can assume that $\omega > 0$. Define $F : [0, \infty) \to L(X)$ by

$$F(t) = \begin{cases} I, & t = 0, \\ (1/t)R(1/t, A), & t \in (0, \delta), \\ 0, & t \geq \delta, \end{cases}$$

for some $\delta \in (0, 1/\omega)$. We choose $\delta > 0$ small enough so that

$$\begin{aligned} \|F(t)^k\| &\leq (1/t^k)\|R(1/t, A)^k\| \\ &\leq M/\left[t^k (t^{-1} - \omega)^k \right] \\ &= M/(1 - \omega t)^k \\ &\leq M e^{k(\omega+1)t}, \quad \forall t \in (0, \delta),\ k \in \mathbb{N}. \end{aligned}$$

We also have

$$\lim_{t \to 0^+} t^{-1}[F(t)x - x] = \lim_{t \to 0^+} (1/t)R(1/t, A)Ax = Ax, \quad \forall x \in D(A).$$

Thus, all the assumptions of Theorem 9.39 are fulfilled and so (9.10.58) holds. $\square$

Another consequence of the Chernoff product formula is the so-called Trotter product formula corresponding to perturbed semigroups:

Corollary 9.42. *Let* $A \in G(M, \omega)$, $M \geq 1$, $\omega \in \mathbb{R}$, *and* $B \in L(X)$. *If* $\{T(t);\ t \geq 0\}$ *is the* C_0*-semigroup generated by* A, $S(t) = e^{tB}$, $t \geq 0$, *and* $\{U(t);\ t \geq 0\}$ *is the* C_0*-semigroup generated by* $A + B$, *then*

$$U(t)x = \lim_{n \to \infty} \left[T(t/n)S(t/n) \right]^n x, \tag{9.10.59}$$

for all $x \in X$, *uniformly for* t *in compact subintervals of* $[0, \infty)$.

Proof. By Theorem 9.35, $A + B \in G(M, \omega + M\|B\|)$. Making use of the previous renorming procedure (see the proof of Theorem 9.30), we can assume $M = 1$. So, defining $F(t) = T(t)S(t)$, $t \geq 0$, we have

$$
\begin{aligned}
\|F(t)^k\| &\leq \|T(t)\|^k \|S(t)\|^k \\
&\leq e^{k\omega t} e^{k\|B\|t} \\
&= e^{k(\omega + \|B\|)t}, \ \forall t \geq 0,
\end{aligned}
$$

and for all $x \in D(A + B) = D(A)$

$$
\begin{aligned}
\lim_{t \to 0^+} t^{-1}[F(t)x - x] &= \lim_{t \to 0^+} T(t)\frac{S(t)x - x}{t} + \lim_{t \to 0^+}\frac{T(t)x - x}{t}. \\
&= Bx + Ax.
\end{aligned}
$$

Therefore, Theorem 9.39 is again applicable and (9.10.59) follows. $\square$

Remark 9.43. The Trotter product formula is valid, under appropriate conditions, for two general C_0-semigroups (see, e.g., [13, p. 154]).

9.11 The Inhomogeneous Cauchy Problem

Consider the Cauchy (initial value) problem

$$
u'(t) = Au(t) + f(t), \quad t \in [0, r]; \quad u(0) = x, \qquad (CP)
$$

where A is the generator of a C_0-semigroup $\{T(t); t \geq 0\} \subset L(X)$, f is a given function from $[0, r]$ to X, $r \in (0, \infty)$. The case $f \equiv 0$ was discussed before.

Definition 9.44. *A function $u : [0, r] \to X$ is a classical solution of problem (CP) if u is continuous on $[0, r]$ and continuously differentiable on $(0, r]$, $u(t) \in D(A)$ for all $t \in (0, r]$, $u(0) = x$, and u satisfies equation $(CP)_1$ for all $t \in (0, r]$.*

Remark 9.45. If $f \in C([0, r]; X)$ and u is a classical solution of problem (CP), then for $0 < s < t \leq r$ we have

$$
\begin{aligned}
\frac{d}{ds}[T(t - s)u(s)] &= -T(t - s)Au(s) + T(t - s)u'(s) \\
&= -T(t - s)Au(s) + T(t - s)Au(s) \\
&\quad + T(t - s)f(s) \\
&= T(t - s)f(s).
\end{aligned}
$$

Therefore, by integration over $[0, t]$ one obtains

$$u(t) = T(t)x + \int_0^t T(t-s)f(s)\,ds, \quad t \in [0, r], \qquad (9.11.60)$$

showing that u is unique (since A generates a unique C_0-semigroup; see Theorem 9.12).

Note also that the integral term in the right-hand side of Eq. (9.11.60) makes sense for $f \in L^1(0, r; X)$, since (see Theorem 9.7)

$$\|T(t-s)f(s)\| \le Me^{\omega(t-s)}\|f(s)\|, \quad 0 \le s \le t \le r.$$

This observation leads to the introduction of a new concept of solution for the Cauchy problem (CP).

Definition 9.46. *Let $x \in X$, $f \in L^1(0, r; X)$, and let A be the generator of a C_0-semigroup $\{T(t); t \ge 0\} \subset L(X)$. The function $u \in C([0, r]; X)$ given by*

$$u(t) = T(t)x + \int_0^t T(t-s)f(s)\,ds \quad \forall t \in [0, r] \qquad (9.11.61)$$

is called a mild solution of problem (CP).

Obviously, if A is the generator of a C_0-semigroup $\{T(t); t \ge 0\}$, then for each $(x, f) \in X \times L^1(0, r; X)$ problem (CP) has a unique mild solution (since the C_0-semigroup generated by A is unique). Formula (9.11.61) above is often called the *variation of constants formula*. Under certain conditions on x and f it gives a classical solution of problem (CP). The following theorem is one such example.

Theorem 9.47. *Let $A : D(A) \subset X \to X$ be the infinitesimal generator of a C_0-semigroup, say $\{T(t); t \ge 0\}$, and let $x \in D(A)$ and $f \in C^1([0, r]; X)$. Then problem (CP) has a unique classical solution (given by (9.11.61)).*

Proof. Uniqueness is already known (see the remark above). To prove existence it suffices to show that

$$v(t) = \int_0^t T(t-s)f(s)\,ds$$

satisfies equation $(CP)_1$ on $(0, r]$ (see also Theorem 9.10). Indeed,

$$v(t) = \int_0^t T(s)f(t-s)\,ds$$

and so there exists

$$
\begin{aligned}
v'(t) &= T(t)f(0) + \int_0^t T(s)f'(t-s)\,ds \\
&= T(t)f(0) + \int_0^t T(t-s)f'(s)\,ds \quad \forall t \in (0,r].
\end{aligned}
$$

On the other hand, for each $t \in (0,r)$ and $h > 0$ small enough, we have

$$
\begin{aligned}
h^{-1}[T(h) - I]v(t) &= h^{-1}\int_{0}^t T(t+h-s)f(s)\,ds - h^{-1}v(t) \\
&= h^{-1}[v(t+h) - v(t)] - h^{-1} \\
&\quad \times \int_t^{t+h} T(t+h-s)f(s)\,ds
\end{aligned}
$$

which converges to $v'(t) - f(t)$ as $h \to 0^+$. Therefore, $v(t) \in D(A)$ and

$$
Av(t) = v'(t) - f(t), \quad \forall t \in (0,r).
$$

In fact, f can be extended to the right of $t = r$ as a continuously differentiable function, so $v(r) \in D(A)$ and there exists $v'(r) = Av(r) + f(r)$. Even more, there exists $v'(0) = f(0)$ so the function $u(t) = T(t)x + v(t)$ is continuously differentiable on $[0,r]$ and satisfies equation $(CP)_1$ for all $t \in [0,r]$. $\qquad\square$

Remark 9.48. From the proof above we see that (under the conditions of Theorem 9.47)

$$
u'(t) = T(t)x + T(t)f(0) + \int_0^t T(t-s)f'(s)\,ds \quad \forall t \in [0,r]. \tag{9.11.62}
$$

Remark 9.49. Let A be the infinitesimal generator of a C_0-semigroup $\{T(t); t \geq 0\}$ and let $(x,f) \in X \times L^1(0,r;X)$. If u is the corresponding mild solution of problem (CP), then it is the uniform limit of a sequence of C^1-solutions (hence classical solutions) of (CP). Indeed, let $((x_n, f_n))$ be a sequence in $D(A) \times C^1([0,r]; X)$ which approximates (x,f) in $X \times L^1(0,r;X)$. For each (x_n, f_n) there exists a unique C^1-solution u_n of problem (CP) with $x := x_n$ and $f := f_n$ given by the variation of constants formula:

$$
u_n(t) = T(t)x_n + \int_0^t T(t-s)f_n(s)\,ds.
$$

By standard arguments one gets for all $t \in [0, r]$

$$\|u_n(t) - u(t)\| \leq \|T(t)(x_n - x)\|$$
$$+ \int_0^t \|T(t-s)\| \cdot \|f_n(s) - f(s)\| \, ds$$
$$\leq Me^{\omega t} \|x_n - x\|$$
$$+ \int_0^t Me^{\omega(t-s)} \|f_n(s) - f(s)\| \, ds$$
$$\leq Me^{\omega r} \left(\|x_n - x\| + \int_0^r \|f_n(s) - f(s)\| \, ds \right).$$

Therefore, $u_n \to u$ in $C([0, r]; X)$.

Remark 9.50. The semigroup approach can be used to solve Cauchy problems for semilinear evolution equations. Specifically, let us consider the following problem,

$$u'(t) = Au(t) + f(t, u(t)), \ t \in [0, r]; \ u(0) = x \in X, \qquad (NCP)$$

where A is the infinitesimal generator of a C_0-semigroup $\{T(t); t \geq 0\} \subset L(X)$ and $f : [0, r] \times X \to X$ is continuous and satisfies the Lipschitz condition

$$\|f(t, x_1) - f(t, x_2)\| \leq L\|x_1 - x_2\|, \ \ (t, x_1), (t, x_2) \in [0, r] \times X.$$

Here L is a positive constant. One can consider the following "mild" form for (NCP)

$$u(t) = T(t)x + \int_0^t T(t-s)f(s, u(s)) \, ds, \ \ t \in [0, r]. \qquad (9.11.63)$$

If u is a classical solution of problem (NCP), then it satisfies (9.11.63). One can prove the existence of a solution $u \in Y := C([0, r]; X)$ of (9.11.63) by using the Banach Contraction Principle. For this purpose, let us consider the Bielecki norm[6] on Y:

$$\|g\|_B = \sup_{0 \leq t \leq r} e^{-\beta t} \|g\|, \ \ g \in Y,$$

where β is a large positive constant. This Bielecki norm is equivalent to the usual sup-norm of Y, so Y is a Banach space with respect to $\| \cdot \|_B$. Define on Y an operator Q by

$$(Qu)(t) = T(t)x + \int_0^t T(t-s)f(s, u(s)) \, ds, \ \ t \in [0, r], \ u \in Y.$$

---

[6]Adam Bielecki, Polish mathematician, 1910–2003.

Obviously, Q maps Y into itself, and for all $t \in [0, r]$ and $u_1, u_2 \in Y$ we have

$$\|(Qu_1)(t) - (Qu_2)(t)\| \leq LM \int_0^t e^{\omega(t-s)} \|u_1(s) - u_2(s)\| \, ds$$

$$= LMe^{\omega t} \int_0^t e^{(\beta-\omega)s} e^{-\beta s} \|u_1(s) - u_2(s)\| \, ds$$

$$\leq LMe^{\omega t} \|u_1 - u_2\|_B \int_0^t e^{(\beta-\omega)s} ds$$

$$= \frac{LM}{\beta - \omega} \|u_1 - u_2\|_B (e^{\beta t} - e^{\omega t})$$

$$\leq \frac{LM}{\beta - \omega} \|u_1 - u_2\|_B e^{\beta t}.$$

Thus,

$$e^{-\beta t} \|(Qu_1)(t) - (Qu_2)(t)\| \leq \frac{LM}{\beta - \omega} \|u_1 - u_2\|_B,$$

$$t \in [0, r], \ u_1, u_2 \in Y,$$

which implies

$$\|Qu_1 - Qu_2\|_B \leq \frac{LM}{\beta - \omega} \|u_1 - u_2\|_B, \ u_1, u_2 \in Y.$$

So, for $\beta > LM + \omega$, Q is a contraction and the Banach Contraction Principle ensures the existence of a unique fixed point u of Q. This u is the unique solution in Y of Eq. (9.11.63), which can be called a mild solution of the given semilinear Cauchy problem. In general, a mild solution is not a classical one. However, under appropriate conditions on x and f it is.

9.12 Applications

In this section we illustrate the above theory with some applications.

9.12.1 The Heat Equation

Consider the heat (diffusion) equation

$$u_t = u_{xx} + f(t, x), \quad t \in [0, r], \ x \in (0, 1), \tag{9.12.64}$$

with Dirichlet boundary conditions

$$u(t,0) = 0 = u(t,1), \quad t \in [0,r], \tag{9.12.65}$$

and initial condition

$$u(0,x) = u_0(x), \quad x \in (0,1), \tag{9.12.66}$$

where $u_0 \in L^2(0,1)$, $f \in L^1(0,r; L^2(0,1))$, and $u = u(t,x)$ is the unknown function representing the temperature (or density in the case of a general diffusion process). We have denoted $u_t := \frac{\partial u}{\partial t}$ and $u_{xx} := \frac{\partial^2 u}{\partial x^2}$. In order to solve problem (9.12.64)–(9.12.66), we choose $X = L^2(0,1)$ as the basic space equipped with the usual scalar product

$$\langle p, q \rangle = \int_0^1 p(x)q(x)\, dx,$$

and the corresponding (Hilbertian) norm. Define $A : D(A) \subset X \to X$ by

$$D(A) = H^2(0,1) \cap H_0^1(0,1), \quad Av = v'' = \frac{d^2v}{dx^2}.$$

So, regarding $u = u(t,x)$ as an X-valued function of $t \in [0,r]$, problem (9.12.64)–(9.12.66) can be expressed as the Cauchy problem in X

$$\frac{d}{dt}u(t,\cdot) = Au(t,\cdot) + f(t,\cdot), \ t \in [0,r]; \ \ u(0,\cdot) = u_0. \tag{9.12.67}$$

Note that the boundary conditions (9.12.65) are incorporated into the definition of $D(A)$. It turns out that A is the generator of a C_0-semigroup of contractions, say $\{T(t) : X \to X; t \geq 0\}$, so there is a unique mild solution u of problem (9.12.64)–(9.12.66) given by the variation of constants formula (see (9.11.61))

$$u(t,\cdot) = T(t)u_0(\cdot) + \int_0^t T(t-s)f(s,\cdot)\, ds, \ \ t \in [0,r]. \tag{9.12.68}$$

In order to show that A is a generator of a C_0-semigroup of contractions one could use the Hille–Yosida Theorem. A better option is to use the Lumer–Phillips Theorem. In fact, as X is a Hilbert (hence reflexive) space, it suffices to prove that A is an m-dissipative operator (cf. Theorem 9.27). This means that we do not need to check the density condition on $D(A)$ (that actually follows by the density of $C_0^\infty(0,1)$ in X and the obvious inclusion relation $C_0^\infty(0,1) \subset D(A)$).

As the dissipativeness of A follows trivially, let us prove that for any $\lambda > 0$ we have $R(\lambda I - A) = X$. In other words, for any $\lambda > 0$, $g \in X$, there exists a solution $v \in H^2(0,1)$ of the following boundary value problem

$$\lambda v - v'' = g, \quad v(0) = 0 = v(1).$$

But this follows easily by imposing the boundary conditions to the general solution of the above differential equation.

One could also use Theorem 9.29 and the fact that A is a self-adjoint operator.

According to Theorem 9.47 (see also its proof), if $u_0 \in D(A) = H_0^1(0,1) \cap H^2(0,1)$ and $f \in C^1([0,r]; X)$ then $u \in C^1([0,r]; X)$. Moreover, since u satisfies the heat equation it follows that $u \in C([0,r]; H^2(0,1))$. Note that the condition $u_0 \in D(A)$ incorporates the compatibility of u_0 with the boundary conditions: $u_0(0) = u_0(1) = 0$. It is also worth pointing out that higher regularity of u can be obtained under additional conditions on u_0 and f.

The above discussion can be extended to more dimensions. Specifically, let $\Omega \subset \mathbb{R}^n$, $n \geq 2$, be a bounded domain with a sufficiently smooth boundary $\partial\Omega$. Consider the n-dimensional heat equation

$$u_t = \Delta u + f(t,x), \quad t \in [0,r], \ x \in \Omega,$$

and associate with it the homogeneous Dirichlet boundary condition

$$u = 0 \ \text{ on } \ \partial\Omega,$$

and the initial condition

$$u(0,x) = u_0(x), \quad x \in \Omega.$$

We have denoted by Δ the classical Laplacian with respect to x. Let $X = L^2(\Omega)$ and let $A = \Delta$ with $D(A) = H_0^1(\Omega) \cap H^2(\Omega)$. So the above initial-boundary value problem can be viewed as a Cauchy problem in X. The fact that A is a dissipative operator follows from Green's formula, and its m-dissipativity can be derived by using the Lax–Milgram Theorem. The reader is encouraged to continue the discussion and derive existence, uniqueness, and regularity of the solution to the above problem. The reader could also consider the case of the homogeneous Neumann or Robin boundary condition and investigate it along the same lines.

9.12.2   The Wave Equation

Consider in a first stage the one-dimensional wave equation

$$u_{tt} - u_{xx} = f(t, x), \quad t \geq 0, \ x \in (0, 1), \tag{9.12.69}$$

with the homogeneous Dirichlet boundary conditions,

$$u(t, 0) = 0 = u(t, 1), \quad t \geq 0, \tag{9.12.70}$$

and initial conditions

$$u(0, x) = u_0(x), \quad u_t(0, x) = v_0(x), \quad x \in (0, 1). \tag{9.12.71}$$

Recall that this problem describes the evolution of the displacement $u(t, x)$ of an elastic string fixed at both its ends ($x = 0$ and $x = 1$), where $f(t, x)$ represents an external force.

Denoting $v = u_t$, problem (9.12.69)–(9.12.71) can be equivalently written as

$$\begin{cases} \frac{\partial}{\partial t}[u, v] = [v, u_{xx} + f], & t \geq 0, \ x \in (0, 1), \\ u(t, 0) = u(t, 1) = 0, & t \geq 0, \\ [u, v](0, x) = [u_0(x), v_0(x)], & x \in (0, 1). \end{cases}$$

Let $X = H_0^1(0, 1) \times L^2(0, 1)$ (the so-called phase space) which is a real Hilbert space with the scalar product

$$\langle [p_1, q_1], [p_2, q_2] \rangle = \int_0^1 p_1' p_2' \, dx + \int_0^1 q_1 q_2 \, dx,$$

and the induced norm. Define $A : D(A) \subset X \to X$ by

$$D(A) = [H_0^1(0, 1) \cap H^2(0, 1)] \times H_0^1(0, 1), \quad A[p, q] = [q, p''].$$

Thus the above problem can be expressed as the following Cauchy problem in X

$$\begin{cases} \frac{d}{dt}[u(t, \cdot), v(t, \cdot)] = A[u(t, \cdot), v(t, \cdot)] + [0, f(t, \cdot)], & t \geq 0, \\ [u(0, \cdot), v(0, \cdot)] = [u_0, v_0]. \end{cases}$$

Denote this Cauchy problem by (CP). In order to derive existence results for (CP), we are going to show in what follows that A is the generator of a C_0-group of isometries. For this purpose, we can use Corollary 9.34.

First, noting that $C_0^\infty(0,1)$ is dense in $H_0^1(0,1)$ as well as in $L^2(0,1)$, and $C_0^\infty(0,1) \times C_0^\infty(0,1) \subset D(A)$, we infer that the closure of $D(A)$ in X equals X. It is also easily seen that A is a closed operator. So we need only to show that condition $(kk)^*$ of Corollary 9.34 is fulfilled. Let $\lambda \in (-\infty, 0) \cup (0, \infty)$ and let $[g, h]$ be an arbitrary pair in X. We claim that there exists a unique $[p, q] \in D(A)$ such that

$$\lambda[p, q] - A[p, q] = [g, h], \qquad (9.12.72)$$

or, equivalently, there exists a unique $p \in H_0^1(0,1) \cap H^2(0,1)$ satisfying the equation

$$\lambda^2 p = p'' + h + \lambda g.$$

We know from the preceding discussion on the heat equation that the last assertion is true. We also have $q = \lambda p - g \in H_0^1(0,1)$ which concludes the proof of our claim. Hence $\lambda I - A$ is invertible.

Now, multiplying Eq. (9.12.72) by $[p, q]$ and taking into account the definition of A we get

$$\lambda \|[p, q]\|_X^2 - \underbrace{\left(\int_0^1 p'q' \, dx + \int_0^1 p''q \, dx \right)}_{=0} = \langle [g, h], [p, q] \rangle,$$

which implies

$$
\begin{aligned}
|\lambda| \cdot \|[p, q]\|_X^2 &= |\langle [g, h], [p, q] \rangle| \\
&\leq \|[g, h]\|_X \cdot \|[p, q]\|_X.
\end{aligned}
$$

Therefore,

$$|\lambda| \cdot \|(\lambda I - A)^{-1}[g, h]\|_X \leq \|[g, h]\|_X \quad \forall [g, h] \in X,$$
$$\lambda \in (-\infty, 0) \cup (0, \infty),$$

and so

$$\|(\lambda I - A)^{-1}\| \leq \frac{1}{|\lambda|} \quad \forall \lambda \in (-\infty, 0) \cup (0, \infty).$$

Thus, according to Corollary 9.34, A generates a group of isometries, say $\{G(t); t \in \mathbb{R}\} \subset L(X)$. Therefore, for all $[u_0, v_0] \in X$ and $f \in L_{\text{loc}}^1([0, \infty); X)$ there exists a unique mild solution $[u, v]$ of (CP) given by the variation of constants formula

$$[u(t, \cdot), v(t, \cdot)] = G(t)[u_0, v_0] + \int_0^t G(t - s)[0, f(s, \cdot)] \, ds, \quad t \geq 0,$$

hence $u \in C([0, \infty); H_0^1(0,1))$. This u can be viewed as a generalized solution of problem (9.12.69)–(9.12.71). If $[u_0, v_0] \in D(A) = [H_0^1(0,1) \cap H^2(0,1)] \times H_0^1(0,1)$ and $f \in C^1([0, \infty); L^2(0,1))$, then $[u, v] \in C^1([0, \infty); X)$ (cf. Theorem 9.47). It follows that $u \in C^2([0, \infty); L^2(0,1)) \cap C^1([0, \infty); H_0^1(0,1)) \cap C([0, \infty); H^2(0,1))$ and u is a classical solution of problem (9.12.69)–(9.12.71).

The above discussion can be extended to the n-dimensional case

$$\begin{cases} u_{tt} - \Delta u = f(t, x), & t \geq 0, \ x \in \Omega, \\ u(t, x) = 0, & t \geq 0, \ x \in \partial\Omega, \\ u(0, x) = u_0(x), & x \in \Omega, \end{cases}$$

where $\Omega \subset \mathbb{R}^n$, $n \geq 2$, is a bounded domain with sufficiently smooth boundary $\partial\Omega$, and Δ is the Laplacian with respect to x. In this case, using the substitution $v = u_t$ again, the above initial-boundary value problem can similarly be expressed as a Cauchy problem in the phase space $X = H_0^1(\Omega) \times L^2(\Omega)$, associated with the operator $A : D(A) \subset X \to X$ defined by

$$D(A) = \left[H_0^1(\Omega) \cap H^2(\Omega)\right] \times H_0^1(\Omega), \quad A[p, q] = [q, \Delta p].$$

One can again use Corollary 9.34 to prove that A generates a C_0-group of isometries on X. In particular, to show that Eq. (9.12.72) has a solution in $D(A)$ we need to use Green's formula (instead of integration by parts) and Lax–Milgram. The rest follows similarly. The case of the homogeneous Neumann or Robin boundary condition can be addressed in a similar manner.

9.12.3 The Transport Equation

Let a be a given vector in $\mathbb{R}^n$, $n \geq 1$. Consider the equation

$$u_t + a \cdot \nabla u = f(t, x), \quad t \geq 0, \ x \in \mathbb{R}^n, \tag{9.12.73}$$

with the initial condition

$$u(0, x) = u_0(x), \quad x \in \mathbb{R}^n, \tag{9.12.74}$$

where ∇u means the gradient of u with respect to x, and $a \cdot b$ denotes the usual scalar product of $a, b \in \mathbb{R}^n$. Equation (9.12.73) is known as the transport equation. The case $a = 0$ is trivial, so in what follows we assume $a \neq 0$ (i.e., $a = (a_1, \ldots, a_n)$ contains nonzero components).

Let us choose $X = L^p(\mathbb{R}^n)$, $p \in (1, \infty)$, equipped with the usual norm. If $f \equiv 0$ and u_0 is a smooth function, then the solution of problem (9.12.73) and (9.12.74) is given by

$$u(t, x) = u_0(x - ta), \quad t \geq 0, \ x \in \mathbb{R}^n.$$

This formula leads us to the definition of the semigroup $\{T(t) : X \to X; t \geq 0\}$,

$$(T(t)v)(x) = v(x - ta), \quad v \in X, \ x \in \mathbb{R}^n, \ t \geq 0.$$

It is easily seen that this is a semigroup of isometries, of class C_0:

$$\lim_{t \to 0^+} \|T(t)v - v\|_X^p = \lim_{t \to 0^+} \int_{\mathbb{R}^n} |v(x - ta) - v(x)|^p \, dx$$
$$= 0, \quad \forall v \in X,$$

by virtue of the Lebesgue Dominated Convergence Theorem.

In order to determine its infinitesimal generator $A : D(A) \subset X \to X$, consider Eq. (9.12.73) with $f \equiv 0$ and deduce $Av = -a \cdot \nabla v$ for all $v \in D(A)$. This follows from the fact that the right derivative of $t \mapsto T(t)v$ at $t = 0$ is equal to Av. Indeed, if $v \in C_0^\infty(\mathbb{R}^n)$ (which is dense in X), then $v \in D(A)$ and

$$\lim_{h \to 0^+} \|h^{-1}[T(h)v - v] + a \cdot \nabla v\|_X^p$$
$$= \lim_{h \to 0^+} \int_{\mathbb{R}^n} |h^{-1}[v(x - ha) - v(x)] + a \cdot \nabla v(x)|^p \, dx$$
$$= 0,$$

by virtue of the Mean Value Theorem (which insures uniform convergence as $h \to 0^+$ under the above integral). Since the range of A must be a subset of X, the maximal domain of A is

$$D(A)$$
$$= \{v \in X; \ \frac{\partial v}{\partial x_i} \in X \ \text{ for all } i \in \{1, \ldots, n\} \ \text{ for which } a_i \neq 0\},$$

where $\frac{\partial v}{\partial x_i}$ denotes the partial derivative of v with respect to x_i in the sense of distributions. Since $C_0^\infty(\mathbb{R}^n)$ is dense in X and $C_0^\infty(\mathbb{R}^n) \subset D(A)$ it follows that $D(A)$ is dense in X. Obviously, A is a closed operator. We can use Theorem 9.29 to prove that A is a generator (the generator of $\{T(t) : X \to X; t \geq 0\}$). Indeed, for all $u \in D(A) \setminus \{0\}$

and $u^* = J(u) = \|u\|_X^{2-p}|u|^{p-2}u$ (here J denotes the duality mapping of X), we have

$$
\begin{aligned}
u^*(Au) &= \|u\|_X^{2-p}\int_{\mathbb{R}^n} Au \cdot |u|^{p-2}u\,dx \\
&= -\|u\|_X^{2-p}\sum_{i=1}^n a_i \int_{\mathbb{R}^n} |u|^{p-2}u\frac{\partial u}{\partial x_i}\,dx \\
&= -\frac{1}{p}\|u\|_X^{2-p}\sum_{i=1}^n a_i \int_{\mathbb{R}^n} \frac{\partial}{\partial x_i}|u|^p\,dx \\
&= 0,
\end{aligned}
$$

so A is dissipative. To derive the last equality, we have used the fact that the function $g(x_i) = \int_{\mathbb{R}^{n-1}} |u|^p\,dx_1 \ldots dx_{i-1}dx_{i+1}\ldots dx_n$ belongs to $W^{1,1}(\mathbb{R})$ so $g(x_i) \longrightarrow 0$ as $|x_i| \longrightarrow \infty$ (prove it, or see [6, Corollary 8.9, p. 214]). Let $X^* = L^q(\mathbb{R})$ be the dual of X (i.e., $\frac{1}{p} + \frac{1}{q} = 1$). The adjoint A^* of A is defined by

$$
D(A^*) = \{w \in X^*; \frac{\partial w}{\partial x_i} \in X^* \ \forall i \in \{1,\ldots,n\} \ \text{ for which } \ a_i \neq 0\},
$$

$$
A^*w = a \cdot \nabla w.
$$

By a computation similar to that performed above for operator A, we infer that operator A^* is also dissipative. Thus, according to Theorem 9.29, A is m-dissipative, hence it is indeed the generator of $\{T(t) : X \to X; t \geq 0\}$. In fact, this semigroup extends to a C_0-group of isometries,

$$
(T(t)v)(x) = v(x - ta) \quad x \in \mathbb{R}^n, \ t \in \mathbb{R},
$$

with generator A (see Sect. 9.4).

Therefore, for all $u_0 \in X = L^p(\mathbb{R}^n)$ and $f \in L^1(0,\infty; X)$, problem (9.12.73) and (9.12.74) has a unique mild solution u,

$$
\begin{aligned}
u(t,x) &= (T(t)u_0)(x) + \int_0^t (T(t-s)f(s,\cdot))(x)\,ds, \\
&= u_0(x - ta) + \int_0^t f(s, x - (t-s)a)\,ds, \quad \forall t \geq 0.
\end{aligned}
$$

If $u_0 \in D(A)$ and $f \in C^1([0,\infty); X)$, then $u \in C^1([0,\infty); X)$ and u is a classical solution, with the additional property $a \cdot \nabla u \in C([0,\infty); X)$.

Remark 9.51. If $n = 1$ and $a = -1$, then the above group $\{T(t); t \in \mathbb{R}\}$ is a group of translations defined on $X = L^p(\mathbb{R})$.

Remark 9.52. Since the above operator A generates a C_0-group of isometries, it follows by Corollary 9.34 that $\mathbb{R} \setminus \{0\} \subset \rho(A)$. Therefore for all $\lambda \in \mathbb{R} \setminus \{0\}$ and $g \in X = L^p(\mathbb{R}^n)$ there exists a unique solution $u \in D(A)$ satisfying the equation

$$\lambda u(x) + a \cdot \nabla u(x) = g(x), \quad x \in \mathbb{R}^n.$$

9.12.4 The Telegraph System

For an electrical long line we have the following PDE system, called the telegraph system (see, e.g., [36, p. 320])

$$\begin{cases} Li_t + v_x + Ri = e(t, x), \\ Cv_t + i_x + Gv = 0, \quad t \geq 0, \ x \in (0, 1), \end{cases}$$

with the boundary conditions (Ohm's law at both ends of the line)

$$v(t, 0) + R_0 i(t, 0) = 0, \quad R_1 i(t, 1) = v(t, 1), \quad t \geq 0,$$

and initial conditions

$$i(0, x) = i_0(x), \quad v(0, x) = v_0(x), \quad x \in (0, 1),$$

where $i = i(t, x)$ is the current flowing in the line and $v = v(t, x)$ represents the voltage across the line; $R \geq 0$, $R_0 > 0$, $R_1 > 0$, $L > 0$, $C > 0$, $G \geq 0$ are constants representing resistances, inductance, capacitance, and conductance, respectively; $e = e(t, x)$ is the voltage per unit length impressed along the line in series with it.

We regard the unknown pair $[i, v]$ as a function of $t \geq 0$ with values in $X = L^2(0, 1) \times L^2(0, 1)$. Consider in X the scalar product

$$\langle [f_1, g_1], [f_2, g_2] \rangle = L \int_0^1 f_1 f_2 \, dx + C \int_0^1 g_1 g_2 \, dx,$$

and the norm induced by this scalar product, so X is a real Hilbert space. Define $A : D(A) \subset X \to X$ by

$$D(A) = \{[f, g] \in X; \ f', g' \in L^2(0, 1),$$
$$g(0) + R_0 f(0) = 0, \ R_1 f(1) = g(1)\},$$
$$A[f, g] = \left[-\frac{1}{L}(g' + Rf), -\frac{1}{C}(f' + Gg) \right].$$

Operator A is densely defined, since $C_0^\infty(0,1) \times C_0^\infty(0,1) \subset D(A)$ and is dense in X. It is also easily seen that A is a closed operator: it suffices to note that the derivative is a closed operator in $L^2(0,1)$ and that convergence in $H^1(0,1)$ implies convergence in $C[0,1]$ (cf. Arzelà–Ascoli). It turns out that A is an m-dissipative operator (thus confirming the fact that A is densely defined and closed, cf. Theorems 9.10 and 9.27). Indeed, for all $[f,g] \in D(A)$ we have

$$
\begin{aligned}
\langle A[f,g],[f,g]\rangle &= \langle[-\frac{1}{L}(g'+Rf),-\frac{1}{C}(f'+Gg)],[f,g]\rangle \\
&= -\int_0^1 f(g'+Rf)\,dx - \int_0^1 g(f'+Gg)\,dx \\
&= -\int_0^1 (fg)'\,dx - R\int_0^1 f^2\,dx - G\int_0^1 g^2\,dx \\
&\leq -\int_0^1 (fg)'\,dx \\
&= f(0)g(0) - f(1)g(1) \\
&= -R_0 f(0)^2 - R_1 f(1)^2 \\
&\leq 0,
\end{aligned}
$$

that is to say, A is dissipative (with respect to the scalar product $\langle\cdot,\cdot\rangle$). Let us now prove that $R(\lambda I - A) = X$ for all $\lambda > 0$, i.e., for all $\lambda > 0$ and $[h,k] \in X$ there exists a solution $[f,g] \in D(A)$ of the equation

$$\lambda[f,g] - A[f,g] = [h,k]. \tag{9.12.75}$$

Equation (9.12.75) can be written as the following boundary value problem

$$
\begin{cases}
f' + (C\lambda + G)g = Ck, \\
g' + (L\lambda + R)f = Lh, \\
g(0) + R_0 f(0) = 0, \quad R_1 f(1) = g(1).
\end{cases}
$$

We compute the general solution $[f,g]$ of the above differential system (see the solution of Exercise 9.13) and then impose upon it the above boundary conditions to deduce that there exists a unique $[f,g] \in D(A)$ satisfying the problem. The details are left to the reader. Thus, A is m-dissipative, so it generates a C_0-contraction semigroup on X, say $\{T(t) : X \to X; t \geq 0\}$ (cf. Theorem 9.27). Therefore, for all $[i_0, v_0] \in X$ and $e \in L^1_{\text{loc}}([0,\infty); L^2(0,1))$ there exists a unique mild solution $[i, v] \in C([0, \infty); X)$ of the Cauchy problem

$$\begin{cases} \frac{d}{dt}[i(t, \cdot), v(t, \cdot)] = A[i(t, \cdot), v(t, \cdot)] + [e(t, \cdot), 0], \quad t \geq 0, \\ [i(0, \cdot), v(0, \cdot)] = [i_0, v_0], \end{cases}$$

which is the representation in X of our initial-boundary value problem formulated above. This mild solution can be written explicitly in terms of $T(t)$, i_0, v_0, and e, by using the usual variation of constants formula. If $[i_0, v_0] \in D(A)$ and $e \in C^1([0, \infty); L^2(0, 1))$, then $[i, v]$ is a classical solution, $[i, v] \in C^1([0, \infty); X) \cap C([0, \infty); H^1(0, 1)^2)$. It is worth pointing out that the condition $[i_0, v_0] \in D(A)$ implies compatibility of the initial data with the boundary conditions and, as a by-product of this compatibility plus smoothness of function e, we obtain a classical solution $[i, v]$ with the above properties. In particular i, v are continuous on $[0, \infty) \times [0, 1]$ and satisfy the boundary conditions for all $t \geq 0$.

Remark 9.53. All the above applications can be extended to the semilinear case, as pointed out in Remark 9.50.

Comment. This chapter represents a short introduction into the theory of semigroups of linear operators, including its implications to linear evolution equations and some applications. Some subjects in the field have not been addressed, e.g., semigroups of compact operators, differentiable semigroups, analytic semigroups, dual semigroups, etc. For more information about linear operator semigroups and their applications, the reader is referred to [7], [12], [19], [21], [39], [49], [51]. For more details on the regularity of solutions to linear evolution equations, including significant examples from the theory of linear partial differential equations, see [6], [19], [39], [49].

9.13 Exercises

1. Compute $T(t) = e^{tA}$, $t \in \mathbb{R}$, where

   $(i)\ A = \begin{bmatrix} 1 & 1 \\ -1 & -1 \end{bmatrix};\quad (ii)\ A = \begin{bmatrix} 0 & 1 \\ -1 & 0 \end{bmatrix};\quad (iii)\ A = \begin{bmatrix} -1 & -1 \\ 2 & -4 \end{bmatrix}.$

2. If A is an $n \times n$ complex matrix, then the following equivalences hold true:

(a) $\sup_{t\geq 0} \|e^{tA}\| < \infty \iff$ all eigenvalues λ of A satisfy $\operatorname{Re}\lambda \leq 0$ and whenever $\operatorname{Re}\lambda = 0$, then λ is a simple eigenvalue;

(b) $\lim_{t\to\infty} \|e^{tA}\| = 0 \iff$ all eigenvalues λ of A satisfy $\operatorname{Re}\lambda < 0$.

3. Let $(X, \|\cdot\|)$ be a Banach space and let $A \in L(X)$. Consider in X the Cauchy problem

$$\begin{cases} u'(t) = Au(t), & t \in \mathbb{R}, \\ u(0) = u_0. \end{cases}$$

Show that if $u_0 \neq 0$ then $u(t) \neq 0$ for all $t \in \mathbb{R}$.

4. Let $(X, \|\cdot\|)$ be a Banach space. Show that for every C_0-semigroup $\{T(t) : X \to X; t \geq 0\}$ the X-valued function $(t, x) \mapsto T(t)x$ is continuous on $[0, \infty) \times X$.

5. Let X denote the space of all functions $f : \mathbb{R} \to \mathbb{R}$ which are continuous and bounded, equipped with the sup-norm. For some $\lambda > 0$ and $\delta > 0$ define $G(t) : X \to X$ by

$$(G(t)f)(x) = e^{-\lambda t} \sum_{k=0}^{\infty} \frac{(\lambda t)^k}{k!} f(x - k\delta), \quad t \geq 0, \ f \in X, \ x \in \mathbb{R}.$$

(a) Prove that $\{G(t) : X \to X; t \geq 0\}$ is a uniformly continuous group and determine its infinitesimal generator;

(b) Show that

$$\|G(t)\| = \begin{cases} 1 & \text{if } t \geq 0, \\ e^{-2\lambda t} & \text{if } t < 0. \end{cases}$$

6. Let X be the real Banach space of all functions $f : \mathbb{R} \to \mathbb{R}$ that are continuous on $\mathbb{R}$ and p-periodic with some period $p > 0$, equipped with the sup-norm

$$\|f\| = \sup_{0\leq s\leq p} |f(s)| \quad \forall f \in X.$$

Define

$$(T(t)f)(s) = f(t+s), \quad t, s \in \mathbb{R}, \ f \in X.$$

Show that $\{T(t) : X \to X; t \in \mathbb{R}\}$ is a C_0-group of isometries, i.e., $\|T(t)\| = 1$, $t \in \mathbb{R}$. Find its infinitesimal generator.

7. Let $M = (m_{ij})$ be a $k \times k$ matrix with real entries. Denote $X = L^p(\mathbb{R}^k)$, where $p \in [1, \infty)$. For $t \in \mathbb{R}$ define $G(t) : X \to X$ by

$$(G(t)f)(x) = f(e^{-tM}x), \quad f \in X, \text{ a.a. } x \in \mathbb{R}^k.$$

(a) Show that $\{G(t) : X \to X; t \geq 0\}$ is a C_0-group and determine its infinitesimal generator;

(b) If $\sum_{i=1}^{k} m_{ii} = 0$, then $\|G(t)\| = 1$ for all $t \in \mathbb{R}$.

8. Let X be the real Banach space of all functions $f : [0, \infty) \to \mathbb{R}$ that are bounded and uniformly continuous on $[0, \infty)$, equipped with the usual sup-norm. Define

$$(T(t)f)(s) = \begin{cases} f(s-t) & \text{for } s-t \geq 0, \\ f(0) & \text{for } s-t < 0. \end{cases}$$

Show that $\{T(t) : X \to X; t \geq 0\}$ is a C_0-semigroup and determine its infinitesimal generator.

9. For a given $1 \leq p < \infty$, consider the real Banach space $X = l^p$ of all sequences $(x_n)_{n \in \mathbb{N}}$ in $\mathbb{R}$ satisfying $\sum_{n=1}^{\infty} |x_n|^p < \infty$, equipped with the usual norm

$$\|(x_n)\|_p = \left(\sum_{n=1}^{\infty} |x_n|^p \right)^{1/p} \quad \forall (x_n)_{n \in \mathbb{N}} \in X.$$

Let $(c_n)_{n \in \mathbb{N}}$ be a sequence of positive numbers. Define $T(t) : X \to X$ by

$$T(t)(x_n)_{n \in \mathbb{N}} = (e^{-c_n t} x_n)_{n \in \mathbb{N}} \quad \forall (x_n)_{n \in \mathbb{N}} \in X, \ t \geq 0.$$

(a) Show that $\{T(t) : X \to X; t \geq 0\}$ is a C_0-semigroup of contractions;

(b) Determine its infinitesimal generator;

(c) Prove that $\{T(t) : X \to X; t \geq 0\}$ is uniformly continuous if and only if (c_n) is bounded.

10. Let $H = L^2(0, 1)$ be equipped with the usual scalar product and the corresponding induced norm. Define $A : D(A) \subset H \to H$ by

$$D(A) = \{v \in H^1(0, 1); v(0) = 0\}, \quad Av = -v' \quad \forall v \in D(A).$$

Show that A generates a C_0-semigroup of contractions $\{T(t) : H \to H; t \geq 0\}$. Find the explicit form of this semigroup and show that, for $u_0 \in H$, $u(t, x) = (T(t)u_0)(x)$ satisfies the transport equation $u_t + u_x = 0$ in $\Omega = (0, \infty) \times (0, 1)$ in the sense of distributions.

11. Consider the initial-boundary value problem

$$\begin{cases} u_t - u_{xx} + au = f(t, x), & t > 0, \ x \in (0, 1), \\ u(t, 0) = 0, \ u_x(t, 1) + \alpha u(t, 1) = 0, & t > 0, \\ u(0, x) = u_0(x), & x \in (0, 1), \end{cases}$$

where $a \in \mathbb{R}$, $\alpha > 0$, $u_0 \in L^2(0, 1)$, $f \in L^1_{loc}[0, \infty)$. Solve this problem using the semigroup approach. Solve the more general problem obtained by replacing the term au in the above equation by $h(u)$, where $h : \mathbb{R} \to \mathbb{R}$ is a Lipschitz function.

12. Consider the initial-boundary value problem

$$\begin{cases} u_{tt} - u_{xx} = f(t, x), & t > 0, \ x \in (0, 1), \\ u(t, 0) = 0, \ u_x(t, 1) = 0, & t > 0, \\ u(0, x) = u_0(x), \end{cases}$$

where $u_0 \in H^1(0, 1)$, $u_0(0) = 0$, $f \in L^1_{loc}[0, \infty)$. Solve this problem using the semigroup approach.

13. Consider the telegraph differential system

$$\begin{cases} Li_t + v_x + Ri = e(t, x), \\ Cv_t + i_x + Gv = 0, & t \geq 0, \ x \in (0, 1), \end{cases}$$

with the following boundary conditions

$$v(t.0) + R_0 i(t, 0) = 0, \ -i(t, 1) + C_1 v_t(t, 1) + D_1 v(t, 1)$$
$$= e_1(t), \ t > 0,$$

and initial conditions

$$i(0, x) = i_0(x), \ v(0, x) = v_0(x), \ x \in (0, 1),$$

where $C > 0$, $C_1 > 0$, $L > 0$, $D_1 \geq 0$, $G \geq 0$, $R \geq 0$, $R_0 \geq 0$, and e, e_1 are given functions.

(a) Solve the above problem by using the semigroup approach;
(b) What can you say about existence in the case when D_1, G, R are Lipschitz functions from $\mathbb{R}$ into itself?

Chapter 10

Solving Linear Evolution Equations by the Fourier Method

In Chap. 9 we used the linear semigroup approach to solve inhomogeneous linear evolution equations. For the same purpose, we use here the Fourier method. More precisely, under appropriate conditions on the linear operators governing such equations, we find the solutions in the form of Fourier series expansions. This approach is based in an essential way on the results discussed in Chap. 8.

10.1 First Order Linear Evolution Equations

Consider the Cauchy problem

$$u'(t) + Qu(t) = f(t), \ \ 0 < t < T, \qquad (E)$$

$$u(0) = u_0, \qquad (IC)$$

where Q satisfies the set of conditions **(a)** originally presented in Chap. 8:

© Springer Nature Switzerland AG 2019

G. Moroşanu, *Functional Analysis for the Applied Sciences*,
Universitext, https://doi.org/10.1007/978-3-030-27153-4_10

(a) $Q : D(Q) \subset H \to H$ is a linear, densely defined, self-adjoint, strongly positive operator, where $(H, (\cdot, \cdot), \| \cdot \|)$ is a real, infinite dimensional, separable Hilbert space.

We also assume that the energetic space H_E defined in Chap. 8 satisfies

(b) H_E is compactly embedded into H,

so that Theorem 8.16 holds true. The notation in the statement of that theorem will be also used in what follows.

If Q satisfies (a) then $-Q$ generates a C_0-semigroup of contractions (see Theorem 9.29) and so for any $u_0 \in H$ and $f \in L^1(0, T; H)$ there exists a unique mild solution $u = u(t)$ of problem (E), (IC) given by the variation of constants formula. If $u_0 \in D(Q)$ and $f \in C^1([0, T]; H)$, then u is a classical solution (cf. Theorem 9.47). The Fourier method we are going to discuss next offers more possibilities to investigate the regularity of solutions and provides good approximations of solutions in terms of eigenfunctions of the operator Q. Let us start with a specific result.

Theorem 10.1. *Assume that* (a) *and* (b) *above are fulfilled. Then for all $u_0 \in H$ and $f \in L^2(0, T; H)$ there exists a unique function $u \in C([0, T]; H) \cap C((0, T]; H_E) \cap L^2(0, T; H_E)$ with $\sqrt{t} u' \in L^2(0, T; H)$ which satisfies (IC) and Eq. (E) for a.a. $t \in (0, T)$. This function u (called a strong solution of problem (E), (IC)) is expressed as the Fourier series expansion*

$$u(t) = \sum_{n=1}^{\infty} u_n(t) e_n, \qquad (10.1.1)$$

where $\{e_n\}_{n=1}^{\infty}$ is the orthonormal basis in H provided by Theorem 8.16, and $u_n(t) = (u(t), e_n)$, $n = 1, 2, \ldots$ If $u_0 \in H_E$ and $f \in L^2(0, T; H)$, then $u \in H^1(0, T; H) \cap C([0, T]; H_E)$, $u(t) \in D(Q)$ for a.a. $t \in (0, T)$, and $Qu \in L^2(0, T; H)$.

Proof. Assume first that $u_0 \in H$ and $f \in L^2(0, T; H)$. As mentioned before, we already know that problem (E), (IC) has a unique mild solution u given by the variation of constants formula. A strong solution is clearly a mild one so the uniqueness part of the theorem is obvious.

In fact, uniqueness also follows by a simple direct proof. If $y = y(t)$ denotes the difference of two strong solutions of problem (E), (IC), then $y(0) = 0$ and

$$y'(t) + Qy(t) = 0 \quad \text{for a.a. } t \in (0, T).$$

Multiplying this equation by $y(t)$ and taking into account the positivity of Q we obtain

$$\frac{1}{2}\frac{d}{dt}\|y(t)\|^2 = (y'(t), y(t)) \leq 0 \quad \text{for a.a. } t \in (0, T),$$

which shows that the function $t \mapsto \|y(t)\|$ is nonincreasing on $[0, T]$. Since $y(0) = 0$ it follows that y is the null function, i.e., the two strong solutions coincide.

We could show that, under our assumptions, the mild solution u is in fact a strong solution by a limiting procedure applied to a sequence of strong solutions $u_n \in C^1([0, T]; H)$ (given by Theorem 9.47) corresponding to sequences $u_{0n} \in D(Q)$ and $f_n \in C^1([0, T]; H)$ which satisfy $\|u_{0n} - u_0\| \to 0$, $\|f_n - f\|_{L^2(0,T; H)} \to 0$. However, we shall provide here the existence proof using the Fourier method. Specifically, we seek a solution in the form (10.1.1) where the u_n's are unknown real valued functions. For u_0 we have the Fourier expansion

$$u_0 = \sum_{n=1}^{\infty} u_{0n}e_n \quad \text{with } u_{0n} = (u_0, e_n), \quad \|u_0\|^2 = \sum_{n=1}^{\infty} u_{0n}^2.$$

Similarly, for a.a. $t \in (0, T)$,

$$f(t) = \sum_{n=1}^{\infty} f_n(t)e_n \quad \text{with } f_n(t) = (f(t), e_n), \quad \|f(t)\|^2 = \sum_{n=1}^{\infty} f_n(t)^2.$$

Denoting $s_k(t) = \sum_{n=1}^{k} f_n(t)e_n$, we can see that

$$\|s_k(t)\|^2 = \sum_{n=1}^{k} f_n(t)^2 \leq \|f(t)\|^2 \quad \forall k \in \mathbb{N}, \quad \text{a.a. } t \in (0, T),$$

so by the Lebesgue Dominated Convergence Theorem $s_k \to f$ strongly in $L^2(0, T; H)$. Now we impose conditions on u (given by (10.1.1)) to formally satisfy Eq. (E),

$$\sum_{n=1}^{\infty} u_n'(t)e_n + \sum_{n=1}^{\infty} u_n(t)\lambda_n e_n = \sum_{n=1}^{\infty} f_n(t)e_n,$$

and (IC),

$$\sum_{n=1}^{\infty} u_n(0)e_n = \sum_{n=1}^{\infty} u_{0n}e_n,$$

so identifying the coefficients of the e_n's yields

$$u_n'(t) + \lambda_n u_n(t) = f_n(t) \quad \text{for all } n \in \mathbb{N} \text{ and a.a. } t \in (0,T), \quad (10.1.2)$$

$$u_n(0) = u_{0n}, \quad n \in \mathbb{N}, \quad (10.1.3)$$

hence

$$u_n(t) = e^{-\lambda_n t} u_{0n} + \int_0^t e^{-\lambda_n(t-s)} f_n(s)\, ds \quad \forall t \in [0,T], \ n \in \mathbb{N}.$$

Therefore, $u_n \in H^1(0,T)$ and, since $\lambda_n \geq \lambda_1 > 0$, we easily obtain by Hölder's inequality

$$u_n(t)^2 \leq 2\left(u_{0n}^2 + T \int_0^T f_n(s)^2 ds\right) \quad \forall t \in [0,T], \ n \in \mathbb{N}. \quad (10.1.4)$$

Since u_{0n}^2 and $\int_0^T f_n(s)^2 ds$ are terms of convergent series, it follows from (10.1.4), by the Weierstrass M-test, that the series $\sum_{n=1}^{\infty} u_n(t)^2$ is uniformly convergent in $[0,T]$ and consequently so is the series (10.1.1) and its sum u is in $C([0,T]; H)$.

Next, we multiply Eq. (10.1.2) by $tu_n'(t)$ and then integrate the resulting equation over $[0,T]$ to obtain $\forall n \in \mathbb{N}$

$$\int_0^T t u_n'(t)^2 dt + \frac{\lambda_n}{2} T u_n(T)^2$$

$$= \frac{\lambda_n}{2} \int_0^T u_n(t)^2 dt + \int_0^T t f_n(t) u_n'(t)\, dt$$

$$\leq \frac{\lambda_n}{2} \int_0^T u_n(t)^2 dt + \frac{1}{2} \int_0^T t u_n'(t)^2 dt + \frac{1}{2} \int_0^T t f_n(t)^2 dt. \quad (10.1.5)$$

On the other hand, multiplying (10.1.2) by $u_n(t)$ and then integrating over $[0,T]$ we obtain

$$\frac{1}{2}\left(u_n^2(T) - u_{0n}^2\right) + \lambda_n \int_0^T u_n(t)^2 dt = \int_0^T f_n(t) u_n(t)\, dt$$

$$\leq \frac{1}{2} \int_0^T f_n(t)^2 dt$$

$$+ \frac{1}{2} \int_0^T u_n(t)^2 dt,$$

for all $n \in \mathbb{N}$, so

$$\sum_{n=1}^{\infty} \lambda_n \int_0^T u_n(t)^2 dt < \infty, \tag{10.1.6}$$

hence

$$\sum_{n=1}^{\infty} \left(u(t), \lambda_n^{-1/2} e_n\right)_E^2 = \sum_{n=1}^{\infty} \left(u(t), \lambda_n^{-1/2} Q e_n\right)^2 = \sum_{n=1}^{\infty} \lambda_n u_n(t)^2$$

is convergent for a.a. $t \in (0, T)$, and $t \mapsto \|u(t)\|_E^2 = \sum_{n=1}^{\infty} \lambda_n u_n(t)^2$ is summable on $(0, T)$, i.e., $u \in L^2(0, T; H_E)$.

From (10.1.5) and (10.1.6) we infer that

$$\sum_{n=1}^{\infty} \int_0^T t u_n'(t)^2 dt < \infty,$$

so $\sqrt{t} u' \in L^2(0, T; H)$. We also have the inequality (similar to (10.1.5))

$$\frac{\lambda_n}{2} t u_n(t)^2 \le -\frac{1}{2} \int_0^t s u_n(s)^2 ds + \frac{\lambda_n}{2} \int_0^T u_n(s)^2 ds + \frac{1}{2} \int_0^T s f_n(s)^2 ds,$$

for all $t \in [0, T]$, $n \in \mathbb{N}$, which combined with (10.1.6) implies (by the Weierstrass M-test) that $\sum_{n=1}^{\infty} \lambda_n t u_n(t)^2$ is uniformly convergent in $[0, T]$ so $\sqrt{t} u \in C([0, T]; H_E)$. This shows that $u \in C((0, T]; H_E)$. Now, passing to the limit in $L^2(0, T; H)$ as $k \to \infty$ in the equation

$$\sum_{n=1}^{k} f_n(t) e_n = \sum_{n=1}^{k} u_n'(t) e_n + \sum_{n=1}^{k} u_n(t) \lambda_n e_n,$$

$$= \sum_{n=1}^{k} u_n'(t) e_n + Q\left(\sum_{n=1}^{k} u_n(t) e_n\right)$$

we conclude that u satisfies Eq. (E) for a.a. $t \in (0, T)$. This uses the fact that Q is a closed operator. It is also obvious that $u(0) = u_0$.

Now, let us assume that $u_0 \in H_E$ and $f \in L^2(0, T; H)$. Multiplying Eq. (10.1.2) by $u_n'(t)$ we obtain

$$u_n'(t)^2 + \frac{\lambda_n}{2} \frac{d}{dt}\left(u_n(t)^2\right)$$
$$= f_n(t) \cdot u_n'(t) \quad \text{for a.a. } t \in (0, T), \ \forall n \in \mathbb{N}. \tag{10.1.7}$$

It follows, by integration over $[0, T]$, that

$$\int_0^T u_n'(t)^2 dt + \frac{\lambda_n}{2} \left(u_n(T)^2 - u_{0n}^2\right) = \int_0^T f_n(t) \cdot u_n'(t)\, dt$$

$$\leq \frac{1}{2} \int_0^T f_n(t)^2 dt + \frac{1}{2} \int_0^T u_n'(t)^2 dt, \tag{10.1.8}$$

for all $n \in \mathbb{N}$. Since $u_0 \in H_E$ (i.e., $\sum_{n=1}^\infty \lambda_n u_{0n}^2 < \infty$), the last inequality implies

$$\sum_{n=1}^\infty \int_0^T u_n'(t)^2 dt < \infty,$$

hence $\sum_{n=1}^\infty u_n'(t) e_n$ is convergent in $L^2(0, T; H)$ and, obviously, its sum is $u' \in L^2(0, T; H)$.

Integration over $[0, t]$ of (10.1.7) leads to an inequality similar to (10.1.8) which implies that $\sum_{n=1}^\infty \lambda_n u_n(t)^2$ is uniformly convergent in $[0, T]$ and so $u \in C([0, T]; H_E)$. As u', $f \in L^2(0, T; H)$ we derive from Eq. (E) that $Qu \in L^2(0, T; H)$. $\qquad\square$

Remark 10.2. For further regularity results see, e.g., [22, Chapter 7].

We continue with a result on the existence of a periodic solution of Eq. (E).

Theorem 10.3. *Assume that* **(a)** *and* **(b)** *are fulfilled and* $f \in L^2$ $(0, T; H)$. *Then, there exists a unique function* $u \in H^1(0, T; H) \cap C([0, T]; H_E)$ *satisfying Eq.* (E) *for a.a.* $t \in (0, T)$ *and* $u(0) = u(T)$, *and* u *is given by Eq.* (10.1.1), *where*

$$u_n(t) = d_n e^{-\lambda_n t} + \int_0^t e^{-\lambda_n(t-s)} f_n(s)\, ds,$$

with

$$d_n = \left(1 - e^{-\lambda_n T}\right)^{-1} \int_0^T e^{-\lambda_n(T-s)} f_n(s)\, ds, \quad n = 1, 2, \dots$$

Proof. By Theorem 10.1, for all $u_0 \in H$ there is a unique strong solution $u = u(t, u_0)$ of problem (E), (IC) which belongs to $C([0, T]; H) \cap C((0, T]; H_E) \cap L^2(0, T; H_E)$ with $\sqrt{t}\, u'(t, u_0) \in L^2(0, T; H)$. For two vectors $u_0, v_0 \in H$ we have

$$\frac{d}{dt}[u(t, u_0) - u(t, v_0)] + Q[u(t, u_0) - u(t, v_0)] = 0 \quad \text{for a.a. } t \in (0, T).$$

If we multiply this equation by $u(t, u_0) - u(t, v_0)$ and use the strong positivity of Q (with some constant $c > 0$), we get

$$\frac{1}{2}\frac{d}{dt}\|u(t, u_0) - u(t, v_0)\|^2$$
$$+ c\|u(t, u_0) - u(t, v_0)\|^2 \leq 0 \quad \text{for a.a. } t \in (0, T),$$

or, equivalently,

$$\frac{d}{dt}\left(e^{2ct}\|u(t, u_0) - u(t, v_0)\|^2\right) \leq 0 \quad \text{for a.a. } t \in (0, T)$$

which shows that the function $t \mapsto e^{ct}\|u(t, u_0) - u(t, v_0)\|$ is nonincreasing and hence

$$\|u(t, u_0) - u(t, v_0)\| \leq e^{-ct}\|u_0 - v_0\| \quad \forall t \in [0, T]. \tag{10.1.9}$$

Now let us consider the so-called Poincaré operator $P : H \to H$ defined by

$$Pu_0 = u(T; u_0) \quad \forall u_0 \in H.$$

From (10.1.9) we see that P is a contraction:

$$\|Pu_0 - Pv_0\| \leq e^{-cT}\|u_0 - v_0\| \quad \forall u_0, v_0 \in H.$$

By the Banach Contraction Principle (see Chap. 2) it follows that P has a unique fixed point $u_0^* \in H$, i.e., $Pu_0^* = u_0^*$. In other words, $u(T, u_0^*) = u_0^*$, which is to say, $u(t, u_0^*)$ is the unique periodic solution of Eq. (E). Since $u_0^* = u(T, u_0^*)$ we deduce from the first part of Theorem 10.1 that $u_0^* \in H_E$. Therefore, by the second part of Theorem 10.1, it follows that $u(t, u_0^*) \in H^1(0, T; H) \cap C([0, T]; H_E)$. Clearly $u(t, u_0^*)$ is the sum of a Fourier series of the form (10.1.1) which is convergent in $C([0, T]; H_E)$ since $u_0^* \in H_E$. From the periodicity condition $u_0^* = u(T, u_0^*)$ we infer

$$u_n(0) = u_n(T) \quad \forall n \in \mathbb{N}, \tag{10.1.10}$$

where the u_n's are solutions of (10.1.2), i.e.,

$$u_n(t) = d_n e^{-\lambda_n t} + \int_0^t e^{-\lambda_n(t-s)} f_n(s)\, ds \quad \forall t \in [0, T], \ n \in \mathbb{N},$$

Taking into account (10.1.10) we can easily find

$$d_n = \left(1 - e^{-\lambda_n T}\right)^{-1} \int_0^T e^{-\lambda_n(T-s)} f_n(s)\, ds, \quad n \in \mathbb{N}. \qquad \square$$

10.2　Second Order Linear Evolution Equations

In this section we keep the notation and assumptions used in the previous section. Consider the Cauchy problem

$$u''(t) + Qu(t) = f(t), \quad 0 < t < T, \tag{e}$$

$$u(0) = u_0, \; u'(0) = u_1. \tag{ic}$$

Theorem 10.4. *Assume that conditions* **(a)** *and* **(b)** *are fulfilled. Then for all $u_0 \in D(Q)$ (i.e., $Qu_0 \in H$), $u_1 \in H_E$ and $f \in L^2(0,T; H_E)$ there exists a unique function $u \in C^1([0,T]; H_E) \cap H^2(0,T; H)$ which satisfies (ic) and (e) for a.a. $t \in (0,T)$, and $Qu \in C([0,T]; H)$. If, in addition, $f \in C([0,T]; H)$ then $u'' \in C([0,T]; H)$. Alternatively, if $u_0 \in D(Q)$, $u_1 \in H_E$ and $f \in H^1(0,T; H)$ then $u \in C^1([0,T]; H_E) \cap C^2([0,T]; H)$ (hence $Qu \in C([0,T]; H)$). In both cases the solution u is given by a Fourier series expansion of the form (10.1.1).*

Proof. Let us first prove uniqueness. Let $y \in H^1(0,T; H)$ be the difference of two solutions of problem (e), (ic). Then, $y(0) = 0$, $y'(0) = 0$, and

$$y''(t) + Qy(t) = 0 \quad \text{for a.a } t \in (0,T).$$

We multiply this equation by $y'(t)$ to obtain

$$(y''(t), y'(t)) + (Qy(t), y'(t)) = 0,$$

so, as Q is self-adjoint, we can write

$$\frac{d}{dt}[\|y'(t)\|^2 + (Qy(t), y(t))] = 0,$$

for a.a. $t \in (0,T)$. This shows that y is the null function (since $y(0) = 0$, $y'(0) = 0$ and Q is strongly positive), so the solution is indeed unique (if it exists).

In order to prove existence, we seek a solution u to problem (e), (ic) in the form (10.1.1). Requiring this series to formally satisfy (e) and (ic) we find

$$u_n''(t) + \lambda_n u_n(t) = f_n(t) \quad \forall n \in \mathbb{N} \text{ and a.a. } t \in (0,T), \tag{10.2.11}$$

$$u_n(0) = u_{0n}, \; u_n'(0) = u_{1n} \quad \forall n \in \mathbb{N}, \tag{10.2.12}$$

where $f_n(t)$, u_{0n} and u_{1n} are the Fourier coefficients of $f(t)$, u_0 and u_1, respectively. For each $n \in \mathbb{N}$ problem (10.2.11) and (10.2.12) has the solution

$$u_n(t) = u_{0n}\cos(\sqrt{\lambda_n}t) + \frac{u_{1n}}{\sqrt{\lambda_n}}\sin(\sqrt{\lambda_n}t)$$

$$+ \frac{1}{\sqrt{\lambda_n}}\int_0^t \sin\left(\sqrt{\lambda_n}(t-s)\right)f_n(s)\,ds\,, \qquad (10.2.13)$$

for all $t \in [0, T]$. Therefore,

$$u_n'(t) = -\sqrt{\lambda_n}u_{0n}\sin(\sqrt{\lambda_n}t) + u_{1n}\cos(\sqrt{\lambda_n}t)$$

$$+ \underbrace{\int_0^t \cos\left(\sqrt{\lambda_n}(t-s)\right)f_n(s)\,ds}_{=\int_0^t \cos\left(\sqrt{\lambda_n}s\right)f_n(t-s)\,ds}\,, \qquad (10.2.14)$$

and

$$u_n''(t) = -\lambda_n u_{0n}\cos(\sqrt{\lambda_n}t) - \sqrt{\lambda_n}u_{1n}\sin(\sqrt{\lambda_n}t)$$

$$+ f_n(t) - \sqrt{\lambda_n}\int_0^t \sin\left(\sqrt{\lambda_n}(t-s)\right)f_n(s)\,ds\,, \qquad (10.2.15)$$

or, equivalently,

$$u_n''(t) = -\lambda_n u_{0n}\cos(\sqrt{\lambda_n}t) - \sqrt{\lambda_n}u_{1n}\sin(\sqrt{\lambda_n}t)$$

$$+ \underbrace{\int_0^t \cos\left(\sqrt{\lambda_n}(s)\right)f_n'(t-s)\,ds}_{=\int_0^t \cos\left(\sqrt{\lambda_n}(t-s)\right)f_n'(s)\,ds}\,. \qquad (10.2.16)$$

From (10.2.13)–(10.2.16) we deduce (where C_1, C_2, C_3, C_4 are constants)

$$u_n(t)^2 \le C_1\left(u_{0n}^2 + \frac{1}{\lambda_n}u_{1n}^2 + \frac{1}{\lambda_n}\int_0^T f_n(s)^2 ds\right), \qquad (10.2.17)$$

$$u_n'(t)^2 \le C_2\left(\lambda_n u_{0n}^2 + u_{1n}^2 + \int_0^T f_n(s)^2 ds\right), \qquad (10.2.18)$$

$$u_n''(t)^2 \le C_3\left(\lambda_n^2 u_{0n}^2 + \lambda_n u_{1n}^2 + f_n(t)^2 + \lambda_n\int_0^T f_n(s)^2 ds\right), \qquad (10.2.19)$$

and

$$u_n''(t)^2 \leq C_4(\lambda_n^2 u_{0n}^2 + \lambda_n u_{1n}^2 + \int_0^T f_n'(s)^2 ds). \qquad (10.2.20)$$

Assume $u_0 \in D(Q)$, $u_1 \in H_E$ and $f \in L^2(0,T; H_E)$. Then

$$\sum_{n=1}^{\infty} \lambda_n^2 \|u_{0n}\|^2 < \infty, \quad \sum_{n=1}^{\infty} \lambda_n \|u_{1n}\|^2 < \infty,$$

$$\sum_{n=1}^{\infty} \lambda_n \int_0^T \|f(t)\|^2 dt < \infty. \qquad (10.2.21)$$

It follows from (10.2.17)–(10.2.19) and (10.2.21) that the series (10.1.1) is convergent in different spaces and its sum u satisfies

$$u \in C^1([0,T]; H_E) \cap H^2(0,T; H), \quad Qu \in C([0,T]; H).$$

If, in addition, $f \in C([0,T]; H)$ then, according to (10.2.20), $u'' \in C([0,T]; H)$.

If $u_0 \in D(Q)$, $u_1 \in H_E$ and $f \in H^1(0,T; H)$ then, according to (10.2.17), (10.2.18), (10.2.20), and (10.2.21), $u \in C^1([0,T]; H_E) \cap C^2([0,T]; H)$ (hence $Qu \in C([0,T]; H)$).

Finally, it is easily seen (as in the proof of Theorem 10.1) that in both cases u, expressed as the sum of the series (10.1.1), satisfies (e), (ic). $\qquad \square$

Remark 10.5. Obviously, further regularity results can be stated under different conditions on u_0, u_1 and f.

On the other hand, using the semigroup approach, one can derive the existence of a solution to problem (e), (ic) which comes from the mild solution for the Cauchy problem associated with a first order differential equation in the product space $X = V \times H$ equipped with the scalar product

$$([v_1, h_1], [v_2, h_2])_X = (v_1, v_2)_E + (h_1, h_2) \quad \forall [v_1, h_1], [v_2, h_2] \in X.$$

Obviously, X is a real Hilbert space. Define $A : D(A) \subset X \to X$ by

$$D(A) = D(Q) \times H_E, \quad A[v, h] = [h, -Qv] \quad \forall [v, h] \in D(A).$$

It is easily seen that A is linear, densely defined, closed, and dissipative. In fact, for all $[v, h] \in D(A)$, we have

$$
\begin{aligned}
(A[v, h], [v, h])_X &= ([h, -Qv], [v, h])_X \\
&= (h, Qv) - (Qv, h) \\
&= 0.
\end{aligned}
$$

Thus, according to Remark 9.26, A is a dissipative operator. We also have $A^* = -A$, so A^* is also dissipative. By Theorem 9.29 it follows that A is m-dissipative, so (according to the Lumer–Phillips Theorem) it generates a C_0-semigroup of contractions, say $\{S(t) : X \to X; t \geq 0\}$.

Problem (e), (ic) can be expressed as the following Cauchy problem in X

$$
\frac{d}{dt}[u(t), w(t)]
$$
$$
= A[u(t), w(t)] + [0, f(t)], \quad 0 < t < T; \quad [u, w](0) = [u_0, u_1].
$$
$$(10.2.22)$$

According to Sect. 9.11, for $[u_0, u_1] \in X$ and $f \in L^1(0, T; H)$ this problem has a unique mild solution $[u, w] \in C([0, T]; X)$,

$$
[u(t), w(t)] = S(t)[u_0, u_1] + \int_0^t S(t - s)[0, f(s)]\, ds, \quad t \in [0, T].
$$
$$(10.2.23)$$

The first component $u = u(t)$ can be called a mild solution of problem (e), (ic). In fact, $w(t) = u'(t)$. In order to show this, we approximate $[u_0, u_1] \in X$ by $[u_0^k, u_1^k] \in D(Q) \times H_E$, and $f \in L^1(0, T; H)$ by $f^k \in H^1(0, T; H)$. Denote by $[u^k, w^k] (= [u^k, (u^k)']$ the solution of problem (10.2.22) with $[u_0, u_1] := [u_0^k, u_1^k]$ and $f := f^k$ which is a strong solution belonging to $C^1([0, T]; H_E) \cap C^2([0, T]; H)$ (cf. Theorem 10.4). Obviously,

$$
[u^k(t), (u^k)'(t)]
$$
$$
= S(t)[u_0^k, u_1^k] + \int_0^t S(t - s)[0, f^k(s)]\, ds, \quad t \in [0, T]. \quad (10.2.24)
$$

As $\{S(t) : X \to X; t \geq 0\}$ is a semigroup of contractions, we have for all $t \in [0, T]$

$$
\|[u^k(t) - u^m(t), (u^k)'(t) - (u^m)'(t)]\|_X \leq \|[u_0^k - u_0^k, u_1^k - u_1^m]\|_X
$$
$$
+ \int_0^T \|f^k(s) - f^m(s)\|\, ds,
$$

hence u^k converges in $C([0,T]; H_E)$ to some $u \in C([0,T] H_E)$, and $(u^k)'$ converges in $C([0,T]; H)$ to $w = u' \in C([0,T]; H)$. Passing to the limit in (10.2.24) we reobtain (10.2.23) with $w = u'$. So the mild solution u belongs to $C([0,T]; H_E) \cap C^1([0,T]; H)$. Since u is a limit of strong solutions u^k that admit Fourier series expansions (as stated in Theorem 10.4), we can easily show that u is the sum of the Fourier series (10.1.1), where $u_n(t) = (u(t), e_n)$ for $n = 1, 2, \ldots$

10.3   Examples

Let $\emptyset \neq \Omega \subset \mathbb{R}^N$, $N \geq 2$, be a bounded domain with smooth boundary $\partial\Omega$. Consider the following problem (associated with the heat equation)

$$
\begin{cases}
u_t - \Delta u = f(t,x), & t \geq 0, \ x \in \Omega, \\
u(t,x) = 0, & t \geq 0, \ x \in \partial\Omega, \\
u(0,x) = u_0(x), & x \in \Omega.
\end{cases}
\tag{10.3.25}
$$

This problem can be solved by the Fourier method using the results presented in Chap. 8 and in Sect. 10.1 above. Thus, the Fourier method provides an approach for solving the above initial-boundary value problem which is complementary to the semigroup approach. Specifically, consider $H = L^2(\Omega)$ equipped with the usual scalar product and Hilbertian norm, $Q = -\Delta$ with $D(Q) = H_0^1(\Omega) \cap H^2(\Omega)$, and $H_E = H_0^1(\Omega)$ (the corresponding energetic space) with

$$
(p,q)_E = \int_\Omega \nabla p \cdot \nabla q \, dx, \quad \|p\|_E^2 = (p,p)_E.
$$

By Theorem 8.16 there exist an increasing sequence $(\lambda_n)_{n \geq 1}$ in $(0, \infty)$ converging to ∞ and an orthonormal basis $\{e_n\}_{n=1}^\infty$ in $H = L^2(\Omega)$ such that

$$
-\Delta e_n = \lambda e_n \quad \text{in } \Omega, \ \forall n \geq 1.
$$

Thus, Theorem 10.1 is applicable to problem (10.3.25) which is of the form (E), (IC) with the above choices. In particular, under suitable conditions, the solution of (10.3.25) is given by

$$
u(t,x) = \sum_{n=1}^\infty u_n(t) e_n(x),
\tag{10.3.26}
$$

where the u_n's are solutions of

$$\begin{cases} u'_n(t) + \lambda_n u_n(t) = f_n(t), & t \geq 0, \\ u_n(0) = u_{0n}, & n = 1, 2, \ldots \end{cases}$$

with

$$f_n(t) = \int_\Omega f(t, \xi) e_n(\xi)\, d\xi, \quad u_{0n} = \int_\Omega u_0(\xi) e_n(\xi)\, d\xi, \quad n = 1, 2 \ldots$$

Theorem 10.3 is also applicable to problem (10.3.25).

Theorem 10.4 can be illustrated with the following problem (associated with the wave equation):

$$\begin{cases} u_{tt} - \Delta u = f(t, x), & t \geq 0,\ x \in \Omega, \\ u(t, x) = 0, & t \geq 0,\ x \in \partial\Omega, \\ u(0, x) = u_0(x), & x \in \Omega. \end{cases} \qquad (10.3.27)$$

The cases of the boundary conditions of Neumann or Robin type can also be analyzed along the same lines.

10.4 Exercises

1. Consider the following initial-boundary value problem:

$$\begin{cases} u_t - u_{xx} = f(t, x), & t \in (0, T),\ x \in (0, 1), \\ u(t, 0) = 0,\ u_x(t, 1) = 0, & t \in [0, T], \\ u_0(x) = u_0(x), & x \in (0, 1). \end{cases}$$

Denote $H = L^2(0, 1)$. Assume that H is equipped with the usual scalar product $(\cdot, \cdot)$ and the induced norm $\|\cdot\|$ (hence H is a real Hilbert space which is infinite dimensional and separable). Define $Q : D(Q) \subset H \to H$ by

$$D(Q) = \{v \in H^2(0, 1);\ v(0) = 0, v'(1) = 0\},$$
$$Qv = -v'' \quad \forall v \in D(Q).$$

The above problem can be expressed as a Cauchy problem in H:

$$\begin{cases} u'(t) + Qu(t) = f(t), & 0 < t < T, \\ u(0) = u_0, \end{cases} \qquad (CP)$$

where $u(t) := u(t, \cdot) \in H$.

(i)   Show that Q satisfies condition **(a)** of Theorem 10.1 (i.e., Q is densely defined, self-adjoint, and strongly positive);

(ii)   Find all the eigenpairs of Q and construct a corresponding orthonormal basis $\{e_n\}_{n=1}^\infty$ of H;

(iii) Determine the energetic space H_E, show that H_E is compactly embedded in H, and determine an orthonormal basis of H_E;

(iv)   Find the explicit Fourier series solution $u(t,x) = \sum_{n=1}^\infty u_n(t)e_n(x)$ for

$$u_0(x) = x(1-x), \quad f(t,x) = (t+1)x.$$

2. Consider a homogeneous thin metal rod occupying an interval $[0,l]$, $l > 0$. The temperature at time $t = 0$ of the rod is constant: $u = u_0$ for $x \in [0,l]$. The temperatures at the ends of the rod are kept constant in time: $u(t,0) = u_1$, $u(t,l) = u_2$, $t \in [0,T]$, where $T > 0$ is a given time instant.

   Find the temperature distribution $u = u(t,x)$ on the rod, if there is no external heat source distributed along the rod.

3. Consider the following initial-boundary value problem:

$$\begin{cases} u_t - u_{xx} = f(t,x), & t \in (0,T), x \in (0,1), \\ -u_x(t,0) + \alpha u(t,0) = 0, \ u_x(t,1) = 0, & t \in [0,T], \\ u_0(x) = u_0(x), & x \in (0,1), \end{cases}$$

   where α is a given positive number. Denote as before $H = L^2(0,1)$ and define $Q : D(Q) \subset H \to H$ by

$$D(Q) = \{v \in H^2(0,1); \ -v'(0) + \alpha v(0) = 0, \ v'(1) = 0\},$$
$$Qv = -v'' \quad \forall v \in D(Q).$$

   Thus, the above problem can be expressed as a Cauchy problem in H:

$$\begin{cases} u'(t) + Qu(t) = f(t), & 0 < t < T, \\ u(0) = u_0, \end{cases} \tag{CP}$$

   where $u(t) := u(t,\cdot) \in H$.

Show that Q satisfies the conditions **(a)** and **(b)** of Theorem 10.1 (thus ensuring existence, uniqueness, and regularity of solutions to the given problem).

4. Repeat Exercise 10.1 above, replacing the boundary conditions by the following (Neumann) boundary conditions

$$u_x(t,0) = 0, \ \ u_x(t,1) = 0, \ \ t \in [0,T].$$

5. Let $(H, (\cdot, \cdot), \|\cdot\|)$ be a real Hilbert space and let $A : D(A) \subset H \to H$ be a linear and positive operator, i.e., $(Ap, p) \geq 0 \ \forall p \in D(A)$, where I is the identity operator on H. Assume that $Q = A + \alpha I$ satisfies both conditions **(a)** and **(b)** of Theorem 10.1, where α is a positive constant.

   (a) Solve the following Cauchy problem:

$$\begin{cases} u'(t) + Au(t) = f(t), & 0 < t < T, \\ u(0) = u_0, \end{cases} \quad (CP)$$

   for some given $u_0 \in H$ and $f \in L^2(0, T; H)$.

   (b) Show that, given T and f, if α is small enough, then there exists $u_0 \in H$ such that $u(T)$ is close to u_0, i.e., $\|u(T) - u_0\|$ is small, where u is the solution of (CP) corresponding to u_0 and f.

6. Let $\Omega = (0, a) \times (0, b) \subset \mathbb{R}^2$, $a, b \in (0, \infty)$. Consider the initial-boundary value problem

$$\begin{cases} u_t - \Delta u = f(t, x), & (t, x) \in (0, T) \times \Omega, \\ u(t, x) = 0, & (t, x) \in [0, T] \times \partial\Omega, \\ u(0, x) = u_0(x), & x \in \Omega. \end{cases}$$

   Find the general Fourier series expansion of the solution $u = u(t, x)$ of the above problem for $u_0 \in H = L^2(\Omega)$ and $f \in L^2((0, T) \times \Omega)$, and determine an explicit expansion for $u_0(x) = c$ and $f(t, x) = tx_1 x_2$, where c is a real constant.

7. Repeat the previous exercise with Neumann conditions on $\partial\Omega$ (instead of the preceding Dirichlet boundary conditions). Consider also combinations of Dirichlet and Neumann conditions on different sides of the rectangle Ω.

8. Solve the following initial-boundary value problem:

$$\begin{cases} u_t - u_{xx} = \alpha\delta(x-1) + \beta\delta(x-2), & (t,x) \in (0,\infty) \times (0,3), \\ u(t,0) = 0, \ u(t,3) = 0, & t \geq 0, \\ u(0,x) = 0, & x \in [0,3], \end{cases}$$

where α, β are real constants, and $\delta(x-1)$, $\delta(x-2)$ are the usual Dirac distributions in $D'(0,3)$, also denoted δ_1, δ_2.

9. Consider an elastic string of length $l > 0$, held fixed at both ends $x = 0$ and $x = l$. Find the displacement $u = u(t,x)$ in the string, which is set in motion from its straight equilibrium position, with the initial velocity v_0 defined by

$$v_0(x) = \begin{cases} Ax, & 0 \leq x \leq l/2, \\ A(l-x), & l/2 \leq x \leq l, \end{cases}$$

where A is a positive constant.

10. Consider an elastic string of length $l > 0$, held fixed at the end $x = 0$, while the end $x = l$ is free. Find the displacement $u = u(t,x)$ in the string, if it is set in motion at $t = 0$ from the initial configuration described by a function $u_0(x)$, with zero initial velocity. Discuss the regularity of u with respect to u_0.

11. Solve the initial-boundary value problem

$$\begin{cases} u_{tt} - u_{xx} + u = 0, & (t,x) \in (0,\infty) \times (0,1), \\ u_x(t,0) = 0, \ u_x(t,1) = 0, & t \geq 0, \\ u(0,x) = u_0(x), \ u_t(0,x) = 0, & x \in [0,1] \end{cases}$$

using the Fourier method.

12. Consider a guitar string of length $l > 0$, fixed at both ends $x = 0$ and $x = l$. Assume that the string is at rest at the time instant $t = 0$ and is set to motion by a force $f = c\delta(x - l/2)$ exerted on the midpoint of the string, where c is a real constant and $\delta(x - l/2)$ is the Dirac distribution (also denoted $\delta_{l/2}$). Determine the displacement $u(t,x)$ of the string for $t > 0$ and $x \in [0,l]$ using the Fourier method.

13. Let $\Omega = (0, a) \times (0, b) \subset \mathbb{R}^2$, $a, b \in (0, \infty)$. Solve the initial-boundary value problem

$$\begin{cases} u_{tt} - \Delta u = f(t, x), & (t, x) \in (0, T) \times \Omega, \\ u(t, x) = 0, & (t, x) \in [0, T] \times \partial\Omega, \\ u(0, x) = u_0(x), \ u_t(0, x) = 0, & x \in \Omega, \end{cases}$$

where

$$u_0(x) = x_1(a - x_1) \sin\left(\frac{3\pi x_2}{b}\right), \ f(t, x) = te^{x_1 x_2}.$$

Chapter 11

Integral Equations

This chapter is an introduction to the theory of linear Volterra and Fredholm equations. Some aspects related to certain nonlinear extensions are also addressed.

11.1 Volterra Equations

We begin with scalar, linear Volterra equations.[1] There are two kinds of such equations that are most relevant to applications, namely

$$f(t) = \int_a^t k(t,s)x(s)\,ds, \quad a \le t \le b, \tag{11.1.1}$$

and

$$x(t) = f(t) + \int_a^t k(t,s)x(s)\,ds, \quad a \le t \le b, \tag{11.1.2}$$

where $a, b \in \mathbb{R}$, $a < b$, $f \in C[a,b] := C([a,b]; \mathbb{R})$, $k \in C(\Delta) := C(\Delta; \mathbb{R})$ (called the **kernel**), with $\Delta = \{(t,s) \in \mathbb{R}^2; a \le s \le t \le b\}$; and $x = x(t)$ denotes the unknown function which is sought in the space $C[a,b]$.

Equation (11.1.1) is known as the *Volterra equation of the first kind*, while (11.1.2) as the *Volterra equation of the second kind*. In the following we examine Eq. (11.1.2). We will show later that Eq. (11.1.1) reduces to (11.1.2) under suitable conditions.

---

[1]Vito Volterra, Italian mathematician and physicist, 1860–1940.

© Springer Nature Switzerland AG 2019
G. Moroşanu, *Functional Analysis for the Applied Sciences*,
Universitext, https://doi.org/10.1007/978-3-030-27153-4_11

Theorem 11.1 (Existence and Uniqueness). *Under the above conditions there exists a unique solution $x \in C[a,b]$ to Eq. (11.1.2).*

We present below three different proofs.

Proof 1: Denote $K = \sup_{(t,s)\in\Delta} |k(t,s)|$ which is finite since Δ is a compact subset of $\mathbb{R}^2$. Assume in a first stage that

$$K(b-a) < 1. \tag{11.1.3}$$

Consider $X = C[a,b]$ equipped with the usual sup-norm, $\|g\| = \sup_{a\leq t\leq b} |g(t)|$, and the corresponding metric, $d(g_1,g_2) = \|g_1 - g_2\|$. Define $T : X \to X$ by

$$(Tg)(t) = f(t) + \int_a^t k(t,s)g(s)\,ds, \quad t \in [a,b], \ g \in X. \tag{11.1.4}$$

It is clear from (11.1.4) that T maps X into itself. We have

$$
\begin{aligned}
|(Tg_1)(t) - (Tg_2)(t)| &= \left| \int_a^t k(t,s)[g_1(s) - g_2(s)]\,ds \right| \\
&\leq \int_a^t |k(t,s)| \cdot |g_1(s) - g_2(s)|\,ds \\
&\leq K(b-a)\|g_1 - g_2\|,
\end{aligned}
$$

for all $g_1, g_2 \in X$, and all $t \in [a,b]$. Hence,

$$d(Tg_1, Tg_2) \leq K(b-a)d(g_1,g_2),$$

i.e., T is a contraction (cf. (11.1.3)). By the Banach Contraction Principle (see Chap. 2), T has a unique fixed point $x \in X$ which is clearly the unique solution of Eq. (11.1.2).

If condition (11.1.3) is not fulfilled, then we consider a subdivision of $[a,b]$, say,

$$a = t_0 < t_1 < t_2 < \cdots < t_{N-1} < t_N = b,$$

where $t_j = a + jh$ for $j = 1,2,\ldots,N$, $h = (b-a)/N$, with N large enough such that $Kh < 1$. In particular, $K(t_1 - t_0) = Kh < 1$, so it follows from above that Eq. (11.1.2) has a unique solution $x_1 = x_1(t)$ on the interval $[t_0,t_1] = [a,t_1]$, i.e.,

$$x_1(t) = f(t) + \int_a^t k(t,s)x_1(s)\,ds, \quad t \in [a,t_1].$$

Now consider the equation

$$x(t) = f(t) + \underbrace{\int_a^{t_1} k(t,s)x_1(s)\,ds}_{=:f_1(t)\in C[t_1,t_2]} + \int_{t_1}^t k(t,s)x(s)\,ds, \quad t \in [t_1, t_2].$$

Since $K(t_2 - t_1) = Kh < 1$, it follows by the above argument that this equation has a unique solution $x_2 \in C[t_1, t_2]$, and obviously $x_2(t_1) = x_1(t_1)$. Similarly, there exists a unique function $x_3 \in C[t_2, t_3]$ satisfying for all $t \in [t_2, t_3]$ the equation

$$x_3(t) = f(t) + \int_a^{t_1} k(t,s)x_1(s)\,ds + \int_{t_1}^{t_2} k(t,s)x_2(s)\,ds$$

$$+ \int_{t_2}^t k(t,s)x_3(s)\,ds,$$

and $x_3(t_2) = x_2(t_2)$. Continuing this procedure we obtain a solution $x \in C[t_0, t_N] = C[a, b]$ of Eq. (11.1.2) defined by $x(t) = x_j(t)$ for $t \in [t_{j-1}, t_j]$, $j = 1, 2, \ldots, N$. The solution x is obviously unique.

Proof 2. Again, consider the operator T defined by (11.1.4), where X is the same as above. It is easily seen that

$$|(Tg_1)(t) - (Tg_2)(t)| \le K\|g_1 - g_2\|(t - a) \quad \forall t \in [a.b], \; g_1, g_2 \in X.$$

Consequently for $T^2 = T \circ T$ we obtain the estimate

$$\begin{aligned}
|(T^2 g_1)(t) - (T^2 g_2)(t)| &\le \int_a^t |k(t,s)| \cdot |(Tg_1)(s) - (Tg_2)(s)|\,ds \\
&\le K^2\|g_1 - g_2\| \int_a^t (s - a)\,ds \\
&= \frac{K^2(t-a)^2}{2!}\|g_1 - g_2\|.
\end{aligned}$$

It can be shown by induction that

$$\begin{aligned}
|(T^k g_1)(t) - (T^k g_2)(t)| &\le \frac{K^k(t-a)^k}{k!}\|g_1 - g_2\| \\
&\le \frac{K^k(b-a)^k}{k!}\|g_1 - g_2\|,
\end{aligned}$$

for all $t \in [a, b]$, $g_1, g_2 \in X$, $k = 1, 2, \ldots$. We now take the supremum to find that

$$d(T^k g_1, T^k g_2) \leq \frac{K^k (b-a)^k}{k!} \|g_1 - g_2\| \quad \forall g_1, g_2 \in X, \ k = 1, 2, \ldots$$

$$(11.1.5)$$

Since $K^k (b-a)^k / k! \to 0$ as $k \to \infty$, T^k is a contraction for k large enough (cf. (11.1.5)). According to Remark 2.38, T has a unique fixed point $x \in X$ which is the unique solution of (11.1.2).

Proof 3. Let T be the same operator as before, but consider another norm on $X = C[a, b]$, the Bielecki norm which is defined by

$$\|g\|_B = \sup_{a \leq t \leq b} e^{-Lt} |g(t)|,$$

with L a large positive constant such that $K/L < 1$. This is indeed a norm on X which is equivalent to the usual sup-norm. Denote by d_B the metric generated by $\| \cdot \|_B$. We have for all $t \in [a, b]$ and $g_1, g_2 \in X$

$$
\begin{aligned}
|(Tg_1)(t) - (Tg_2)(t)| &\leq \int_a^t |k(t,s)| e^{Ls} e^{-Ls} |g_1(s) - g_2(s)| \, ds \\
&\leq K \|g_1 - g_2\|_B \int_a^t e^{Ls} ds \\
&= \frac{K \|g_1 - g_2\|_B}{L} \left(e^{Lt} - e^{La} \right),
\end{aligned}
$$

so that

$$
\begin{aligned}
e^{-Lt} |(Tg_1)(t) - (Tg_2)(t)| &\leq \frac{K}{L} \|g_1 - g_2\|_B \left(1 - e^{-L(t-a)} \right) \\
&\leq \frac{K}{L} \|g_1 - g_2\|_B.
\end{aligned}
$$

Now take the supremum for $t \in [a, b]$ to find

$$d_B(Tg_1, Tg_2) \leq \frac{K}{L} d_B(g_1, g_2) \quad \forall g_1, g_2 \in X.$$

As $K/L < 1$, T is a contraction with respect to d_B, hence the conclusion of the theorem follows again by the Banach Contraction Principle.

Resolvent Kernel

Assume that the conditions above on f and k are satisfied. For $n \in \mathbb{N}$, $t \in [a, b]$, define

$$x_n(t) = f(t) + \int_a^t k(t,s) x_{n-1}(s)\, ds\,,$$

$$x_0(t) = f(t)\,.$$

Clearly, the $x_n \in X = C[a,b]$ for all n. In fact, the above sequence $(x_n)_{n \geq 0}$ can be expressed as

$$x_n = T x_{n-1}, \quad n \in \mathbb{N}; \ x_0 = f\,,$$

where $T : X \to X$ is the operator defined by (11.1.4). So, (x_n) is the sequence of successive approximations (associated with operator T) which was used in the proof of the Banach Contraction Principle (see Chap. 2). Here we consider a particular starting function, $x_0 = f$. From the proof of the Banach Contraction Principle we know that (x_n) converges in $(C[a,b], \|\cdot\|_B)$ (see also Proof 3 above) to the unique fixed point of T, i.e., (x_n) converges uniformly in $[a,b]$ to the unique solution x of Eq. (11.1.2). On the other hand, we have for all $t \in [a,b]$

$$x_1(t) = f(t) + \int_a^t k(t,s) f(s)\, ds\,,$$

$$x_2(t) = f(t) + \int_a^t k(t,s)\Big[f(s) + \int_a^s k(s,\tau) f(\tau)\, d\tau \Big] ds$$

$$= f(t) + \int_a^t k(t,s) f(s)\, ds + \int_a^t \int_a^s k(t,s) k(s,\tau) f(\tau)\, d\tau\, ds\,.$$

We can interchange the integration to find that the last integral is equal to

$$\int_a^t \Big[\int_\tau^t k(t,s) k(s,\tau)\, ds \Big] f(\tau)\, d\tau\,,$$

so by simply relabeling τ and s we get

$$\int_a^t \underbrace{\Big[\int_s^t k(t,\tau) k(\tau,s)\, d\tau \Big]}_{=:k_2(t,s)} f(s)\, ds\,,$$

and have a new kernel, k_2. In general, if we denote for $n = 2, 3, \ldots$

$$k_n(t, s) := \int_s^t k(t, \tau) k_{n-1}(\tau, s) \, d\tau \,,$$
$$k_1(t, s) := k(t, s) \,,$$

we have for $n = 1, 2, \ldots$

$$x_n(t) = f(t) + \int_a^t \Big[\sum_{j=1}^n k_j(t, s) \Big] f(s) \, ds \,. \tag{11.1.6}$$

Since the k is continuous on the compact set Δ, we have for all $(t, s) \in \Delta$,

$$|k_1(t, s)| \leq K < \infty \,,$$
$$|k_2(t, s)| \leq K^2 (t - s) \,,$$
$$|k_3(t, s)| \leq K^3 \int_s^t |\tau - s| \, d\tau$$
$$= K^3 \frac{(t - s)^2}{2!} \,,$$
$$\vdots$$
$$|k_n(t, s)| \leq K^n \frac{(t - s)^{n-1}}{(n - 1)!}$$
$$\leq K^n \frac{(b - a)^{n-1}}{(n - 1)!} \,.$$

By the Weierstrass M-test the series $\sum_{n=1}^\infty k_n(t, s)$ clearly converges uniformly on Δ since

$$\sum_{n=1}^\infty \frac{K^n (b - a)^{n-1}}{(n - 1)!} < \infty \,.$$

Denote

$$R(t, s) = \sum_{n=1}^\infty k_n(t, s) \,,$$

which is in $C(\Delta)$. Letting $n \to \infty$ in (11.1.6) we deduce that

$$x(t) = f(t) + \int_a^t R(t, s) f(s) \, ds, \quad t \in [a, b] \,. \tag{11.1.7}$$

We call $R(t, s)$ the **resolvent kernel**. It depends on k but is independent of f, so that once we find $R(t, s)$ we have the solution of (11.1.2) for any f (cf. (11.1.7)).

Notice that

$$\sum_{n=2}^{N+1} k_n(t, s) = \int_s^t k(t, \tau) \sum_{n=2}^{N+1} k_{n-1}(\tau, s) \, d\tau \,,$$

which implies

$$-k(t, s) + \sum_{n=1}^{N+1} k_n(t, s) = \int_s^t k(t, \tau) \sum_{n=1}^{N} k_n(\tau, s) \, d\tau \,.$$

Letting $N \to \infty$ we find that R satisfies

$$R(t, s) = k(t, s) + \int_s^t k(t, \tau) R(\tau, s) \, d\tau \quad \forall (t, s) \in \Delta \,,$$

which is a Volterra equation similar to (11.1.2).

Now let us examine Eq. (11.1.1). Assume that

$$f \in C^1[a, b], \text{ and } k, \frac{\partial k}{\partial t} \in C(\Delta), k(t, t) \neq 0 \text{ for all } t \in [a, b]. \quad (H)$$

We also assume $f(a) = 0$ which is a necessary condition for Eq. (11.1.1) to have a solution. If Eq. (11.1.1) has a solution $x \in C[a, b]$, then differentiating (11.1.1) gives

$$
\begin{aligned}
f'(t) &= \frac{d}{dt} \int_a^t k(t, s) x(s) \, ds \quad &(11.1.8) \\
&= k(t, t) x(t) + \int_a^t k_t(t, s) x(s) \, ds, \quad t \in [a, b] \,,
\end{aligned}
$$

which is equivalent to the following integral equation of the second kind,

$$x(t) = \frac{f'(t)}{k(t, t)} + \int_a^t \left[\frac{-k_t(t, s)}{k(t, t)} \right] x(s) \, ds. \quad (11.1.9)$$

So x is also a solution of Eq. (11.1.9). On the other hand, we know from the previous theorem that (11.1.9) has a unique solution $x \in C[a, b]$. This x is also a solution of Eq. (11.1.1). This follows by integrating Eq. (11.1.8) over $[a, t]$ and using the condition $f(a) = 0$. Thus we have proved the following result.

Theorem 11.2. *Under the conditions (H) above, plus $f(a) = 0$, Eq. (11.1.1) has a unique solution $x \in C[a, b]$.*

We continue with the nonlinear Volterra equation

$$x(t) = f(t) + \int_a^t k(t, s, x(s))\, ds, \quad t \in [a, b], \tag{11.1.10}$$

and prove the following general result.

Theorem 11.3. *Assume that $f \in C[a, b]$, $k \in C(D)$, where $D := \Delta \times \mathbb{R} = \{(t, s, v) \in \mathbb{R}^3;\ a \leq s \leq t \leq b,\ v \in \mathbb{R}\}$, and there exists a $K > 0$ such that*

$$|k(t, s, v) - k(t, s, w)| \leq K|v - w| \quad \forall a \leq s \leq t \leq b,\ v, w \in \mathbb{R}. \tag{11.1.11}$$

Then there exists a unique function $x \in C[a, b]$ which satisfies Eq. (11.1.10) in $[a, b]$.

Proof. Consider $X = C[a, b]$ equipped with the Bielecki norm and define $T : X \to X$ by

$$(Tg)(t) = f(t) + \int_0^t k(t, s, g(s))\, ds \quad \forall t \in [a, b],\ g \in X.$$

The conclusion follows by the Banach Contraction Principle similarly as in Proof 3 of Theorem 11.1. □

Theorem 11.3 gives a *global solution* in the sense that the existence interval is the whole $[a, b]$. Obviously this is a generalization of Theorem 11.1. Indeed, to obtain Theorem 11.1 it is enough to assume that k is linear in the third variable, i.e., $k := k(t, s)v$, $a \leq s \leq t \leq b$, $v \in \mathbb{R}$, with $k \in C(\Delta)$ so that the Lipschitz condition (11.1.11) is automatically satisfied.

Now let us examine a case when the resulting solution is only a *local* one, i.e., its domain may not be the whole $[a, b]$.

Theorem 11.4. *Assume that $f \in C[a, b]$, $k = k(t, s, v) \in C(D)$, where $D := \Delta \times [x_0 - c, x_0 + c] = \{(t, s, v) \in \mathbb{R}^3;\ a \leq s \leq t \leq b,\ |v - x_0| \leq c\}$, with $x_0 \in \mathbb{R}$ and $c \in (0, \infty)$. If in addition there exists a $K > 0$ such that*

$$|k(t, s, v) - k(t, s, w)| \leq K|v - w| \quad \forall (t, s, v), (t, s, w) \in D, \tag{11.1.12}$$

and for some $d \in [0, c)$

$$|f(t) - x_0| \leq d \quad \forall t \in [a, b], \qquad (11.1.13)$$

then there exists a unique function $x \in C[a, a + \delta]$ which satisfies Eq. (11.1.10) in $[a, a + \delta]$, where

$$\delta = \min\{b - a, (c - d)/M\}, \quad M = \sup\{|k(t, s, v)|; \ (t, s, v) \in D\}.$$

(M is assumed to be positive since the case $M = 0$ is trivial).

Proof. Consider the space $C[a, a + \delta]$ with the usual sup-norm and the metric d generated by it. Denote

$$Y = \{g \in C[a, a + \delta]; \ |g(t) - x_0| \leq c \ \forall t \in [a, a + \delta]\}.$$

Clearly (Y, d) is a complete metric space (since Y is a closed subset of $(C[a, a + \delta], d)$). As usual, define an operator T by

$$(Tg)(t) = f(t) + \int_a^t k(t, s, g(s)) \, ds, \quad t \in [a, a + \delta], \ g \in Y.$$

Let us show that T takes Y into itself. Indeed, for all $g \in Y$ and $t \in [a, a + \delta]$ we have (see (11.1.13))

$$
\begin{aligned}
|(Tg)(t) - x_0| &\leq |f(t) - x_0| + \int_a^t |k(t, s, g(s))| \, ds \\
&\leq d + M(t - a) \\
&\leq d + M\delta \\
&\leq c,
\end{aligned}
$$

which proves the assertion. By arguments similar to those used in Proof 2 of Theorem 11.1 we deduce that T^k is a contraction on (Y, d) for k large enough. So T has a unique fixed point $x \in Y$ which is the unique solution of Eq. (11.1.10) in $[a, a + \delta]$. $\qquad\square$

Another existence and uniqueness result is obtained if k is defined on a different domain,

$$\tilde{D} = \{(t, s, v) \in \mathbb{R}^3; \ a \leq s \leq t \leq b, \ |v - f(s)| \leq c\}, \ c \in (0, \infty),$$

which is a compact subset of $\mathbb{R}^3$. The following result makes that precise.

Theorem 11.5. *Assume* $f \in C[a, b]$ *and* $k = k(t, s, v) \in C(\tilde{D})$, *with* $M = \sup_{\tilde{D}} |k| > 0$. *If, in addition, there exists a* $K > 0$ *such that*

$$|k(t, s, v) - k(t, s, w)| \leq K|v - w| \quad \forall (t, s, v), (t, s, w) \in \tilde{D}, \quad (11.1.14)$$

then there exists a unique function $x \in C[a, a + \delta]$ *which satisfies Eq.* (11.1.10) *in* $[a, a + \delta]$, *where* $\delta = \min \{b - a, c/M\}$.

Proof. The proof is similar to that of Theorem 11.4 above. Here the domain of operator T is conveniently chosen as

$$\tilde{Y} = \{g \in C[a, a + \delta]; \; |g(t) - f(t)| \leq c \; \forall t \in [a, a + \delta]\},$$

which is the closed ball in $(C[a, a + \delta], d)$ centered at f (restricted to $[a, a+\delta]$) of radius c. Obviously, T is well defined on $\tilde{Y}$ and takes $\tilde{Y}$ into itself. It is also easily seen that T^k is a contraction for some sufficiently large $k \in \mathbb{N}$. This completes the proof (see Remark 2.38). $\square$

Comments

1. If in Theorem 11.4 we assume $d = 0$ (i.e., $f \equiv x_0$) and k is independent of t, i.e., $k(t, s, v) = h(s, v)$, then we reobtain a well-known existence and uniqueness result for the Cauchy problem

   $$x'(t) = h(t, x(t)), \quad x(a) = x_0 .$$

   See the introductory part of Sect. 2.5. The same result can also be derived from Theorem 11.5.

2. If all the conditions of Theorem 11.4 are fulfilled, except for the Lipschitz condition (11.1.12), then local existence still holds, but without uniqueness. Indeed, $k = k(t, s, v)$ can be approximated uniformly on D by a sequence of smooth functions (hence Lipschitzian, even in all variables), say $(k_n)_{n \in \mathbb{N}}$. To obtain such a sequence we can use, for instance, Friedrichs' mollification with $\varepsilon = 1/n$ (see Chap. 5). In fact, by a classical result, $k = k(t, s, v)$ can even be approximated by polynomials in t, s, v. According to Theorem 11.4, for each $n \in \mathbb{N}$ there exists a unique function x_n which satisfies the equation

   $$x_n(t) = f(t) + \int_a^t k_n(t, s, x_n(s)) \, ds \; \forall t \in [a, a + \hat{\delta}], \quad (11.1.15)$$

where $\hat{\delta} = \min\{b-a, (c-d)/\hat{M}\}$, with $\hat{M}$ being the least upper bound of $\{\sup_D |k_n|\}_{n\in\mathbb{N}}$, i.e., $\hat{M} = \sup_{(t,s,v)\in D, n\in\mathbb{N}} |k_n(t,s,v)|$ (which is finite since $k_n \to k$ uniformly in D). Of course, $\hat{\delta}$ is less than the δ given by Theorem 11.4. It is easily seen that (x_n) satisfies the conditions of the Arzelà–Ascoli Criterion (see Chap. 2), so there exists a subsequence $(x_{n_j})_{j\in\mathbb{N}}$ which converges uniformly on $[a, a+\hat{\delta}]$ to a function $x \in C[a, a+\hat{\delta}]$. Letting $j \to \infty$ in (11.1.15) with $n := n_j$, we infer that this x satisfies Eq. (11.1.10) in $[a, a+\hat{\delta}]$.

Similar remarks are valid for Theorem 11.5.

3. Qualitative problems, such as continuability of local solutions, existence on the half-axis $[a, \infty)$, behavior of solutions at the end of their existence intervals, etc., are avoided here. For details in this respect we refer the reader to [9], where Volterra equations in L^2-spaces and abstract Volterra equations are also addressed.

4. All the above remarks apply to linear and nonlinear Volterra equations in $\mathbb{R}^k$, $k \in \mathbb{N}$, $k \geq 2$, with obvious slight changes.

11.2 Fredholm Equations

In the following $\mathbb{K}$ is either $\mathbb{R}$ or $\mathbb{C}$. Consider in $\mathbb{K}$ the integral equation

$$x(t) = f(t) + \int_a^b k(t,s)x(s)\,ds, \quad t \in [a,b], \tag{11.2.16}$$

where $a, b \in \mathbb{R}$, $a < b$, $f \in C([a,b]; \mathbb{K})$ and $k \in C([a,b] \times [a,b]; \mathbb{K})$. Here we prefer $\mathbb{K}$ instead of $\mathbb{R}$ since some specific aspects are better described in this framework. Equation (11.2.16) is known as the *Fredholm equation* (it is sometimes called the Fredholm equation of the second kind). It involves a fixed interval of integration and is fundamentally different from Eq. (11.1.2) (the corresponding Volterra analogue). A first remark that confirms this assertion is that, while the corresponding Volterra equation (of the second kind) always has a (unique, continuous) solution in $[a, b]$, Eq. (11.2.16) may have no solution in some cases. For instance, assuming that there exists a solution $x \in C[0, 1] := C([0, 1]; \mathbb{R})$ of the equation (see [9, p. 41])

$$x(t) = t + \int_0^1 k(t,s)x(s)\,ds, \quad t \in [0,1], \tag{11.2.17}$$

where

$$k(t, s) = \begin{cases} \pi^2 s(1 - t) & s \le t, \\ \pi^2 t(1 - s) & t \le s, \end{cases}$$

it follows by differentiating Eq. (11.2.17) twice that x should satisfy the problem

$$\begin{cases} x''(t) + \pi^2 x(t) = 0, & t \in [0, 1], \\ x(0) = 0, \ x(1) = 1. \end{cases}$$

On the other hand, it is easily seen that actually this problem has no solution. Therefore Eq. (11.2.17) has no solution. It is worth pointing out, however, that under the above assumptions, Eq. (11.2.16) has a unique solution in $C[a, b]$ whenever the sup-norm of $|k|$ is sufficiently small, more precisely if $(b - a) \sup_{[a,b] \times [a,b]} |k| < 1$. This result follows readily by the Banach Contraction Principle. In fact, the existence question can be discussed in the space $L^2(a, b; \mathbb{K})$, which is a larger framework. Specifically, let us assume $f \in L^2(a, b; \mathbb{K}), k \in L^2(Q; \mathbb{K})$, where $Q = (a, b) \times (a, b)$.

The solution x of Eq. (11.2.16) will be sought in $L^2(a, b; \mathbb{K})$ which is a Hilbert space with respect to the usual scalar product and norm,

$$\langle g_1, g_2 \rangle_{L^2} = \int_a^b g_1(t) \cdot \overline{g_2(t)} \, dt, \quad \|g\|_{L^2}^2 = \langle g, g \rangle.$$

Of course, if we find a solution $x \in L^2(a, b; \mathbb{K})$ of Eq. (11.2.16) with $f \in C([a, b]; \mathbb{K})$, $k \in C([a, b] \times [a, b]; \mathbb{K})$, then obviously $x \in C([a, b]; \mathbb{K})$. We have the following result.

Theorem 11.6. *If $f \in L^2(a, b; \mathbb{K})$, $-\infty < a < b < +\infty$, $k \in L^2(Q; \mathbb{K})$ and $\iint_Q |k(t, s)|^2 dt \, ds < 1$, where $Q = (a, b) \times (a, b)$, then there exists a unique function $x \in L^2(a, b; \mathbb{K})$ satisfying the equation*

$$x(t) = f(t) + \int_a^b k(t, s) x(s) \, ds,$$

almost everywhere in (a, b).

Proof. Let T be the operator defined by

$$(Tg)(t) = f(t) + \int_a^b k(t, s) g(s) \, ds \quad \forall g \in L^2(a, b; \mathbb{K})$$

and for a.a. $t \in (a, b)$.

It is easily seen that T takes $L^2(a, b; \mathbb{K})$ into itself. Moreover, $\|k\|_{L^2(Q;\,\mathbb{K})} < 1$, and so T is a contraction with respect to the metric generated by $\| \cdot \|_{L^2}$. Hence it has a unique fixed point $x \in L^2(a, b; \mathbb{K})$ which is the unique L^2-solution of the equation

$$x(t) = f(t) + \int_a^b k(t, s)x(s)\, ds\,. \qquad \qquad \square$$

Remark 11.7. Using a procedure similar to that used for the Volterra Eq. (11.1.2), we find that the solution given by Theorem 11.6 can be represented by the formula

$$x(t) = f(t) + \int_a^b R(t, s)f(s)\, ds \quad \text{for a.a. } t \in (a, b)\,,$$

where the *resolvent kernel* R is given by

$$R(t, s) = \sum_{i=1}^\infty k_i(t, s)\,, \qquad \qquad (11.2.18)$$

with

$$k_1(t, s) := k(t, s), \quad k_m(t, s) = \int_a^b k(t, \tau)k_{m-1}(\tau, s)\, d\tau \quad \forall m \geq 2\,.$$

The series in (11.2.18) converges in $L^2(Q; \mathbb{K})$ and almost everywhere on Q. We encourage the reader to check the details.

Remark 11.8. Theorem 11.6 can be extended to the *nonlinear Fredholm equation*

$$x(t) = f(t) + \int_a^b k(t, s, x(s))\, ds, \quad t \in [a, b]\,. \qquad (11.2.19)$$

Indeed, if $f \in L^2(a, b; \mathbb{K})$, $k : Q \times \mathbb{K} \to \mathbb{K}$ is Lebesgue measurable, $k(\cdot, \cdot, 0) \in L^2(Q; \mathbb{K})$, and

$$|k(t, s, v) - k(t, s, w)| \leq \alpha(t, s)|v - w|$$
$$\text{for all } v, w \in \mathbb{K} \text{ and a.a. } (t, s) \in Q\,,$$

for a given $\alpha \in L^2(Q)$ with $\|\alpha\|_{L^2(Q)} < 1$, then there exists a unique $x \in L^2(a, b; \mathbb{K})$ which satisfies Eq. (11.2.19) almost everywhere in (a, b). As usual, the conclusion follows by the Banach Contraction Principle. Let us just notice that for every $g \in L^2(a, b; \mathbb{K})$ the function $(t, s) \mapsto k(t, s, g(s))$ belongs to $L^2(Q; \mathbb{K})$, since

$$|k(t, s, g(s))| \leq |k(t, s, 0)| + \alpha(t, s)|g(s)| \quad \text{for a.a. } (t, s) \in Q\,.$$

Remark 11.9. In the case of Fredholm equations, the concept of a local solution does not make sense since the integral term involves the values $x(t)$ for a.a. $t \in (a, b)$. This shows once more that the Fredholm equations are fundamentally different from the Volterra equations of the second kind.

On the other hand, the reader may be wondering whether Eq. (11.2.16) still has solutions when the condition $\|k\|_{L^2(Q;\mathbb{K})} < 1$ is no longer satisfied. A complete answer is given by the Fredholm alternative (see Remark 7.11). In our specific case $H = L^2(a, b; \mathbb{K})$ and $A : H \to H$ is defined by

$$(Ag)(t) = \int_a^b k(t, s)g(s)\, ds \ \forall g \in H \ \text{ and for a.a. } t \in (a, b)\,. \quad (11.2.20)$$

Clearly, $A \in L(H)$. Moreover, we have the following lemma:

Lemma 11.10. *If* $k \in L^2(Q; \mathbb{K})$, *then operator* $A : H \to H$ *defined by* (11.2.20) *is compact.*

Proof. Assume first that $k \in C([a, b] \times [a, b]; \mathbb{K})$. In order to show that A is compact in this case, we shall make use of the Arzelà–Ascoli Criterion (see Chap. 2 and notice that the criterion is valid with $\mathbb{K}$ instead of $\mathbb{R}^k$). Let $B(0, r)$, $r \in (0, \infty)$, be a ball in H. Then the set $\mathcal{F} = \{Ag; g \in B(0, r)\}$ is a bounded subset of $C([a, b]; \mathbb{K})$:

$$
\begin{aligned}
|(Ag)(t)| &\leq \int_a^b |k(t, s)| \cdot |g(s)|\, ds \\
&\leq \Big(\int_a^b |k(t, s)|^2 ds \Big)^{1/2} \|g\|_{L^2} \\
&\leq r(b - a)^{1/2} \sup_Q |k| < \infty,
\end{aligned}
$$

for all $g \in B(0, r)$ and all $t \in [a, b]$. Set $\mathcal{F}$ is also equicontinuous since k is uniformly continuous on $[a, b] \times [a, b]$, so (by the Arzelà–Ascoli criterion) $\mathcal{F}$ is relatively compact in $C([a, b]; \mathbb{K})$, hence also in $H = L^2(a, b; \mathbb{K})$. Therefore, A is indeed a compact operator.

Now, assume $k \in L^2(Q; \mathbb{K})$. Then there is a sequence (k_n) in $C([a, b] \times [a, b]; \mathbb{K})$ such that $\|k_n - k\|_{L^2(Q;\mathbb{K})} \to 0$ as $n \to \infty$ (one can use, for instance, the density of $C_0^\infty(Q)$ in $L^2(Q)$, see Theorem 5.8). Let us associate with each k_n the operator $A_n \in L(H)$ defined by

$$(A_n g)(t) = \int_a^b k_n(t, s)g(s)\, ds \ \forall g \in H, \ t \in [a, b]\,,$$

which is compact, by the above argument. A straightforward computation shows that $\|A_n - A\|_{L(H)} \le \|k_n - k\|_{L^2(Q;\mathbb{K})}$ for all n, hence $\|A_n - A\|_{L(H)} \to 0$ as $n \to \infty$. It follows by Theorem 4.11 that A is compact. $\qquad\qquad\qquad\qquad\qquad\qquad\qquad\qquad\qquad\qquad\qquad\square$

Consider (in $\mathbb{K}$) the equation

$$x(t) = f(t) + \lambda \int_a^b k(t,s)x(s)\,ds, \quad t \in [a,b], \qquad (11.2.21)$$

where $\lambda \in \mathbb{K}$, $f \in L^2(a,b;\mathbb{K})$, $k \in L^2(Q;\mathbb{K})$, $Q = (a,b) \times (a,b)$. According to Theorem 11.6, Eq. (11.2.21) has a unique solution in $L^2(a,b;\mathbb{K})$ provided that $|\lambda|$ is sufficiently small. More precisely, this happens if

$$|\lambda| \cdot \|k\|_{L^2(Q;\mathbb{K})} < 1. \qquad (11.2.22)$$

We shall show in what follows that there are solutions for Eq. (11.2.21) even if λ does not satisfy condition (11.2.22). Using the above notation we can write Eq. (11.2.21) as an abstract equation in $H = L^2(a,b;\mathbb{K})$, namely

$$x = f + \lambda A x. \qquad (11.2.23)$$

Note that A^*, the adjoint of A, is given by

$$(A^*h)(t) = \int_a^b \overline{k(s,t)} \cdot h(s)\,ds \quad \forall h \in H.$$

Note also that $(\lambda A)^* = \bar{\lambda} A^*$.

According to Lemma 11.10 and Theorem 8.4, operator A has a countable set of eigenvalues with 0 being the only possible accumulation point; moreover, for any eigenvalue $\nu \ne 0$ of A, dim $N(I - \lambda A) < \infty$, where $\lambda = 1/\nu$. Of course, similar assertions hold for A^*, in particular dim $N(I - \bar{\lambda}A^*) < \infty$. In fact, we can prove that, if $\nu \ne 0$ is an eigenvalue of A, then

$$\dim N(I - \lambda A) = \dim N(I - \bar{\lambda} A^*), \quad \text{where } \lambda = 1/\nu. \qquad (11.2.24)$$

First of all, note that $\bar{\nu}$ is an eigenvalue of A^* (cf. Theorem 7.10), so dim $N(I - \bar{\lambda}A^*) \ge 1$. Let $\{\phi_1, \phi_2, \dots, \phi_m\}$ and $\{\psi_1, \psi_2, \dots, \psi_n\}$ be orthonormal bases in $N(I-\lambda A)$ and $N(I-\bar{\lambda}A^*)$, respectively. Assume

by way of contradiction that $m < n$. Let B be the operator associated
with the kernel

$$K(t,s) = k(t,s) - \sum_{j=1}^{m} \overline{\phi_j(s)} \cdot \psi_j(t),$$

and let $\phi, \psi \in H$ be solutions of the equations

$$\phi(t) = \lambda(B\phi)(t)$$
$$= \lambda \int_a^b k(t,s)\phi(s)\,ds - \lambda \sum_{j=1}^{m} \psi_j(t) \int_a^b \overline{\phi_j(s)} \cdot \phi(s)\,ds,$$

$$(11.2.25)$$

$$\psi(t) = \bar{\lambda}(B^*\psi)(t)$$
$$= \bar{\lambda} \int_a^b \overline{k(s,t)}\psi(s)\,ds - \bar{\lambda} \sum_{j=1}^{m} \phi_j(t) \int_a^b \overline{\psi_j(s)} \cdot \psi(s)\,ds.$$

$$(11.2.26)$$

Multiplying Eq. (11.2.25) by $\overline{\psi_k(t)}$ and then integrating over $[a,b]$ the
resulting equation yields

$$(\phi, \psi_k)_{L^2} = \int_a^b \underbrace{\left[\lambda \int_a^b k(t,s) \cdot \overline{\psi_k(t)}\,dt \right]}_{= \overline{\psi_k(s)}} \phi(s)\,ds - \lambda(\phi, \phi_k)_{L^2}$$
$$= (\phi, \psi_k)_{L^2} - \lambda(\phi, \phi_k)_{L^2},$$

hence

$$(\phi, \phi_k)_{L^2} = 0, \quad k = 1, 2, \ldots, m. \qquad (11.2.27)$$

From (11.2.25) and (11.2.27) we deduce that $\phi \in N(I - \lambda A)$. Thus
$\phi = \sum_{i=1}^{m} c_i \phi_i$ with some $c_i \in \mathbb{K}$, $i = 1, 2, \ldots, m$. This combined
with (11.2.27) yields $\phi = 0$, hence Eq. (11.2.25) has only the null
solution. On the other hand, Eq. (11.2.26) is satisfied by ψ_k for all
$k \in \{m+1, \ldots, n\}$. Indeed, since $(\psi_k, \psi_j)_{L^2} = 0$ for $j \in \{1, \ldots, m\}$,
$k \in \{m+1, \ldots, n\}$, Eq. (11.2.26) with $\psi = \psi_k$, $k = m+1, \ldots, n$,
can be written as $\psi_k = \bar{\lambda}A^*\psi_k$, $k = m+1, \ldots, n$. This means that,
$N(I - \bar{\lambda}B^*) = N(I - (\lambda B)^*) \neq \{0\}$, while $N(I - \lambda B) = \{0\}$, which
contradicts Theorem 7.10. Therefore, $m \geq n$. The converse inequality

follows from the fact that $(\bar{\lambda}A^*)^* = \lambda A$, so the proof of (11.2.24) is complete.

Notice that in the case of Eq. (11.2.21) above the Fredholm Alternative (see Remark 7.11) has the following specific form:

Theorem 11.11 (Fredholm Alternative). *Assume $\lambda \in \mathbb{K}$, $f \in H = L^2(a,b; \mathbb{K})$, $k \in L^2(Q; \mathbb{K})$, where $Q = (a,b) \times (a,b)$, and let $A : H \to H$ be the operator defined by*

$$(Ag)(t) = \int_a^b k(t,s)g(s)\, ds \quad \forall g \in H \quad \text{and for a.a. } t \in (a,b).$$

Then, one of the following holds:

- $N(I - \lambda A) = \{0\}$ *(if and only if $N(I - \bar{\lambda}A^*) = \{0\}$) and in this case the equation*

$$x(t) = f(t) + \lambda \int_a^b k(t,s)x(s)\, ds, \quad t \in [a,b] \qquad (F)$$

  has a unique solution for all $f \in H$,

- $\dim N(I - \lambda A) = \dim N(I - \bar{\lambda}A^*) = m$ *with $1 \leq m < \infty$ and in this case Eq. (F) is solvable if and only if*

$$(f, \psi)_{L^2} = \int_a^b f(t) \cdot \overline{\psi(t)}\, dt = 0 \quad \forall \psi \in \ker(I - \bar{\lambda}A^*),$$

  (equivalently, $(f, \psi_k)_{L^2} = 0$, $k \in \{1, 2, \ldots, m\}$, where the ψ_k's form an orthonormal basis in $N(I - \bar{\lambda}A^*)$).

Remark 11.12. Since the set $S = \{\lambda \in \mathbb{K};\, N(I - \lambda A) = \{0\}\}$ is countable it follows by Theorem 11.11 that there exist "many" λ's which do not satisfy condition (11.2.22), but for which Eq. (F) has a (unique) solution for all $f \in H = L^2(a,b; \mathbb{K})$. Even for $\lambda \in S$ Eq. (F) is solvable if and only if $f \perp N(I - \bar{\lambda}A^*)$.

The Case of Hermitian Kernels: Schmidt's Formula

In addition to the conditions

$$f \in H = L^2(a,b; \mathbb{K}), \quad k \in L^2(Q; \mathbb{K}), \quad Q = (a,b) \times (a,b),$$

we have used before, let us assume that k is Hermitian, i.e.,

$$k(t,s) = \overline{k(s,t)}, \quad \text{for a.a. } (t,s) \in Q.$$

Then obviously $A = A^*$. According to Proposition 8.5 every eigenvalue of A is real.

Next, we try to use the Hilbert–Schmidt Theorem to investigate the Fredholm equation in its abstract form (11.2.23), i.e.

$$x = f + \lambda Ax, \tag{11.2.23}$$

In fact, in the following A in (11.2.23) may be any linear, symmetric, compact operator from an infinite dimensional, separable Hilbert space $(H, (\cdot,\cdot), \|\cdot\|)$ into itself, and $f \in H$.

As a first step, let us assume that $N(A) = \{0\}$, i.e., zero is not an eigenvalue of A. Thus the Hilbert–Schmidt Theorem (Theorem 8.7) is applicable to A (see also Lemma 11.10). Denote by $\lambda_1, \lambda_2, \ldots, \lambda_n, \ldots$ the eigenvalues of A given by this theorem and by $u_1, u_2, \ldots, u_n, \ldots$ the corresponding eigenvectors, i.e., $Au_n = \lambda_n u_n$, $n = 1, 2 \ldots$. According to the proof of the Hilbert–Schmidt Theorem, each eigenvalue is taken into account k-times, where k means its multiplicity (the dimension of the corresponding eigenspace). The system $\{u_n\}_{n\geq 1}$ is an orthonormal basis in H.

For $\lambda \in \mathbb{K} \setminus \{0\}$ we distinguish two cases

(i) $N(I - \lambda A) = \{0\}$, i.e., $1/\lambda$ is not an eigenvalue of A;

(ii) $N(I - \lambda A) \neq \{0\}$, i.e., $1/\lambda$ is an eigenvalue of A.

Let us first discuss the case (i). By Remark 7.11 Eq. (11.2.23) has a unique solution x for each $f \in H$. By formula (8.2.11) from the proof of Theorem 8.7 (the Hilbert–Schmidt Theorem) we have

$$Ax = \sum_{n=1}^{\infty} \lambda_n(x, u_n)u_n. \tag{11.2.28}$$

On the other hand, using Eq. (11.2.23) and the fact that A is symmetric, we get

$$(x, u_n) = (f, u_n) + \lambda\lambda_n(x, u_n), \quad n = 1, 2, \ldots,$$

hence

$$(x, u_n) = \frac{1}{1 - \lambda\lambda_n}(f, u_n), \quad n = 1, 2, \ldots \tag{11.2.29}$$

Now, from (11.2.23), (11.2.28), and (11.2.29) we can derive the following formula for the solution x of Eq. (11.2.23) (known as Schmidt's formula)

$$x = f + \lambda \sum_{n=1}^{\infty} \frac{\lambda_n}{1 - \lambda\lambda_n}(f, u_n)u_n. \qquad (11.2.30)$$

Now, let us discuss the case (ii), i.e., when $1/\lambda$ is an eigenvalue of operator A, say $1/\lambda = \lambda_k$ for some $k \in \mathbb{N}$. Obviously, formula (11.2.30) does not make sense in this case.

Denote $H_0 := N(I - \lambda A) = N(\lambda_k I - A)$, $H_1 := H_0^{\perp}$, so that $H = H_0 \oplus H_1$. By Theorem 8.4, H_0 is finite dimensional. Denote $m := \dim H_0 \in \mathbb{N}$. Let $B_0 = \{v_1, v_2, \ldots, v_m\}$ be a basis of H_0. As H is a separable space, so is H_1.

Taking into account the fact that A is symmetric, it is easily seen that A maps H_1 into itself. Clearly, the restriction $A_1 = A|_{H_1}$ is symmetric and $A_1 \in K(H_1)$, i.e., A_1 is compact in H_1 which is a Hilbert subspace of H with the same $(\cdot, \cdot)$ and $\|\cdot\|$. Obviously, $N(A_1) = \{0\}$ so the Hilbert–Schmidt Theorem is applicable to H_1 and A_1 and shows the existence of a sequence of (real) eigenvalues of A_1 (hence of A), which does not include λ_k, and of a corresponding orthonormal basis in H_1, with

$$A_1 u_n = A u_n = \lambda_n u_n, \quad n \in \mathbb{N}, n \neq k.$$

According to the previous analysis corresponding to the case (i), Eq. (11.2.23) has a (unique) solution $x = x_1$ in H_1 (i.e., $x_1 - \lambda A_1 x_1 = f$) if and only if $f \in H_1$, and (see (11.2.30))

$$x_1 = f + \lambda \sum_{\lambda_n \neq \lambda_k} \frac{\lambda_n}{1 - \lambda\lambda_n}(f, u_n)u_n.$$

If we consider (11.2.23) in H, then for $f \in H_1$ and for all $y \in H_0$,

$$x = f + \lambda \sum_{\lambda_n \neq \lambda_k} \frac{\lambda_n}{1 - \lambda\lambda_n}(f, u_n)u_n + y$$

is a solution of Eq. (11.2.23). Consequently, the formula

$$x = f + \lambda \sum_{\lambda_n \neq \lambda_k} \frac{\lambda_n}{1 - \lambda\lambda_n}(f, u_n)u_n + \sum_{i=1}^{m} c_i v_i, \qquad (11.2.31)$$

with $c_1, \ldots, c_m \in \mathbb{K}$, gives all solutions of Eq. (11.2.23).

We now turn our attention to the case when $N(A) \neq \{0\}$.

Denoting $Y_0 = N(A)$ and $Y_1 = Y_0^\perp$, we can write $H = Y_0 \oplus Y_1$. We can assume that Y_0 is a proper subspace of H, otherwise $A = 0$ which is a trivial case. It is easy to see that A takes Y_1 to itself. Obviously, Y_1 is a Hilbert subspace of H with respect to the same $(\cdot, \cdot)$ and $\| \cdot \|$, and the restriction $\tilde{A} = A|_{Y_1}$ is symmetric, compact, and $N(\tilde{A}) = \{0\}$. If Y_1 is infinite dimensional, then the Hilbert–Schmidt Theorem is applicable to Y_1 and $\tilde{A}$. In order to solve Eq. (11.2.23) we use the decompositions $x = x_0 + x_1$, $f = f_0 + f_1$, where $x_0, f_0 \in Y_0$ and $x_1, f_1 \in Y_1$. Thus (11.2.23) becomes

$$x_0 - f_0 = -x_1 + f_1 + \lambda A x_1,$$

hence both sides are equal to zero, so $x_0 = f_0$ and

$$x_1 = f_1 + \lambda \tilde{A} x_1. \tag{11.2.32}$$

Clearly, for every $f \in H$, $f = f_0 + f_1$, x is a solution of Eq. (11.2.23) if and only if $x = f_0 + x_1$, where $x_1 \in Y_1$ satisfies Eq. (11.2.32).

It is worth pointing out that Eq. (11.2.32), with $\tilde{A} : Y_1 \to Y_1$, $N(\tilde{A}) = \{0\}$, is in the situation we had before, so one can similarly discuss the solvability of (11.2.32) in terms of the eigenvectors of $\tilde{A}$ (i.e., the eigenvectors of A corresponding to nonzero eigenvalues).

If it turns out that Y_1 is finite dimensional, then Eq. (11.2.32) reduces to a linear algebraic system which can be solved by using elementary algebraic computations.

Example. Let $H = L^2(-\pi, \pi)$ with the usual scalar product and norm. Consider the usual orthonormal basis in H, i.e. (see Chap. 6),

$$u_0 = \frac{1}{\sqrt{2\pi}}, \quad u_{2k-1}(t) = \frac{1}{\sqrt{\pi}} \cos(kt),$$

$$u_{2k}(t) = \frac{1}{\sqrt{\pi}} \sin(kt), \quad k = 1, 2, \ldots$$

For a given $m \in \mathbb{N}$, define

$$k(t, s) = \sum_{n=m}^{\infty} \frac{1}{n^2} u_n(t) u_n(s), \quad (t, s) \in Q = (-\pi, \pi) \times (-\pi, \pi).$$

Clearly, $k \in C(\bar{Q}) \subset L^2(Q)$. If A is the operator defined by (11.2.20), where $a = -\pi$, $b = \pi$, with this kernel (which is symmetric, hence Hermitian), then $Ag = 0$ for every g which is a linear combination of $u_0, u_1, \ldots, u_{m-1}$. Therefore

$$\mathrm{Span}\{u_0, u_1, \ldots, u_{m-1}\} \subset N(A).$$

On the other hand, if $Af = 0$, where f is a member of H, i.e., $f = \sum_{k=0}^{\infty}(f, u_k)_{L^2} u_k$ (which is the Fourier expansion of f), then

$$
\begin{aligned}
0 &= (Af, f)_{L^2}\\
&= \Big(\sum_{n=m}^{\infty} \frac{1}{n^2}(f, u_n)_{L^2} u_n, \sum_{k=0}^{\infty}(f, u_k)_{L^2} u_k\Big)_{L^2}\\
&= \sum_{n=m}^{\infty} \frac{1}{n^2}(f, u_n)_{L^2}^2,
\end{aligned}
$$

hence $(f, u_n)_{L^2} = 0$ for all $n \geq m$ and so $f = \sum_{k=0}^{m-1}(f, u_k)_{L^2} u_k$, i.e., $f \in \mathrm{Span}\{u_0, u_1, \ldots, u_{m-1}\}$. Therefore,

$$N(A) = \mathrm{Span}\{u_0, u_1, \ldots, u_{m-1}\}.$$

On the other hand, if we choose, for example,

$$k(t, s) = 1 + \sum_{n=1}^{\infty} \frac{1}{n^2} u_n(t) u_n(s), \quad (t, s) \in Q,$$

then the corresponding operator A satisfies the condition $N(A) = \{0\}$. The solvability of the Fredholm equation $x = f + \lambda Ax$, with A, associated with the k's, defined above, is left to the reader.

Comments.

1. If in the equation

$$x(t) = f(t) + \lambda \int_a^b k(t, s) x(s) \, ds, \quad t \in [a, b],$$

(which is (11.2.21) above) we assume $f \in C[a, b]$ and $k \in C([a, b] \times [a, b])$, then $x \in C[a, b]$. Moreover, if f and k are more regular, then so is x.

2. The above theory also works if $[a, b]$ is replaced by a bounded domain $D \subset \mathbb{R}^N$ or by the boundary of such a domain. It is well known that the main elliptic boundary value problems (Dirichlet, Neumann, Robin) can be reduced, by using potentials, to Fredholm equations that live on the boundary of the corresponding domains. Thus the above theory can be used to solve such problems.

3. The following nonlinear extension of the Fredholm equation, known as the Hammerstein equation,

$$x(t) = f(t) + \int_D k(t, s)g(s, x(s)) \, ds \quad \text{for a.a. } t \in D,$$

   where g is a nonlinear function, is also heavily discussed in the literature (see [20], [9], [26]).

11.3   Exercises

1. Calculate the resolvent kernels of the following Volterra equations and then find the corresponding solutions:

   (a) $x(t) = e^{t^2} + \int_0^t e^{t^2 - s^2} x(s) \, ds, \quad t \geq 0$;

   (b) $x(t) = e^t \sin t + \int_0^t \frac{2 + \cos t}{2 + \cos s} x(s) \, ds, \quad t \geq 0$;

   (c) $x(t) = t + \int_0^t (t - s)x(s) \, ds, \quad t \geq 0$.

2. Solve the following integral equations by converting them into Cauchy problems for differential equations:

   (a) $x(t) = t - \frac{t^3}{6} + \int_0^t (t - s + 1)x(s) \, ds, \quad t \geq 0$;

   (b) $x(t) = t^3 + 1 - \int_0^t (t - s)x(s) \, ds, \quad t \geq 0$;

   (c) $x(t) = 3t - \int_0^t e^{t-s} x(s) \, ds, \quad t \geq 0$.

3. Solve the following Volterra equations of the first kind:

   (a) $\int_0^t (1 - t^2 + s^2) \cdot x(s) \, ds = \frac{t^2}{2}, \quad t \geq 0$;

   (b) $\int_0^t \cos(t - s) \cdot x(s) \, ds = 2t(t + 1), \quad t \geq 0$;

   (c) $\int_0^t e^{t+s} \cdot x(s) \, ds = t \cos t, \quad t \geq 0$.

4. Let $h \in C[0, b]$, where $b \in (0, \infty)$. Define $k(t, s) = h(t - s)$, $0 \le s \le t \le b$. Show that the resolvent kernel $R(t, s)$ associated with $k(t, s)$ depends only on $t - s$.

5. Let $a, b \in \mathbb{R}$, $a < b$. Let $f, x \in C[a, b]$, $k \in C(\Delta)$ be nonnegative functions, where $\Delta = \{(t, s) \in \mathbb{R}^2; a \le s \le t \le b\}$. If

$$x(t) \le f(t) + \int_a^t k(t, s)x(s) \, ds, \quad t \in [a, b],$$

then

$$x(t) \le f(t) + \int_a^t R(t, s)f(s) \, ds, \quad t \in [a, b],$$

where $R(t, s)$ is the resolvent kernel associated with $k(t, s)$.

6. Consider in $C([0, \pi]; \mathbb{K})$ the equation

$$x(t) = \lambda \int_0^\pi (\sin t \cdot \cos s)x(s) \, ds, \quad t \in [0, \pi].$$

Show that for any $\lambda \in \mathbb{K}$ the equation has only the null solution.

7. Let $a, b \in (0, \infty)$. Define $D = \{(t, s); 0 \le t \le a, \ 0 \le s \le b\}$, $Q = \{(t, s, \xi, \eta); 0 \le \xi \le t \le a, \ 0 \le \eta \le s \le b\}$. Consider the integral equation

$$x(t, s) = f(t, s) + \int_0^t \int_0^s k(t, s, \xi, \eta) \, d\xi d\eta, \quad (t, s) \in D. \quad (E)$$

Assume $k \in C(Q) := C(Q; \mathbb{R})$.

Show that for each $f \in C(D) := C(D; \mathbb{R})$ there exists a unique function $x = x(t, s) \in C(D)$ satisfying Eq. (E) for all $(t, s) \in D$.

8. Consider the problem

$$\begin{cases} x'(t) = f(t) + \int_0^t k(t, s)x(s) \, ds, & t \in (0, T), \\ x(0) = x_0, \end{cases}$$

where $x_0 \in \mathbb{R}$, $T \in (0, \infty)$, $f \in L^1(0, T)$, $k \in C(\Delta)$, and $\Delta = \{(t, s) \in \mathbb{R}^2; 0 \le s \le t \le T\}$.

Show that there exists a unique function $x \in W^{1,1}(0, T)$ satisfying the above integro-differential equation for a.a. $t \in (0, T)$ and the initial condition $x(0) = x_0$.

9. Solve the following integral equations, where λ is a real parameter:

(a) $x(t) = \cos t + \lambda \int_0^\pi \sin(t - s) \cdot x(s)\,ds$;

(b) $x(t) = t + \lambda \int_0^{2\pi} |\pi - s| \sin t \cdot x(s)\,ds$;

(c) $x(t) = f(t) + \lambda \int_0^1 (1 - 3ts) \cdot x(s)\,ds$, $f \in L^2(0,1)$.

10. Consider, in $\mathbb{K}$, the following Fredholm equation with degenerate (separable) kernel:

$$x(t) = f(t) + \lambda \int_a^b \underbrace{\Big[\sum_{i=1}^n a_i(t)b_i(s) \Big]}_{k(t,s)} x(s)\,ds, \qquad (F)$$

where $\lambda \in \mathbb{K}$, $f, a_i, b_i \in L^2(a,b;\mathbb{K})$, $i = 1, 2, \ldots, n$. One can assume without any loss of generality that the systems $\{a_1, \ldots, a_n\}$, $\{b_1, \ldots, b_n\}$ are linearly independent. Denoting

$$c_i = \int_a^b b_i(s)x(s)\,ds, \quad i = 1, \ldots, n, \qquad (1)$$

we obtain from (F)

$$x(t) = f(t) + \lambda \sum_{i=1}^n c_i a_i(t). \qquad (2)$$

Plugging (2) into (1) we obtain the algebraic system

$$c_i = f_i + \lambda \sum_{j=1}^n k_{ij} c_j, \quad i = 1, \ldots, n, \qquad (3)$$

where

$$f_i = \int_a^b b_i(s)f(s)\,ds, \quad k_{ij} = \int_a^b b_i(s)a_j(s)\,ds, \quad i, j = 1, \ldots, n.$$

Show that the Fredholm alternative for Eq. (F) can be expressed as an equivalent alternative for the algebraic system (3).

11. Let $(H, (\cdot, \cdot), \| \cdot \|)$ be a Hilbert space and let $\{e_1, \ldots, e_m\} \subset H$ be an orthonormal system, where m is a given natural number. Define $A : H \to H$ by

$$Ax = \sum_{k=1}^{m} k(x, e_k)e_k, \quad x \in H.$$

Solve the abstract Fredholm equation

$$x = f + \lambda Ax,$$

where $f \in H$ and $\lambda \in \mathbb{K}$.

12. Consider the functions

$$u_n(t) = \sqrt{2} \cos\left((n + 1/2)\pi t\right), \quad t \in [0, 1], \ n = 0, 1, 2, \ldots$$

It is well known that the system $\{u_n\}_{n=0}^{\infty}$ is an orthonormal basis in $H = L^2(0, 1)$ equipped with the usual scalar product and norm (see the solution to Exercise 8.11). Define the kernel $k(t, s)$ by

$$k(t, s) = \sum_{n=m}^{\infty} \frac{1}{(n + 1)^2} u_n(t)u_n(s), \quad t, s \in [0, 1],$$

where $m \in \{0, 1, 2, \ldots\}$, and the integral operator $A : H \to H$,

$$(Ag)(t) = \int_0^1 k(t, s)g(s)\, ds, \quad g \in H.$$

Discuss the existence for the Fredholm equation

$$x = f + \lambda Ax, \quad f \in H, \ \lambda \in \mathbb{R},$$

in two cases: $m = 0$ and $m \geq 1$.

Chapter 12

Answers to Exercises

This chapter provides solutions to almost all exercises proposed at the end of each chapter. The solutions are labeled with the same numbers used for the corresponding exercises. For easy exercises we shall provide hints or just their final solutions. Answers to very easy exercises are left to the reader.

12.1   Answers to Exercises for Chap. 1

1. Left to the reader.

2. Answers: $X = (C \setminus A) \cup B; \quad X = C \cup (A \setminus B)$.

3. Left to the reader.

4. It is easily seen that the statements (a) and (c) are true, while statement (b) is not true in general as shown by the following counterexample: $A = \{1, 2\}$, $B = \{3\}$, $C = \{3, 4\}$, $D = \{4, 5\}$.

5. Let b be a minimal element of A. Since $a = \min A$, we have $a \leq x$ for all $x \in A$. In particular, $a \leq b \Rightarrow b = a$.

6. Observe that

$$
\begin{aligned}
a_n &= \left(1 - \frac{1}{2}\right) + \left(\frac{1}{2} - \frac{1}{3}\right) + \cdots + \left(\frac{1}{n} - \frac{1}{n+1}\right) \\
&= \left(1 - \frac{1}{n+1}\right).
\end{aligned}
$$

It follows that $\inf A = \frac{1}{2}$ and $\sup A = 1$.

7. Parts (a) and (b) are left to the reader; in order to solve (c), we suggest the partial order

$$z_1 = x_1 + y_1 i \preceq z_2 = x_2 + y_2 i \iff x_1 \leq x_2 \text{ and } y_1 \geq y_2.$$

8. It is easily seen by induction that (a_n) is increasing and $a_n < 2$, for all $n \geq 1$. According to the Monotone Convergence Theorem, (a_n) is convergent and its limit $a \geq a_1 = \sqrt{2}$. Letting $n \to \infty$ in $a_n = \sqrt{2 + a_{n-1}}$ we get $a = \sqrt{2 + a}$, so $a = 2$.

9. Use Zorn's Lemma.

10. Left to the reader.

11. Left to the reader.

12. For part (a), Systems (i), (iii), (v), and (vi) are linearly independent, while (ii) and (iv) are linearly dependent. Part (b) is left to the reader.

13. It is readily seen that B is a basis of X, and the coordinates of $p = p(t)$ with respect to this basis are

$$p(1), \quad \frac{p'(1)}{1!}, \quad \frac{p''(1)}{2!}, \quad \frac{p'''}{3!}.$$

14. Left to the reader.

15. Left to the reader.

16. Left to the reader.

17. Left to the reader.

18. We have
$$F(x) = (Bx, x) = a\|x\|^2 + \|Ax\|^2,$$

where $(\cdot, \cdot)$ and $\|\cdot\|$ denote the usual scalar product of $\mathbb{R}^n$ and the induced norm, respectively. Clearly, F is positive definite if $a > 0$. If $a = 0$, then F is positive definite if and only if $\det A \neq 0$, otherwise F is only positive semidefinite.

19. Left to the reader.

20. Left to the reader.

12.2   Answers to Exercises for Chap. 2

1. Consider first the case of finite unions. Since $A_i \subset \operatorname{Cl} A_i$ we have $\bigcup_{i=1}^n A_i \subset \bigcup_{i=1}^n \operatorname{Cl} A_i$. Here the right-hand side is a closed set, hence $\operatorname{Cl} \left(\bigcup_{i=1}^n A_i \right) \subset \bigcup_{i=1}^n \operatorname{Cl} A_i$. For the converse inclusion, let $x \in \bigcup_{i=1}^n \operatorname{Cl} A_i$. So there exists a $j \in \{1, 2, \ldots, n\}$ such that $x \in \operatorname{Cl} A_j$. Therefore, $x \in \operatorname{Cl} \left(\bigcup_{i=1}^n A_i \right)$ which implies the converse inclusion.

   For the inclusion relation $\bigcup_{i=1}^\infty \operatorname{Cl} A_i \subset \operatorname{Cl} \left(\bigcup_{i=1}^\infty A_i \right)$, one can use a similar argument. This inclusion can be proper: for example, let $(\mathbb{R}, d)$ be our metric space, where $d(x, y) = |x - y|$ for all $x, y \in \mathbb{R}$, and consider also the set $\mathbb{Q}$ of rational numbers, which can be written as $\mathbb{Q} = \{r_1, r_2, \ldots\}$. If $A_i = \{r_i\}$, $i \in \mathbb{N}$, then $\bigcup_{i=1}^\infty \operatorname{Cl} A_i = \mathbb{Q}$ and $\operatorname{Cl} \left(\bigcup_{i=1}^\infty A_i \right) = \mathbb{R}$.

2. Consider $(\mathbb{R}, d)$ with $d(x, y) = |x - y|$ for all $x, y \in \mathbb{R}$, and let $A = \mathbb{Q}$. Then $\operatorname{Cl} A = \mathbb{R}$, $\operatorname{Int} A = \emptyset$, $\operatorname{Int}(\operatorname{Cl} A) = \mathbb{R}$, and $\operatorname{Cl}(\operatorname{Int} A) = \emptyset$, so both answers are **No**.

3. Let A be an arbitrary nonempty subset of (X, d_0). For any $x \in A$ the ball $B(x, 1/2) = \{x\} \subset A$, so A is indeed open in (X, d_0).

4. Let $p \in \operatorname{Cl} A$. Assume by way of contradiction that

   $$\inf \{d(p, x) : x \in A\} =: r > 0.$$

   Obviously, $A \subset X \setminus B(p, r/2)$, the latter being a closed set. It follows that $\operatorname{Cl} A \subset X \setminus B(p, r/2)$, which contradicts the assumption $p \in \operatorname{Cl} A$.

   For the converse implication, assume $\inf \{d(p, x) : x \in A\} = 0$. Then for every $n \in \mathbb{N}$ there exists an $a_n \in A$ such that $d(a_n, p) < 1/n$. So $a_n \to p \Rightarrow p \in \operatorname{Cl} A$.

5. It is easy to see that $(BC(A; Y), \|\cdot\|_{\sup})$ is a normed space, hence a metric space with the metric generated by $\|\cdot\|_{\sup}$. Let $(f_n)_{n \in \mathbb{N}}$ be a Cauchy sequence in $BC(A; Y)$. It is a Cauchy sequence among the bounded functions, so it has a limit f among the bounded functions (in fact, $B(A; Y)$ is similar to $B(A; \mathbb{R})$ which was discussed in Chap. 2, since $(Y, \|\cdot\|)$ is a Banach space). It remains to prove that f is continuous. Let x_0 be an arbitrary

point in A and let ε be a small positive number. We have

$$
\begin{aligned}
\|f(x) - f(x_0)\| \quad &\leq \quad \|f(x) - f_N(x)\| + \|f_N(x) - f_N(x_0)\| \\
&\quad + \|f_N(x_0) - f(x)\| \\
&< \quad \frac{\varepsilon}{3} + \frac{\varepsilon}{3} + \frac{\varepsilon}{3} = \varepsilon,
\end{aligned}
$$

(for $N \in \mathbb{N}$ sufficiently large and) for $\|x - x_0\|$ sufficiently small, hence f is continuous at x_0.

6. Answers:

   (a) $\emptyset$, since all the points of $\mathbb{Z} \times \mathbb{Z}$ are isolated;

   (b) $\mathbb{R}^2$;

   (c) $\mathbb{R} \times \{0\}$;

   (d) $\{(0,0)\} \cup \{(1/m, 0); \ m \in \mathbb{Z}, \ m \neq 0\}$.

   Proof of (c): In order to obtain accumulation points we have to consider as denominators terms of integer sequences that converge to $\pm\infty$. Define

   $$
   h_k := \frac{pn_k + z}{qn_k} = \frac{p}{q} + \frac{z}{qn_k}, \quad k = 1, 2, \ldots,
   $$

   where $p, q, z \in \mathbb{Z}$, $q \neq 0$, $n_k \in \mathbb{N}$, $n_k \to \infty$. Then $(h_k, 1/qn_k)$ converges to $(p/q, 0)$. Since $\mathbb{Q}$ is dense in $\mathbb{R}$ the result follows.

7. Answers: $\partial A = [0, 1]$; $\partial B = B \cup \{0\}$; $\partial C = \{(x, y) \in \mathbb{R}^2; \ x^2 - y^2 = 1\}$.

8. We have

   $$
   B(x, r) \subset \overline{B(x, r)} \ \Rightarrow \ \mathrm{Cl}\, B(x, r) \subset \overline{B(x, r)}.
   $$

   For the converse inclusion, take an arbitrary $y \in \overline{B(x, r)} \backslash B(x, r)$, i.e., $\|y - x\| = r$. Define the sequence (y_n) by

   $$
   y_n = \frac{1}{n}x + \left(1 - \frac{1}{n}\right)y, \quad n = 1, 2, \ldots
   $$

   Since

   $$
   d(y_n, y) = \frac{1}{n}\|y - x\| = \frac{r}{n} \to 0,
   $$

   we see that $y \in \mathrm{Cl}\, B(x, r)$, which concludes the proof.

   In (X, d_0), $B(x, r) = \{x\} \ \Rightarrow \ \mathrm{Cl}\, B(x, r) = \{x\}$, while $\overline{B(x, r)} = X$ if $r \geq 1$.

9. Let (x_n) be a Cauchy sequence in a metric space (X, d), and let y, z be two cluster points of (x_n), i.e., y, z are the limit points of two subsequences of (x_n), say $(y_n), (z_n)$. From

$$d(y, z) \leq d(y, y_n) + d(y_n, z_n) + d(z_n, z)$$

we derive $d(y, z) = 0 \Rightarrow y = z$.

10. Use the decompositions

$$2\pi\sqrt{n^2 + 3n} = 2\pi n + \alpha_n, \quad \pi\sqrt{n^2 + n} = \pi n + \beta_n,$$

where

$$\alpha_n = \frac{6\pi n}{\sqrt{n^2 + 3n} + n} \to 3\pi, \quad \beta_n = \frac{\pi n}{\sqrt{n^2 + n} + n} \to \frac{\pi}{2},$$

to infer that (x_n) is convergent to 0, while (y_n) has two cluster points, $+1$ and -1.

11. Denote $C[0, 1] := C([0, 1]; \mathbb{R})$. Let $d_{\sup}$ be the metric generated by the sup-norm $\|\cdot\|_{\sup}$ on $C[0, 1]$. Let g be an arbitrary element of B. By Weierstrass' Theorem $\inf_{x \in [0,1]} g(x) = g(x_0) > 0$ for some $x_0 \in [0, 1]$. Denoting $c := g(x_0)$, we see that the ball centered at g with radius $c/2$ of the metric space $(C[0, 1], d_{\sup})$ is included in B. Therefore B is open in $(C[0, 1], d_{\sup})$.

To answer the latter question, observe that

$$\tilde{B} := \{f \in C[0, 1] : f(x) \geq 0 \text{ for all } x \in [0, 1]\}$$

is a closed subset of $(C[0, 1], d_{\sup})$, and every element f of $\tilde{B}$ is an accumulation point of B: $f = \lim f_n$, where $f_n(x) = f(x) + 1/n$, $x \in [0, 1]$, $n = 1, 2, \dots$ Therefore $\mathrm{Cl}\, B = \tilde{B}$.

12. Let $d_{\sup}$ be the metric on $BC(\mathbb{R}; \mathbb{R})$ generated by the sup-norm. Obviously, D contains functions whose infimum is zero. For example, $g(x) = e^{-x}$, $x \in \mathbb{R}$, is such a function. Any ball $B(g, \varepsilon) \subset (BC(\mathbb{R}; \mathbb{R}), d_{\sup})$ contains functions with negative values, so g is not an interior point of D, hence D is not open in $(BC(\mathbb{R}; \mathbb{R}), d_{\sup})$.

Obviously, $\mathrm{Int}\, D = \{f \in BC(\mathbb{R}; \mathbb{R}); \inf_{\mathbb{R}} f > 0\}$. The closure of D is $\{f \in BC(\mathbb{R}; \mathbb{R}); f(x) \geq 0 \, \forall x \in \mathbb{R}\}$. The proof is the same as in the previous exercise.

13. Clearly, such an open cover does exist since $(0, 1]$ is not a closed set, hence not compact. Indeed, for example the collection $\{(1/n, 2)\}_{n \in \mathbb{N}}$ is an open cover with no finite subcover (easy to check).

14. Let $S \subset (X, d)$ be a discrete subset of a metric space (X, d). If S is finite, then it is clearly compact. Now, let us assume that S is compact and show that it is finite. Assume by way of contradiction that S is infinite and consider an open cover of S, $\{B(x, r_x)\}_{x \in S}$, where $B(x, r_x) \cap S = \{x\}$ for all $x \in S$ (such balls exist since all the points of S are isolated). Any proper sub-collection of $\{B(x, r_x)\}_{x \in S}$ is no longer a cover of S. This contradicts the fact that S is compact. Thus S is compact if and only if it is finite. In fact, this result holds in any topological space.

15. The conclusion follows from the total boundedness of A.

16. Both assertions follow easily by using sequences. As a counterexample, consider the following two subsets of $(\mathbb{R}, |\cdot|)$: $A = [2, +\infty)$ and $B = \{1\}$.

17. The reader is encouraged to first draw the graphs of the f_n's. Denoting by $\|\cdot\|_{\sup}$ the sup-norm of $C[0, 1]$, we have $\|f_n\|_{\sup} = 1$ for all $n \in \mathbb{N}$, so $\mathcal{F}$ is bounded (with respect to the metric $d_{\sup}$, $d_{\sup}(f, g) = \|f - g\|_{\sup}$, $\forall f, g \in C[0, 1]$). On the other hand, $d_{\sup}(f_n, f_m) = 1$ for all $m, n \in \mathbb{N}$, $m \neq n$, so all elements of $\mathcal{F}$ are isolated points (in other words, $\mathcal{F}$ is a discrete set). Therefore, $\mathcal{F}$ is a closed set. Being an infinite discrete set, $\mathcal{F}$ is not compact (one can use the open cover $\{B(f_n, 1/2)\}_{n \in \mathbb{N}}$).

18. We need to analyze the case when A is an infinite set, otherwise the conclusion is obvious.

   Let $(x_n)_{n \in \mathbb{N}}$ be a sequence in A. Since A is totally bounded, it is a subset of a finite union of open balls of radius 1. One of these balls, say $B(a_1, 1)$, contains infinitely many terms of (x_n). Denote $C_1 := B(a_1, 1) \cap A$ and pick a term $x_{n_1} \in C_1$. Obviously, C_1 is also totally bounded, so it is a subset of a finite union of open balls of radius $r = 1/2$. There exists a ball $B(a_2, 1/2)$ such that $C_2 := B(a_2, 1/2) \cap B(a_1, 1) \cap A$ contains infinitely many terms of (x_n). Choose one of these terms, x_{n_2}, with $n_2 > n_1$. Continuing this process, we find $x_{n_j} \in C_j = A \cap \{\cap_{i=1}^{j} B(a_i, 1/i)\}$, with $n_j > n_{j-1}$, $j = 2, 3, \ldots$. Since $x_{n_j} \in B(a_k, 1/k)$ for all $j \geq k$, we have

$$d(x_{n_j}, x_{n_i}) \leq d(x_{n_j}, a_k) + d(x_{n_i}, a_k)) < \frac{2}{k}, \ \forall i, j \geq k,$$

so the subsequence $(x_{n_k})_{k \in \mathbb{N}}$ is a Cauchy sequence in (X, d), hence convergent (to a point in $\mathrm{Cl}\, A$), since (X, d) is complete.

19. (a) It is easily seen that l^1 is a vector space over $\mathbb{R}$ with respect to the usual operations, $\| \cdot \|$ is a norm on l^1, and $(l^1, \| \cdot \|)$ is a Banach space.

    (b) For every $a = (a_n)_{n \in \mathbb{N}} \in A$ we have

    $$\sum_{k=N}^{\infty} |a_k| \leq \frac{1}{N} \sum_{k=N}^{\infty} k |a_k| \leq \frac{1}{N},$$

    so, for all $\varepsilon > 0$, there exists an $N \in \mathbb{N}$ large enough such that

    $$\|(0, 0, \ldots, 0, a_N, a_{N+1}, \ldots)\| < \varepsilon.$$

    This shows that A is totally bounded, since $[-1, +1]^{N-1}$ is. As (l^1, d) is a complete metric space, A is relatively compact (see the previous exercise). It is easily seen that A is closed in (l^1, d), hence A is, in fact, compact.

20. Let us first consider the case $D = [0, 1]$. Denote by A the set of all sequences $a = (a_n)_{n \in \mathbb{N}}$ in $\mathbb{R}$ satisfying $\sum_{n=1}^{\infty} n |a_n| \leq 1$. According to Weierstrass' M-test, for all $a \in A$, the function f_a,

$$f_a(x) = \sum_{n=1}^{\infty} a_n \sin(n\pi x), \ x \in [0, 1],$$

is well defined and belongs to $C[0, 1]$. Indeed, for all $x \in [0, 1]$,

$$|a_n \sin(n\pi x)| \leq |a_n|, \ \forall n \in \mathbb{N}, \ \text{and} \ \sum_{n=1}^{\infty} |a_n| \leq \sum_{n=1}^{\infty} n |a_n| \leq 1,$$

hence the series $\sum_{n=1}^{\infty} a_n \sin(n\pi x)$ is uniformly convergent in $[0, 1]$, so $f_a \in C[0, 1]$. In fact, f_a is continuously differentiable in $[0, 1]$ and

$$f_a'(x) = \sum_{n=1}^{\infty} n a_n \sin(n\pi x), \ x \in [0, 1].$$

It is easily seen that the function $a \mapsto f_a$ is continuous from $A \subset l^1$ to $C[0,1]$. By the previous exercise, A is compact in l^1 so $\mathcal{F} = \{f_a; \, a \in A\}$ is compact in $C[0,1]$ (cf. Theorem 2.19).

Now, if $D = \mathbb{R}$, then $C(\mathbb{R}; \mathbb{R})$ cannot be equipped with the sup-norm. However, the result holds if $C(\mathbb{R}; \mathbb{R})$ is replaced by $BC(\mathbb{R}; \mathbb{R})$ (bounded and continuous functions: $\mathbb{R} \to \mathbb{R}$).

21. We try to apply the Arzelà–Ascoli criterion. First, we have

$$u_n(t) = u_n(s) + \int_s^t u_n'(\tau)\, d\tau \implies |u_n(t)| \leq |u_n(s)|$$
$$+ \int_a^b |u_n'(\tau)|\, d\tau .$$

Integration over $[a,b]$ with respect to s yields

$$(b - a)|u_n(t)| \leq \int_a^b |u_n(s)|\, ds + (b - a) \int_a^b |u_n'(\tau)|\, d\tau ,$$

which shows (by Hölder's inequality) that (u_n) is bounded in $C([a,b]; \mathbb{R})$. On the other hand, we obtain by using Hölder's inequality and the boundedness of (u_n') in $L^p(a,b; \mathbb{R})$ that, for all $a \leq s \leq t \leq b$ and for all $n \in \mathbb{N}$,

$$
\begin{aligned}
|u_n(t) - u_n(s)| &= |\int_s^t u_n'(\tau)\, d\tau| \\
&\leq \int_s^t |u_n'(\tau)|\, d\tau \\
&\leq \left(\int_s^t |u_n'(\tau)|^p\, d\tau \right)^{1/p} (t - s)^{1/q} \\
&\leq \left(\int_a^b |u_n'(\tau)|^p\, d\tau \right)^{1/p} (t - s)^{1/q} \\
&\leq C|t - s|^{1/q} ,
\end{aligned}
$$

where q is the conjugate of p. This shows that (u_n) is equicontinuous, so the result follows by the Arzelà–Ascoli criterion.

22. Proceed by way of contradiction. Assume that $\mathcal{F}$ is not uniformly equicontinuous, i.e., there exists an $\varepsilon_0 > 0$ such that

$$\forall n \in \mathbb{N}, \ \exists x_n, \ y_n \in A, \ d(x_n, y_n) < \frac{1}{n}, \ \forall f \in \mathcal{F},$$

$$\rho(f(x_n, f(y_n)) \geq \varepsilon_0.$$

As A is compact, there exist convergent subsequences $(x_{n_k})_{k \in \mathbb{N}}$, $(y_{n_k})_{k \in \mathbb{N}}$, with the same limit, say $x_0 \in A$. So the above statement contradicts the equicontinuity of $\mathcal{F}$ at $x = x_0$.

23. Apply the Arzelà–Ascoli criterion. Denote by $\| \cdot \|_{\sup}$ the norm of $C[0, 1]$ and by $d_{\sup}$ the induced metric. We have $\|f_a\|_{\sup} \leq 1$ for all $a \in \mathbb{R}$, so $\mathcal{F}$ is bounded in $C[0, 1]$. We also have

$$|f_a'(x)| \leq 1, \ \forall x \in [0, 1], \ a \in \mathbb{R},$$

thus, according to the Mean Value Theorem, $\mathcal{F}$ is (Lipschitz) equicontinuous. It follows by the Arzelà–Ascoli criterion that $\mathcal{F}$ is relatively compact in $C[0, 1]$, but $\mathcal{F}$ is not closed, hence not compact. Indeed, f_a converges uniformly, as $a \to \infty$, to the null function which does not belong to $\mathcal{F}$ (there is no $a \in \mathbb{R}$ such that $f_a \equiv 0$).

24.   (a) Denote

$$y(t) = \int_{t_0}^t b(s)u(s)\, ds.$$

We have

$$\begin{aligned} y'(t) &= b(t)u(t) \\ &\leq a(t)b(t) + b(t)y(t), \ t \in [t_0, T], \end{aligned}$$

which, after multiplication by $e^{-\int_{t_0}^t b(s)\, ds}$, becomes

$$\frac{d}{dt}\left(e^{-\int_{t_0}^t b(s)\, ds} y(t) \right) \leq a(t)b(t)e^{-\int_{t_0}^t b(s)\, ds}, \ t \in [t_0, T].$$

Integrating this inequality over $[t_0, t]$ gives

$$y(t) \leq \int_{t_0}^t a(s)b(s)e^{\int_s^t b(\tau)\, d\tau} ds, \ t \in [t_0, T],$$

which leads to the desired conclusion.

Bellman's lemma follows trivially from this conclusion.

(b) Let $y = y(t)$ be another solution on $[t_0 - \delta, t_0 + \delta]$. Then, for $t \in [t_0, t_0 + \delta]$,

$$\|x(t) - y(t)\| \leq \left\| \int_{t_0}^{t} [f(s, x(s)) - f(s, y(s))] \, ds \right\|$$

$$\leq L \int_{t_0}^{t} \|x(s) - y(s)\| \, ds.$$

According to Bellman's lemma (with $C = 0$), this implies $x(t) = y(t)$, $t \in [t_0, t_0 + \delta]$. The case $t \in [t_0 - \delta, t_0]$ can be reduced to a similar one if we use the change $t = t_0 - \tau$, $\tau \in [0, \delta]$.

25. First consider the case $c \neq 0$. Denote

$$y(t) = |c|^2 + 2 \int_{t_0}^{t} f(s) \cdot |x(s)| \, ds \geq |c|^2.$$

Then

$$|x(t)|^2 \leq y(t), \ t \in I \implies |x(t)| \leq \sqrt{y(t)}, \ t \in I.$$

Thus,

$$\begin{aligned} y'(t) &= 2f(t) \cdot |x(t)| \\ &\leq 2f(t)\sqrt{y(t)}, \ t \in I. \end{aligned}$$

Having in mind that $y(t) \geq |c|^2 > 0$, $\forall t \in I$, we can write

$$\frac{d}{dt}\sqrt{y(t)} \leq f(t), \ t \in I,$$

which gives the desired inequality by integration over $[t_0, T]$.

Now, if $c = 0$ we replace it by $\varepsilon > 0$, so by the above reasoning we obtain

$$|x(t)| \leq |\varepsilon| + \int_{t_0}^{t} f(s) \, ds \quad \forall t \in I,$$

and now let $\varepsilon \to 0$ to finish.

26. According to Peano's Theorem, for any $a, b \in (0, \infty)$ there exists a solution defined on $[-\delta, +\delta]$, where $\delta = \min\{a, b/M\}$, $M = 1 + a^2 + b^2/(1 + b^2)$. The solution is also unique, since the function

defined by the right-hand side of the equation is Lipschitzian with
respect to the second variable. If we fix an arbitrary $a > 0$, we
can choose a sufficiently large $b > 0$ such that

$$\frac{b}{M} \geq a \implies \delta = a.$$

Thus, for any $a > 0$ there exists a unique solution of the given
Cauchy problem, defined on the whole interval $[-a, a]$, so the
solution can (uniquely) be extended to $\mathbb{R}$.

Remark. Note that the function defined by the right-hand side
of the equation, $f(t, v) = 1 + t^2 + v^2/(1 + v^2)$, is Lipschitz continuous with respect to v on $\mathbb{R}^2$. If we consider an arbitrary interval
$[-a, a]$ and use Euler's method of polygonal lines (as in the proof
of Peano's Theorem), we observe that the functions ϕ_ε can be
defined on $[-a, a]$, so we obtain a solution defined on the whole
interval $[-a, a]$. By the Lipschitz continuity of f with respect to
v this solution is also unique. Since a was arbitrarily chosen, the
solution exists on $\mathbb{R}$ and is unique. Another approach towards
solving this exercise is based on the Banach Contraction Principle. Indeed, for an arbitrary but fixed $a > 0$, the operator T
defined by

$$(Tv)(t) = \int_0^t f(s, v(s))\, ds, \ \ t \in [-a, a],$$

maps $C[-a, a] := C([-a, a]; \mathbb{R})$ into itself and T^k is a contraction
(with respect to the usual sup-norm of $C[-a, a]$) for a sufficiently
large k, hence T has a unique fixed point, which is the unique
solution of our Cauchy problem on $[-a, a]$. Hence there exists a
unique solution on $\mathbb{R}$.

27. Apply arguments similar to those used for the previous exercise.

28. Choose $-\infty < t_1 \leq 0 \leq t_2 < +\infty$, and define

$$x(t) = \begin{cases} -(t - t_1)^2, & t < t_1, \\ 0, & t_1 \leq t \leq t_2, \\ (t - t_2)^2, & t > t_2. \end{cases}$$

There are infinitely many pairs (t_1, t_2) that can be chosen in
this way, and all the corresponding x's are solutions of the given

Cauchy problem. Moreover, if we replace one of the restrictions of x to $(-\infty, t_1)$ and $(t_2, +\infty)$ by zero, we obtain further solutions. Of course, the null function is also a solution on $\mathbb{R}$. In fact, these functions are the only solutions of the problem.

29. As $f(t, v) = 1 + t(1 + v^2)$ is Lipschitzian on compact sets (locally Lipschitzian), it follows by Theorem 2.33 that there exists a unique (local) solution, say $x = \phi(t)$, defined on an interval $[0, \delta]$. This solution can be extended (uniquely) to the right. Indeed, let us consider the Cauchy problem

$$x'(t) = 1 + t\big(1 + x(t)^2\big), \ t \geq \delta; \ x(\delta) = \phi(\delta).$$

This problem has a unique solution, say $x = \psi(t)$, on an interval $[\delta, \delta_1]$. Thus, the function defined by

$$x(t) = \begin{cases} \phi(t), & t \in [0, \delta], \\ \psi(t), & t \in (\delta, \delta_1] \end{cases}$$

is the unique solution of our problem on $[0, \delta_1]$. This interval can be further extended. We can prove the existence of a unique solution, $x = x(t)$, defined on a maximal interval $[0, T)$. Let us prove that $T < +\infty$. Assume by contradiction that $T = +\infty$. Therefore,

$$x'(t) = 1 + t\big(1 + x(t)^2\big), \ \forall t \geq 0 \implies \frac{x'(t)}{1 + x(t)^2} \geq t, \ \forall t \geq 0.$$

Integrating over $[0, t]$ we get

$$\arctan x(t) - \arctan x_0 \geq \frac{t^2}{2}, \ \forall t \geq 0,$$

which is impossible.

30. Note that $f : \mathbb{R}^2 \to \mathbb{R}$, $f(t, v) = t^2 + v^2$, is continuous and locally Lipschitzian with respect to v. According to Theorem 2.33, there exists a unique solution $x = x(t)$ on an interval $[-\delta, \delta]$. This solution can be uniquely extended to a maximal interval $(-T_1, T)$. Let us prove that $T < +\infty$. Assume the contrary: $T = +\infty$. Then, for $t \geq 1$, $x'(t) \geq 1 + x(t)^2$, hence

$$\int_1^t \frac{x'(s)}{1 + x(s)^s} \, ds \geq t - 1, \ t \geq 1$$

i.e.,
$$\arctan x(t) \geq \arctan x(1) + t - 1, \ t \geq 1,$$

which is impossible. Thus, $T < +\infty$. On the other hand, we observe that $\tilde{x}(t) = -x(-t)$ is also a solution of the problem. By the uniqueness property, it follows that the solution is an odd function and hence its maximal interval is symmetric with respect to $t = 0$, i.e., $T_1 = T$.

It remains to show that $T > \sqrt{2}/2$. Consider our Cauchy problem on a rectangle $[-a, a] \times [-b, b]$ with $a, b > 0$. From Theorem 2.33 we derive existence and uniqueness on $[-\delta, \delta]$, where $\delta = \min\{a, b/(a^2 + b^2)\}$. Note that, for a given $a > 0$, the maximal value of $b/(a^2 + b^2)$ is $1/(2a)$, being attained for $b = a$. Now, the maximal value of $\min\{a, 1/(2a)\}$ is reached for $a = \sqrt{2}/2$. Summarizing, for $a = b = \sqrt{2}/2$ the corresponding $\delta = \sqrt{2}/2$, so $T > \sqrt{2}/2$.

31. For any $(t_0, x_0) \in \Omega$ one can choose sufficiently small numbers $a, b > 0$ such that

$$D_{a,b} = \{(t, v) \in \Omega; \ |t - t_0| \leq a, \ \|v - x_0\| \leq b\} \subset \Omega.$$

Apply Peano's Theorem to get (local) existence, and for uniqueness just observe that $f = f(t, v)$ is Lipschitzian with respect to v on the compact $D_{a,b}$. See also Theorem 2.33.

32. It is enough to prove existence and uniqueness on every compact subinterval of I containing t_0. Let $[a, b]$ be such a subinterval. Obviously the function $f : [a, b] \times \mathbb{R}^k \to \mathbb{R}^k$, $f(t, v) = A(t)v + b(t)$, is continuous. It is easily seen that f is Lipschitzian with respect to v (actually, with respect to any norm of $\mathbb{R}^k$) since $a_{ij}|_{[a,b]} \in C[a, b]$, $i, j = 1, 2, \ldots, k$. According to Theorem 2.33, there exists a unique solution on the whole interval $[a, b]$.

33. For $n \in \mathbb{N}$ consider the operator $T_n : \overline{B(0, 1)} \to \overline{B(0, 1)}$, defined by

$$T_n x = \left(1 - \frac{1}{n}\right) Tx, \quad x \in \overline{B(0, 1)}.$$

Obviously, for each $n \in \mathbb{N}$, T_n is a contraction on the metric space $(\overline{B(0, 1)}, d_2)$, so it has a unique fixed point $x_n \in \overline{B(0, 1)}$ (cf. Banach's Contraction Principle). To conclude, we need merely to use the (sequential) compactness of $\overline{B(0, 1)}$ and the continuity of T.

34. Apply the Banach Contraction Principle in $C[0,1]$, equipped with the usual sup-norm, to the operator $T : C[0,1] \to C[0,1]$ defined by the right-hand side of the equation. As the function $y \mapsto \cos(\alpha y)$ is Lipschitzian of constant α and T is a contraction, it has a unique fixed point $x \in C[0,1]$, which is the unique solution of the given equation.

35. Let $m \in (0, \infty)$ be arbitrary but fixed. Note that $C([0,m]; X)$ is a Banach space with respect to the sup-norm: $\|y\|_m = \sup_{t \in [0,m]} \|y(t)\|$, $y \in C([0,m]; X)$. Define

$$(Ty)(t) = x_0 + \int_0^t f(s, y(s))\, ds, \ t \in [0, m], \ y \in C([0,m]; X).$$

It is easily seen that T maps $C([0,m]; X)$ into itself and T^k is a contraction on this space for a sufficiently large $k \in \mathbb{N}$ (since $f = f(t,v)$ is Lipschitz continuous with respect to v with Lipschitz constant $L_m = \sup\{|a(t)|; t \in [0,m]\}$). Therefore T has a unique fixed point $x \in C([0,m]; X)$ (see Remark 2.38), which is the unique solution of our Cauchy problem on $[0,m]$. As m was chosen arbitrarily, x can be uniquely extended to $[0, \infty)$. From the original equation we see that $x \in C^1([0,\infty); X)$.

12.3   Answers to Exercises for Chap. 3

1. Assume that $\Omega \subset \mathbb{R}^k$ is measurable. Then $\mathbb{R}^k \setminus \Omega$ is measurable, so for every $\varepsilon > 0$ there exists an open set $D \supset \mathbb{R}^k \setminus \Omega$ such that $m(D \setminus (\mathbb{R}^k \setminus \Omega)) < \varepsilon$ (see Definition 3.1). It follows that $F := \mathbb{R}^k \setminus D$ is closed and $m(\Omega \setminus F) < \varepsilon$.

   The converse implication can be proved similarly.

2. Let $\varepsilon > 0$ be arbitrary, but fixed. Since Ω is measurable, it follows from the previous exercise that there exists a closed set $F \subset \Omega$ such that $m(\Omega \setminus F) < \varepsilon/2$. If it turns out that F is bounded, hence compact, we are done. Assume now that F is unbounded. Observe that $F_n := F \cap \overline{B(0, n)}$ is a compact set for each $n \in \mathbb{N}$, where $\overline{B(0, n)}$ is the closed ball centered at 0 of radius n. Since $F_n \subset F_{n+1}$ and $F_n \subset F \subset \Omega$, it follows that the sequence $(m(F_n))$ is nondecreasing and $m(F_n) \leq m(F) \leq m(\Omega) < \infty$, $n = 1, 2, \ldots$. Therefore, there exists

$\lim_{n\to\infty} m(F_n) \leq m(F) < \infty$. On the other hand, F can be written as a countable union of measurable disjoint sets,

$$F = F_1 \cup \left(\cup_{n=2}^{\infty} (F_n \setminus F_{n-1}) \right),$$

hence

$$
\begin{aligned}
m(F) &= m(F_1) + \sum_{n=2}^{\infty} [m(F_n) - m(F_{n-1})] \\
&= \lim_{n\to\infty} m(F_n) \\
&\leq m(F).
\end{aligned}
$$

Therefore $m(F_n) \to m(F)$. It follows that for a sufficiently large N,

$$m(F \setminus F_N) = m(F) - m(F_N) < \varepsilon/2,$$

where $F_N =: K$ is a compact set. Since $\Omega \setminus K = (\Omega \setminus F) \cup (F \setminus K)$, we conclude that

$$
\begin{aligned}
m(\Omega \setminus K) &= m(\Omega \setminus F) + m(F \setminus K) \\
&< \frac{\varepsilon}{2} + \frac{\varepsilon}{2} = \varepsilon.
\end{aligned}
$$

3. Since $B = (B \setminus A) \cup A$, we have

$$m(B) = m(B \setminus A) + m(A) \implies m(B \setminus A) = 0.$$

As $\Omega \setminus A$ is a subset of the null set $B \setminus A$, it follows that $\Omega \setminus A$ is measurable with $m(\Omega \setminus A) = 0$. From $\Omega = (\Omega \setminus A) \cup A$ we deduce that Ω is measurable, and $m(\Omega) = m(\Omega \setminus A) + m(A) = m(A)$.

4. (a) If $C \subset \mathbb{R}^k$ is a closed cube, then C_ξ is so for any $\xi \in \mathbb{R}^k \setminus \{0\}$ and $v(C_\xi) = v(C)$. Therefore,

$$m_e(\Omega_h) = m_e(\Omega) = m(\Omega).$$

So Ω_h is measurable and $m(\Omega_h) = m(\Omega)$.

(b) Employ similar arguments.

5. Assume in addition that $f \geq 0$. The function $f_h : \mathbb{R}^k \to \mathbb{R}$ defined by

$$f_h(x) = f(x - h), \quad x \in \mathbb{R}^k,$$

is measurable. Indeed, for any $\lambda \in \mathbb{R}$, we have

$$\{x \in \mathbb{R}^k; \, f_h(x) > \lambda\} = \{y + h; \, f(y) > \lambda\},$$

which is measurable (see the previous exercise). Obviously, the property holds for nonnegative simple functions and hence, by using the standard limiting process, one can obtain it for f. For a general $f \in L^1(\mathbb{R}^k)$, one can use the decomposition $f = f^+ - f^-$.

Similar arguments work for the function $x \mapsto f(\alpha x)$.

6. Consider a sequence of partitions P_n of $[a, b]$ with norms tending to 0, as well as the corresponding sequences of the lower and upper Riemann sums, say L_n and U_n, which can be interpreted as Lebesgue integrals of some simple functions l_n and u_n, $l_n \leq f \leq u_n$. As f is Riemann integrable, we have

$$L_n \longrightarrow (R) \int_a^b f(x)\, dx \longleftarrow U_n \quad \text{as } n \to \infty,$$

which leads to the desired conclusion. For more details, see, for example, [50, Theorem 5.52, p. 83].

Now, obviously, D is not Riemann integrable, but it is Lebesgue integrable, since $D = 0$ almost everywhere.

7. Use integration by parts (with $u = 1/(1 + x)$, $v = x^n$), then Lebesgue's Dominated Convergence Theorem.

8. Set

$$f_n(x) = \chi_{[1,n]}(x) x^{-2} \ln x, \quad x \in \mathbb{R}, \; n = 2, 3, \ldots$$

Observe that the f_n's are measurable, $f_n(x) \to \chi_{[1,\infty)}(x) x^{-2} \ln x$ and $0 \leq f_n(x) \leq f_{n+1}(x)$ for each $x \in \mathbb{R}$. By the Monotone Convergence Theorem,

$$\lim_{n \to \infty} \int_{-\infty}^{+\infty} f_n(x)\, dx = \int_1^{\infty} f(x)\, dx \, .$$

On the other hand,

$$\lim_{n\to\infty} \int_{-\infty}^{+\infty} f_n(x)\, dx = \int_1^n x^{-2} \ln x\, dx$$

$$= -\int_1^n \ln x\, d(x^{-1})$$

$$= -x^{-1} \ln x \big|_1^n + \int_1^n x^{-2}\, dx$$

$$= -\frac{\ln n}{n} + 1 - \frac{1}{n} \to 1,$$

which, combined with the previous equality, implies the result.

9. Let $f_n : (0, \infty) \to \mathbb{R}$ be defined by

$$f_n(x) = \chi_{(0,n]}(x)\left(1 + \frac{x}{n}\right)^n e^{-2x}, \quad x > 0, \; n = 1, 2, \ldots$$

We have for each $x > 0$,

$$n \ln\left(1 + \frac{x}{n}\right) \le x \implies \left(1 + \frac{x}{n}\right)^n \le e^x, \; n = 1, 2, \ldots,$$

$$\lim_{n\to\infty} n \ln\left(1 + \frac{x}{n}\right) = x \implies \lim_{n\to\infty} \left(1 + \frac{x}{n}\right)^n = e^x.$$

So, by the Lebesgue Dominated Convergence Theorem,

$$\lim_{n\to\infty} \int_0^n \left(1 + \frac{x}{n}\right)^n e^{-2x}\, dx = \lim_{n\to\infty} \int_0^\infty f_n(x)\, dx$$

$$= \lim_{n\to\infty} \int_0^n e^{-x}\, dx = 1.$$

10. Apply Lebesgue's Dominated Convergence Theorem.

11. (a) Left to the reader;

(b) For each $m \in \mathbb{N}$,

$$f_m(x) = \chi_{[m^{-1}, 1]}(x) f(x), \quad x \in [0, 1],$$

is a simple function. We also have $0 \le f_m \le f_{m+1}$ and $f_m \to f$ as $m \to \infty$. Hence f is measurable and, for $1 \le p < \infty$,

$$\int_0^1 f(x)^p\, dx = \sum_{n=1}^{\infty} (\sqrt{n})^p \left(\frac{1}{n} - \frac{1}{n+1}\right)$$

$$= \sum_{n=1}^{\infty} \frac{n^{(p-2)/2}}{n+1},$$

which is finite for $1 \leq p < 2$ and infinite for $2 \leq p < \infty$. It is obvious that $f \notin L^\infty(0,1)$.

12. Left to the reader.

13. By assumption, there exists

$$\lim_{x \to 0^+} \frac{f(x) - f(0)}{x - 0} = \lim_{x \to 0^+} x^{-1} f(x) = f'_+(0),$$

hence

$$|g(x)| = x^{-1/2} |x^{-1} f(x)| \leq C x^{-1/2} \quad \forall x(0,1],$$

where C is a positive constant. For each $n \in \mathbb{N}$, define

$$h_n(x) = \chi_{[n^{-1},1]}(x) |g(x)|, \quad x \in (0,1],$$

extended as zero on $\mathbb{R} \setminus (0,1]$. Applying the Monotone Convergence Theorem, we find

$$\begin{aligned}
\int_0^1 |g(x)| \, dx &= \lim_{n \to \infty} \int_0^1 h_n(x) \, dx \\
&= \lim_{n \to \infty} \int_{1/n}^1 |g(x)| \, dx \\
&\leq C \int_0^1 x^{-1/2} \, dx = 2C < \infty.
\end{aligned}$$

14. Apply Lebesgue's Dominated Convergence Theorem.

15. The case $q = \infty$ is trivial. For all the other cases, use Hölder's inequality.

16. The case $\|f\|_{L^\infty(\Omega)} = 0 \iff f = 0$ a.e. in Ω is trivial, so let us assume that $\|f\|_{L^\infty(\Omega)} > 0$. Obviously (see also the previous exercise),

$$\|f\|_{L^p(\Omega)} \leq \|f\|_{L^\infty(\Omega)} m(\Omega)^{1/p} \quad \forall p \geq 1,$$

which implies

$$\limsup_{p \to \infty} \|f\|_{L^p(\Omega)} \leq \|f\|_{L^\infty(\Omega)}. \tag{$*$}$$

Now, for $0 < \alpha < \|f\|_{L^\infty(\Omega)}$, set $E = \{x \in \Omega;\ |f(x)| > \alpha\}$. Clearly, $m(E) > 0$ and for $p \geq 1$ we have

$$\int_\Omega |f|^p\, dx \geq \int_E |f|^p\, dx \geq \alpha^p m(E) \implies \liminf_{p \to \infty} \|f\|_{L^p(\Omega)} \geq \alpha.$$

Therefore,

$$\liminf_{p \to \infty} \|f\|_{L^p(\Omega)} \geq \|f\|_{L^\infty(\Omega)},$$

which, combined with $(*)$ above, concludes the proof.

12.4 Answers to Exercises for Chap. 4

1. If G is the graph of a linear operator $A : X \to Y$, i.e., $G = \{[x, Ax];\ x \in X\}$, then necessarily $P_X G$ (the projection of G onto X) is the whole of X and G is a linear subspace of $X \times Y$ (or, equivalently, $P_Y G$ is a linear subspace of Y) whose pairs enjoy the property that the right component is uniquely associated with the first component. Conversely, let G be a linear subspace of $X \times Y$ with $P_X G = X$ satisfying the property:

$$\forall x \in X\ \exists \text{ a unique } y \in P_Y G \text{ such that } [x, y] \in G.$$

Define $A : X \to Y$ by $Ax = y$. It is easy to check that A is a linear operator whose graph is precisely G.

2. It is readily seen that

$$A(rx) = rAx \quad \forall r = \frac{m}{n} \in \mathbb{Q},\ x \in X.$$

Using the density of $\mathbb{Q}$ in $\mathbb{R}$ and the continuity of A we derive

$$A(\alpha x) = \alpha Ax \quad \forall \alpha \in \mathbb{R},\ x \in X,$$

hence A is linear.

3. (i) Denote the norm of $X = C[a, b]$ by $\|\cdot\|_{\mathrm{sup}}$. Obviously,

$$\|Af\|_{\mathrm{sup}} \leq \max(|a|, |b|) \cdot \|f\|_{\mathrm{sup}} \quad \forall f \in X$$
$$\implies \|A\| \leq \max(|a|, |b|).$$

On the other hand, for the constant function $f(t) = 1$, $t \in [a, b]$, we have $\|f\|_{\mathrm{sup}} = 1$ and $\|Af\|_{\mathrm{sup}} = \max(|a|, |b|)$, hence $\|A\| = \max(|a|, |b|)$.

(ii)   Denote the norm of $X = L^p(a, b)$ by $\| \cdot \|_p$. We have

$$\|Af\|_p \leq \max\left(|a|, |b|\right) \cdot \|f\|_p \forall f \in X \implies \|A\| \leq \max\left(|a|, |b|\right).$$

Let us prove that the converse inequality, $\|A\| \geq \max\left(|a|, |b|\right)$, is also satisfied and thus $\|A\| = \max\left(|a|, |b|\right)$. Consider first the case

$$\max\left(|a|, |b|\right) = |b| \implies b > 0.$$

Define the sequence of functions

$$f_n(t) = \begin{cases} 0, & a < t < b - \frac{1}{n}, \\ n^{1/p}, & b - \frac{1}{n} < t < b, \end{cases}$$

for $n \in \mathbb{N}$, $n > 1/b$. Obviously, $\|f_n\|_p = 1$ and

$$\|Af_n\|_p = \left(\int_{b-\frac{1}{n}}^n |t|^p n \, dt\right)^{1/p} \geq b - \frac{1}{n} \quad \forall n \in \mathbb{N}, \ n > \frac{1}{b},$$

which implies $\|A\| \geq b = \max\left(|a|, |b|\right)$. The case

$$\max\left(|a|, |b|\right) = |a| \implies a < 0$$

is similar and is left to the reader.

4. Clearly, A is a linear operator from X into itself. Using Theorem 2.29 (the Arzelà–Ascoli Criterion) we find $A \in K(X)$. We also have

$$\|Af\|_{\sup} \leq \|f\|_{\sup} \int_a^b g(s) \, ds \quad \forall f \in X \implies \|A\|$$

$$\leq \int_a^b g(s) \, ds.$$

Testing with the constant function $f(t) = 1$, $t \in [a, b]$, we see that, in fact, $\|A\| = \int_a^b g(s) \, ds$.

5. Obviously, if A is continuous, then $(*)$ holds true. For the converse implication, it suffices to prove that there exists an $r > 0$ such that $A\left(B_X(0, r)\right) = \{Ax; \ x \in X, \|x\|_X < r\}$ is bounded in $(Y, \| \cdot \|_Y)$. Assume the contrary, i.e., for all $n \in \mathbb{N}$, the set $A\left(B_X(0, 1/n)\right)$ is unbounded. This means there is a sequence (x_n) in X such that $\|x_n\|_X < 1/n$, $\|Ax_n\|_Y > n$, for all $n \in \mathbb{N}$, which contradicts $(*)$.

6. According to Theorem 4.6, $L(X, Y)$ is a Banach space. Denote $S_n = A_1 + A_2 + \cdots + A_n$. For every $\varepsilon > 0$ there exists an $N_\varepsilon \in \mathbb{N}$ such that

$$
\begin{aligned}
\|S_{n+p} - S_n\| &= \|A_{n+1} + A_{n+2} + \cdots + A_{n+p}\| \\
&\leq \|A_{n+1}\| + \|A_{n+2}\| + \cdots + \|A_{n+p}\| \\
&\leq a_{n+1} + a_{n+2} + \cdots + a_{n+p} \\
&< \varepsilon, \quad \forall n > N_\varepsilon, \ p \in \mathbb{N},
\end{aligned}
$$

since the series $\sum_{n=1}^\infty a_n$ is convergent. So $(S_n)_{n \in \mathbb{N}}$ is a Cauchy sequence (in the Banach space $L(X, Y)$), hence it is convergent. This means that $\sum_{n=1}^\infty A_n$ is convergent in $L(X, Y)$.

7. (i)   Use the previous exercise with $Y = X$ and

$$
A_n = \frac{1}{n!} A^n, \quad a_n = \frac{1}{n} \|A\|^n \ \forall n \in \mathbb{N}.
$$

Indeed, we have

$$
\|A_n\| = \frac{1}{n!} \|A^n\| \leq \frac{1}{n!} \|A\|^n = a_n \ \forall n \in \mathbb{N}.
$$

The notation e^A for the sum of this series arises naturally from the similar notation for the classical exponential e^a, $a \in \mathbb{R}$.

(ii)   From classical analysis we know that $(1 - a)^{-1}$ is the sum of the geometric series $1 + a + a^2 + \cdots + a^n + \cdots$ if $|a| < 1$. So we are naturally led to the following geometric series in $L(X)$

$$(\alpha) \qquad I + A + A^2 + \cdots + A^n + \cdots,$$

where I denotes the identity operator. Since $\|A^n\| \leq \|A\|^n$ and $a := \|A\| < 1$, it follows that the series (α) above is convergent in $L(X)$ (see the solution of the previous exercise). Denote its sum by S, i.e., $\|S_n - S\| \to 0$, where $S_n = I + A + A^2 + \cdots + A^n$, $n \in \mathbb{N}$. Note that

$$(I - A)S_n = S_n(I - A) \ \forall n \in \mathbb{N}.$$

Letting $n \to \infty$ in this equality yields $(I - A)S = S(I - A)$, so $I - A$ is invertible and $(I - A)^{-1} = S$ which is an element of $L(X)$.

8. The answer is based on arguments similar to those used in classical analysis for the identity $e^a \cdot e^b = e^{a+b}$ $(a, b \in \mathbb{K})$.

9. (a) This follows directly from Theorem 4.7 (Uniform Boundedness Principle);

   (b) It is clear that T is a linear operator. From (a) we infer that there exists a constant $C > 0$ such that

$$\|T_n x\|_Y \le C\|x\|_X \quad \forall x \in X.$$

Therefore,

$$\|Tx\|_Y \le C\|x\|_X \quad \forall x \in X \implies T \in L(X, Y).$$

   (c) From
$$\|T_n x\|_Y \le \|T_n\| \quad \forall x \in X, \ \|x\| \le 1$$

we find

$$\|Tx\|_Y \le \liminf \|T_n\| \quad \forall x \in X, \ \|x\| \le 1 \implies \|T\|$$
$$\le \liminf \|T_n\|.$$

10. Use Theorem 4.7 (Uniform Boundedness Principle) with

$$(X := X^*, \|\cdot\|_{X^*}), \ (Y := \mathbb{K}, |\cdot|), \ I := S.$$

For $x \in S$ define $T_x : X^* \to \mathbb{K}$ by

$$T_x(f) = f(x), \quad f \in X^*.$$

By the condition on S from the statement of the problem, we have

$$\sup_{x \in S} |T_x(f)| < \infty \quad \forall f \in X^*.$$

So, by Theorem 4.7, there exists a constant $c > 0$ such that

$$|f(x)| \le c\|f\|_{X^*} \quad \forall f \in X^*, \ x \in S \implies \|x\| \le c \quad \forall x \in S,$$

cf. Corollary 4.18.

11. Apply Theorem 4.10 (Closed Graph Theorem). In order to do that, it is sufficient to show that A is a closed operator (equivalently, the graph of A is closed in $X \times Y$). Let $x_n \to x$ in X, $Ax_n \to f$ in X^*. Letting $n \to \infty$ in $(Ax_n)(y) = (Ay)(x_n)$, $y \in X$ yields

$$
\begin{aligned}
f(y) &= (Ay)(x) \\
&= (Ax)(y) \ \ \forall y \in X,
\end{aligned}
$$

hence $f = Ax$.

12. Obviously, $(D(A), \| \cdot \|_X)$ is a Banach space. Using Theorem 4.10 we infer that the restriction of A to $D(A)$ is a linear continuous operator from $(D(A), \| \cdot \|_X)$ into $(Y, \| \cdot \|_Y)$.

13. In order to apply Theorem 4.10, we show that A is a closed operator. To this purpose, consider $x_n \to x$ in X and $Ax_n =: f_n \to f$ in X^*. By the assumption we have

$$
\begin{aligned}
(f_n - Ay)(x_n - y) \geq 0 \ \ \forall y \in X \implies (f - Ay)(x - y) \\
\geq 0 \ \ \forall y \in X.
\end{aligned}
$$

Now take $y = x - tz$, $t \in \mathbb{R}$, $z \in X$ to conclude that $f = Ax$.

14. The identity operator $I : (X, \| \cdot \|_1) \to (X, \| \cdot \|_2)$ is bijective and continuous (due to the inequality which was assumed to be satisfied by the two norms). According to Theorem 4.8 (Open Mapping Theorem), $I^{-1} = I \in L\big((X, \| \cdot \|_2), (X, \| \cdot \|_1)\big)$, hence there exists a constant $C_1 > 0$ such that

$$
\|x\|_1 \leq C_1 \|x\|_2 \ \ \forall x \in X.
$$

This combined with the inequality from the statement of the problem shows the equivalence of the two norms.

15. We shall assume that $f \neq 0$, otherwise its norm is zero in all cases.

(i) For $u = \sum_{i=1}^n \alpha_i u_i \in X$, we have

$$
|f(u)| \leq \sum_{i=1}^n |\alpha_i| \cdot |f_i| \leq \|u\|_\infty \sum_{i=1}^n |f_i|,
$$

hence $\|f\|_{X^*} \leq \sum_{i=1}^{n} |f_i|$. Now, choose a particular u, namely $\tilde{u} = \sum_{i=1}^{n} \tilde{\alpha}_i u_i \in X$, where

$$\tilde{\alpha}_i = \begin{cases} 0, & f_i = 0, \\ |f_i|^{-1} \bar{f}_i, & f_i \neq 0, \end{cases}$$

for $i = 1, 2, \ldots, n$. Here $\bar{f}_i$ denotes the complex conjugate of f_i. Since $\|\tilde{u}\|_\infty = 1$, $f(\tilde{u}) = \sum_{i=1}^{n} |f_i|$ and $\|f\|_{X^*} \leq \sum_{i=1}^{n} |f_i|$ (see above), it follows that

$$\|f\|_{X^*} = \sum_{i=1}^{n} |f_i|.$$

(ii) In this case, for $u = \sum_{i=1}^{n} \alpha_i u_i \in X$, we have

$$|f(u)| \leq \sum_{i=1}^{n} |\alpha_i| \cdot |f_i| \leq \|u\|_1 \cdot \max_{1 \leq i \leq n} |f_i|,$$

hence $\|f\|_{X^*} \leq \max_{1 \leq i \leq n} |f_i|$. Assume that $\max_{1 \leq i \leq n} |f_i|$ (which is a positive number) is achieved for some i_0 and choose the vector $\tilde{u} \in X$ whose coordinates are null, except for $\tilde{\alpha}_{i_0} = |f_{i_0}|^{-1} \bar{f}_{i_0}$. Since $\|\tilde{u}\|_1 = 1$ and $f(\tilde{u}) = |f_{i_0}|$, we infer that

$$\|f\|_{X^*} = \max_{1 \leq i \leq n} |f_i|.$$

(iii) In this case, for $u = \sum_{i=1}^{n} \alpha_i u_i \in X$, we have

$$|f(u)| \leq \sum_{i=1}^{n} |\alpha_i| \cdot |f_i| \leq \|u\|_p \left(\sum_{i=1}^{n} |f_i|^q \right)^{1/q},$$

where q is the conjugate of p (i.e., $1/p + 1/q = 1$), so

$$\|f\|_{X^*} \leq \left(\sum_{i=1}^{n} |f_i|^q \right)^{1/q}.$$

On the other hand, for $u := \tilde{u} = \sum_{i=1}^{n} \tilde{\alpha}_i u_i$, where

$$\tilde{\alpha}_i = \begin{cases} 0, & f_i = 0, \\ \left(\sum_{j=1}^{n} |f_j|^q \right)^{-1+1/q} |f_i|^{q-2} \bar{f}_i, & f_i \neq 0, \end{cases}$$

we have $\|\tilde{u}\|_p = 1$ and $f(\tilde{u}) = \left(\sum_{i=1}^n |f_i|^q \right)^{1/q}$, hence

$$\|f\|_{X^*} = \left(\sum_{i=1}^n |f_i|^q \right)^{1/q}.$$

16. It is easily seen that $\|f\|_{X^*} \leq 1$. In fact, $\|f\|_{X^*} = 1$. Indeed, choosing $u_n : [0.1] \to \mathbb{R}$, $n \in \mathbb{N}$, defined by

$$u_n(t) = \begin{cases} nt, & 0 \leq t \leq 1/n, \\ 1, & 1/n < t \leq 1, \end{cases}$$

we have

$$u_n \in X, \quad \|u_n\|_{\sup} = 1, \quad f(u_n) = 1 - \frac{1}{2n} \quad \forall n \in \mathbb{N},$$

which proves the assertion.

Now, to answer the question from the statement of the problem, observe that $|f(u)| \leq f(|u|)$, $u \in X$, so it is enough to consider only nonnegative functions in the definition of $\|f\|_{X^*}$. Assume by way of contradiction that there exists a function $u \in X$, $u \geq 0$, $\|u\|_{\sup} = 1$ such that $f(u) = \|f\|_{X^*} = 1$, i.e., $\int_0^1 (1 - u(t))\, dt = 0$. As $u \in C[0,1]$ with values in $[0,1]$, this implies $u(t) = 1$ for all $t \in [0,1]$, which contradicts the fact that $u(0) = 0$.

So the answer is "no."

12.5   Answers to Exercises for Chap. 5

1. It is easy to see that u and all its partial derivatives of order $k \leq 2$ are in $C(\Omega)$, so $u \in C^2(\Omega)$. In order to find $\operatorname{supp} u$, notice that $u = 0$ on $\{0\} \times (-1, 1)$ as well as on the graph of the function

$$x_1 = -\frac{1}{x_2} - |x_2| x_2, \quad x_2 \in (-1, 0) \cup (0, 1),$$

and $u \neq 0$ otherwise. Therefore, $\operatorname{supp} u = \operatorname{Cl}\Omega = \mathbb{R} \times [-1, +1]$.

2. Set $\mathcal{F} = \{p_t : C[0,1] \to \mathbb{R}; \, t \in [0,1]\}$, where

$$p_t(f) = |f(t)|, \quad f \in C[0,1].$$

Obviously, p_t is a seminorm for all $t \in [0,1]$. In addition, $\mathcal{F}$ satisfies the axiom of separation: for all $f \in C[0,1]$, $f \neq 0$, there exists a $t \in [0,1]$, such that $f(t) \neq 0 \iff p_t(f) \neq 0$. It is easily seen that convergence with respect to the topology generated by $\mathcal{F}$ means pointwise convergence.

3. It is enough to prove the triangle inequality for d (the other axioms are trivially satisfied). For each $j \in \mathbb{N}$, we have

$$d_j(f,g) \leq d_j(f,h) + d_j(h,g), \quad f,g,h \in C^j(\Omega).$$

This follows from the inequality

$$\frac{|u-v|}{1+|u-v|} \leq \frac{|u-w|}{1+|u-w|} + \frac{|w-v|}{1+|w-v|}, \quad u,v,w \in \mathbb{R},$$

which is a consequence of

$$\frac{\alpha+\beta}{1+\alpha+\beta} \leq \frac{\alpha}{1+\alpha} + \frac{\beta}{1+\beta}, \quad \alpha,\beta \geq 0.$$

The triangle inequality for d follows similarly.

4. Having in mind the typical example of a test function (see Sect. 5.1) we can provide the following example:

$$\phi(x) = \begin{cases} C \exp\left(\frac{1}{(x-2)^2-4}\right), & 0 < x < 4, \\ 0 & \text{otherwise}, \end{cases}$$

where we choose $C = \exp(1/4)$ to obtain $\sup_{\mathbb{R}} \phi = 1$.

5. Any function $\phi = \phi(t) \in C_0^\infty(\mathbb{R})$ with $\int_{\mathbb{R}} \phi(t)\,dt = 0$ can be expressed as the derivative of the function $\phi_1 = \phi_1(t) := \int_{-\infty}^{t} \phi(s)\,ds$ which belongs to $C_0^\infty(\mathbb{R})$. Conversely, if ϕ is the derivative of a function $\phi_1 \in C_0^\infty(\mathbb{R})$, then $\int_{\mathbb{R}} \phi(t)\,dt = 0$. This result can easily be used for the case of k variables to derive the conclusion.

6. If $\phi \in C_0^\infty(\mathbb{R})$ with $\phi(n) = a_n$, $n \in \mathbb{N}$, then $a_n = 0$ for all sufficiently large n (located outside $\text{supp}\,\phi$). Conversely, if $a_n = 0 \; \forall n > n_0$, we can construct the test function $\phi : \mathbb{R} \to \mathbb{R}$,

$$\phi(x) = \begin{cases} a_n \exp\left(\frac{81(x-n)^2}{9(x-n)^2-1}\right), & |x-n| < 1/3, \; n = 1,2,\ldots,n_0, \\ 0, & \text{otherwise}, \end{cases}$$

which satisfies the required properties.

7. Let $r > 0$ be such that $\operatorname{supp} \psi \subset \overline{B(0,r)}$. Then $\operatorname{supp} \phi_n \subset \overline{B(0,r)}$ for all $n \in \mathbb{N}$. We also have for all $n \in \mathbb{N}$ and C a constant

$$|\phi_n(x)| \leq C\,n^m 2^{-n}, \quad x \in \overline{B(0,r)} \implies \phi_n \to 0 \quad \text{uniformly.}$$

The same fact holds for

$$D^\alpha \phi_n(x) = n^{m+|\alpha|} 2^{-m} D^\alpha \psi(nx), \quad x \in \mathbb{R}^k, \ n \in \mathbb{N},$$

for every multi-index $\alpha = (\alpha_1, \ldots, \alpha_k)$.

8. Let $r > 0$ be such that $\operatorname{supp} \psi \subset \overline{B(0,r)}$. Then

$$\operatorname{supp} \phi_n \subset \overline{B(0, r + \|h\|)} \quad \forall n \in \mathbb{N}.$$

Using Taylor's formula we can write

$$\left| \phi_n(x) - \sum_{j=1}^{k} h_j \frac{\partial \psi}{\partial x_j}(x) \right| = \mathcal{O}\!\left(\frac{1}{n}\right), \quad \forall x \in \mathbb{R}^k, \ n \in \mathbb{N},$$

and similar formulas for the $D^\alpha \phi_n$'s showing that

$$\phi_n \to \sum_{j=1}^{k} h_j \frac{\partial \psi}{\partial x_j} \quad \text{in } D(\mathbb{R}^k).$$

The last claim follows trivially from the previous one with h and $-h$.

9. Obviously, for every $n \in \mathbb{N}$ large enough, ϕ_n is well defined and $\operatorname{supp} \phi_n \subset K$, where K is a compact subset of Ω. From Proposition 5.4 we know that $\phi_n \to \phi$ uniformly. Notice that

$$\begin{aligned}
\frac{\partial \phi_n}{\partial x_j}(x) &= \int_\Omega \phi(y) \frac{\partial}{\partial x_j} \omega_{1/n}(x - y)\, dy \\
&= -\int_\Omega \phi(y) \frac{\partial}{\partial y_j} \omega_{1/n}(x - y)\, dy \\
&= \int_\Omega \frac{\partial \phi}{\partial y_j}(y) \omega_{1/n}(x - y)\, dy \to \frac{\partial \phi}{\partial x_j}(x), \quad j = 1, \ldots, k,
\end{aligned}$$

uniformly in Ω as $n \to \infty$. This result extends to $D^\alpha \phi_n$ for every multi-index α,

$$D^\alpha \phi_n \to D^\alpha \phi \quad \text{uniformly in } \Omega,$$

so $\phi_n \to \phi$ in $D(\Omega)$.

10. Left to the reader.

11. Let u be the regular distribution generated by ϕ. Then

$$u(\phi) = \int_\Omega \phi^2 dx = 0,$$

which implies $\phi = 0$ as claimed.

12. For any $\phi \in D(\mathbb{R})$ we have, for some constant C,

$$|\phi(1/i^2) - \phi(0)| \;=\; |\phi'(\theta_i)| \cdot \frac{1}{i^2}, \quad 0 < \theta_i < 1,$$

$$\leq \; C\frac{1}{i^2},$$

which implies that the series defining $u(\phi)$ is absolutely convergent, i.e., u is well defined. It is also easily seen that $u \in D'(\mathbb{R})$.

Now assume by way of contradiction that u is a regular distribution, i.e., there exists $f \in L^1_{\text{loc}}(\mathbb{R})$ such that

$$u(\phi) = \sum_{i=1}^{\infty} \big(\phi(1/i^2) - \phi(0)\big) = \int_{-\infty}^{+\infty} f(t)\phi(t)\, dt \quad \forall \phi \in D(\mathbb{R}).$$

$$(*)$$

Choosing ϕ with support in $\mathbb{R} \setminus \{0, 1, 1/2^2, 1/3^2, \dots\}$ we deduce that $f = 0$ almost everywhere in $\mathbb{R}$ (see Theorem 5.9). So, according to $(*)$, $u(\phi) = 0$ for all $\phi \in D(\mathbb{R})$, which is a contradiction (take, for instance, $\phi = \omega$).

13. Left to the reader.

14. Left to the reader.

15. We have $f'(x) = |x|$, $f'' = 2H - 1$, $f''' = 2\delta$.

The computation of g', g'', g''' is left to the reader.

16. We are intuitively led to consider the usual Friedrichs' approximations of H,

$$F_n(x) \;=\; \int_{-\infty}^{\infty} H(y)\omega_{1/n}(x - y)\, dy$$

$$=\; \int_0^{\infty} \omega_{1/n}(x - y)\, dy, \quad x \in \mathbb{R}.$$

Obviously, for all $n \in \mathbb{N}$, F_n is in $C^\infty(\mathbb{R})$, but $\operatorname{supp} F_n$ is not compact. So we consider (instead of F_n)

$$
\begin{aligned}
H_n(x) &= \int_0^n \omega_{1/n}(x-y)\,dy \\
&= \int_{x-n}^x \omega_{1/n}(t)\,dt, \quad x \in \mathbb{R},\ n \in \mathbb{N},
\end{aligned}
$$

with $\operatorname{supp} H_n = [-1/n, n+1/n]$, i.e., $H_n \in C_0^\infty(\mathbb{R})$ for all $n \in \mathbb{N}$. We are going to prove that

$$
H_n(\phi) \to H(\phi) \quad \forall \phi \in D(\mathbb{R}),
$$

where H_n and H denote the regular distributions associated with H_n and H.

Take an arbitrary $\phi \in D(\mathbb{R})$. Its support, $\operatorname{supp}\phi \subset [-a, a]$ for some $a > 0$. We have

$$
\begin{aligned}
H_n(\phi) &= \int_{-\infty}^\infty \phi(x) H_n(x)\,dx \\
&= \int_{-a}^a \phi(x) \underbrace{\left(\int_{x-n}^x \omega_{1/n}(t)\,dt \right)}_{=:f_n(x)} dx.
\end{aligned}
$$

Since

$$
|f_n(x)| \le |\phi(x)| \int_{\mathbb{R}} \omega_{1/n}(t)\,dt = |\phi(x)|, \quad x \in [-a, a],\ n \in \mathbb{N},
$$

and

$$
f_n(x) \to H(x) \quad \forall x \in [-a, a] \setminus \{0\},
$$

we can apply the Lebesgue Dominated Convergence Theorem to derive

$$
\begin{aligned}
\lim_{n\to\infty} H_n(\phi) &= \int_{-a}^a \phi(x) H(x)\,dx \\
&= \int_{-\infty}^\infty H(x)\phi(x)\,dx \\
&= H(\phi).
\end{aligned}
$$

17. (i) Left to the reader.

(ii) Proceed by way of contradiction. Assuming the existence of $f \in L^1_{\text{loc}}(\mathbb{R}^2)$ that generates u, we have

$$\int_{\mathbb{R}^2} f\phi \, dx = 0 \quad \forall \phi \in D(\mathbb{R}^2),$$

with $\operatorname{supp} \phi \subset \mathbb{R}^2 \setminus \{(x_1, 0); x_1 \in \mathbb{R}\}$.

This implies $f = 0$ a.e. in $\mathbb{R}^2 \implies u = 0$, which is a contradiction.

(iii) Left to the reader.

18. For any $\phi \in D(\Omega)$ the series $\sum_{n=1}^{\infty} a_n \phi(x_n)$ has a finite number of nonzero terms since $\operatorname{supp} \phi$ is a compact subset of Ω, hence $\operatorname{supp} \phi$ contains finitely many points in S.

19. Assume by contradiction that $\delta_{x_n} \to 0$ in $D'(\mathbb{R}^k)$ and $\liminf \|x_n\| < \infty$, i.e., there exists a bounded subsequence $(x_{n_m})_{m \in \mathbb{N}}$. Therefore, a subsequence of $(x_{n_m})_{m \in \mathbb{N}}$, again denoted $(x_{n_m})_{m \in \mathbb{N}}$, converges to some $x^* \in \mathbb{R}^k$ as $m \to \infty$. So

$$\delta_{x_{n_m}}(\phi) = \phi(x_{n_m}) \to \phi(x^*) \quad \text{as } m \to \infty,$$

which implies $\phi(x^*) = 0$ for all $\phi \in D(\mathbb{R}^k)$, which is a contradiction.

20. Left to the reader.

21. Left to the reader.

22. Denote $I_n = \big(n\pi, (n+1)\pi\big)$, $n \in \mathbb{Z}$. If u is a solution of the given equation, then u is a solution of the equation in $D'(I_n)$ for all $n \in \mathbb{Z}$, i.e., for all $\phi \in C_0^\infty(I_k)$ we have

$$((\sin t)u', \phi) = 0 \implies (u', (\sin t)\phi) = 0.$$

So

$$\forall \psi \in D(I_n) \ (u', \psi) = 0 \implies u' = 0 \implies u \text{ is constant on } I_n.$$

Hence,

$$u = \sum_{n \in \mathbb{Z}} c_n \chi_{I_n},$$

where the c_n's are real constants and χ_{I_n} denotes the characteristic function of I_n, $n \in \mathbb{Z}$. In fact, this is the general solution of the given equation. Clearly, $\{\chi_{I_n}; n \in \mathbb{Z}\}$ is an infinite, linearly independent system, hence the claim is confirmed.

Notice that an equivalent form for the general solution of the given equation is

$$u = \sum_{n \in \mathbb{Z}} c_n H(t - n\pi),$$

and $\{H(t - n\pi); n \in \mathbb{Z}\}$ is a linearly independent system of solutions.

23. First of all, solve the third equation for u_1,

$$u_1 = u_3' - u_3 - H.$$

Then solve the first equation for u_2 and use the above equation to find

$$
\begin{aligned}
u_2 &= -u_1' + 4u_1 + H \\
&= -(u_3'' - u_3' - \delta) = 4(u_3' - u_3 - H) + H \\
&= -u_3'' + 5u_3' - 4u_3 - 3H + \delta.
\end{aligned}
$$

Finally, we obtain from the second equation a third order linear differential equation in u_3 which can be solved by the usual method, etc.

24. Recall that $W_0^{1,1}(a, \infty)$ is the closure in $W^{1,1}(a, \infty)$ of $C_0^\infty(a, \infty)$. So, as $u \in W_0^{1,1}(a, \infty)$, there exists a sequence $(u_n)_{n \in \mathbb{N}}$ in $C_0^\infty (a, \infty)$ which converges to u in $W^{1,1}(a, \infty)$. Let $b \in (a, \infty)$ be arbitrary but fixed. We have for all $t \in [a, b]$ and $m, n \in \mathbb{N}$

$$
\begin{aligned}
|u_n(t) - u_m(t)| &\leq \left| \int_a^t \left(u_n'(s) - u_m'(s) \right) ds \right| \\
&\leq \int_a^b |u_n'(s) - u_m'(s)| \, ds \to 0 \text{ as } n, m \to \infty.
\end{aligned}
$$

Therefore, u_n converges in $C[a, b]$ to some $v \in C[a, b]$ and $v(a) = 0$. In fact, v is an absolutely continuous representative of $u|_{[a,b]}$ (cf. Theorem 5.35). Since b was arbitrary v can be extended as a function in $C[a, \infty)$.

25. The embedding of $W^{2,p}(0,1)$ into $C^1[0,1]$ is realized by the map (injection) which associates with each element $u \in W^{2,p}(0,1)$ its representative from $A^{2,p}(0,1) \subset C^1[0,1]$, also denoted by u (see Theorem 5.35). Let $(u_n)_{n \in \mathbb{N}}$ be a bounded sequence in $W^{2,p}(0,1)$. Let us apply the Arzelà–Ascoli criterion to show that (u_n) has a subsequence which is convergent in $C^1[0,1]$. For $t, s \in [0,1]$ and $n \in \mathbb{N}$ we have

$$
\begin{aligned}
|u_n(t)| &= |u_n(s) + \int_s^t u_n'(\tau)\, d\tau| \\
&\leq |u_n(s)| + \int_0^1 |u_n'(\tau)|\, d\tau \\
&= |u_n(s)| + \|u_n'\|_{L^1(0,1)} \\
&\leq |u_n(s)| + \|u_n'\|_{L^p(0,1)} \quad \text{(by Hölder's inequality)}.
\end{aligned}
$$

By integration over $[0,1]$ with respect to s we get

$$
\begin{aligned}
|u_n(t)| &\leq \|u_n\|_{L^1(0,1)} + \|u_n'\|_{L^p(0,1)} \\
&\leq \|u_n\|_{L^p(0,1)} + \|u_n'\|_{L^p(0,1)} \\
&\leq C \quad \text{(by assumption)},
\end{aligned}
$$

where C is some constant. Hence, (u_n) is bounded in $C[0,1]$. We also have for $t, s \in [0,1]$ and $n \in \mathbb{N}$,

$$
\begin{aligned}
|u_n(t) - u_n(s)| &= |\int_s^t u_n'(\tau)\, d\tau| \\
&\leq |\int_s^t |u_n'(\tau)|\, d\tau| \\
&\leq |t-s|^{1/q} \|u_n'\|_{L^p(0,1)} \quad \text{(by Hölder with } q = \tfrac{p}{p-1}) \\
&\leq C\, |t-s|^{1/q},
\end{aligned}
$$

where C is a constant, which shows that (u_n) is equicontinuous. According to the Arzelà–Ascoli criterion, (u_n) has a subsequence $(u_{n_k})_{k \in \mathbb{N}}$ which is convergent in $C[0,1]$ to some $u \in C[0,1]$.

By repeating the above arguments for $(u_{n_k}')_{k \in \mathbb{N}}$ we deduce the existence of a subsequence of $(u_{n_k}')_{k \in \mathbb{N}}$ which converges in $C[0,1]$ and its limit is $u' \in C[0,1]$. Consequently, the original sequence (u_n) has a subsequence which converges in $C^1[0,1]$.

26. We have $\operatorname{supp}\phi \subset [-a, a]$ for some $a > 0$, so $\operatorname{supp}\phi^{(j)} \subset [-a, a]$ for all $j \in \mathbb{N}$.

(i) Let us first discuss the case $p = \infty$. Since ϕ is not the null function, it is easily seen that for each $j \in \{0, 1, \dots\}$ there exists $t_j \in (-a, a)$ such that

$$\sup_{[-a,a]} |\phi^{(j)}| = |\phi^{(j)}(t_j)| > 0.$$

This implies

$$\sup_{\mathbb{R}} |u_n^{(j)}| \le |\phi^{(j)}(t_j)| \quad \forall n \in \mathbb{N}, \ j = 0, 1, \dots$$

Therefore (u_n) is bounded in $W^{m,\infty}(\mathbb{R})$ for all $m \in \mathbb{N}$.

In the case $1 \le p < \infty$ we have

$$\int_{-\infty}^{+\infty} |u_n^{(j)}(t)|^p dt = \int_{-\infty}^{+\infty} |\phi^{(j)}(t+n)|^p dt$$

$$= \int_{-a}^{a} |\phi^{(j)}(t)|^p dt \quad \forall n \in \mathbb{N}, \ j = 0, 1, \dots,$$

which confirms the claim.

(ii) Clearly, for each $j \in \{0, 1, \dots\}$, $(u_n^{(j)})$ converges pointwise to zero.

Let $q = \infty$. Assume by way of contradiction that there exists a subsequence $(u_{n_k})_{k \in \mathbb{N}}$ which converges uniformly to the null function. Let $t_0 \in (-a, a)$ such that $\phi(t_0) \ne 0$. Choose $t_k = t_0 - n_k$, $k \in \mathbb{N}$. We have

$$u_{n_k}(t_k) = \phi(t_k + n_k) = \phi(t_0) \ne 0 \quad \forall k \in \mathbb{N},$$

so $(u_{n_k})_{k \in \mathbb{N}}$ cannot converge uniformly.

If $1 \le q < \infty$, we can write

$$\int_{-\infty}^{+\infty} |u_n(t)|^q dt = \int_{-\infty}^{+\infty} |\phi(t+n)|^q dt$$

$$= \int_{-a}^{a} |\phi(t)|^q dt \ne 0,$$

and thus $(u_{n_k})_{k \in \mathbb{N}}$ cannot converge in $L^q(\mathbb{R})$ (to the null function).

27. According to Theorem 5.21, there exist some sequences $(u_n)_{n \in \mathbb{N}}$, $(v_n)_{n \in \mathbb{N}}$ in $C^1(\bar{\Omega})$ such that

$$u_n \to u, \quad v_n \to v \quad \text{in } H^1(\Omega).$$

Obviously, for each $i \in \{1, 2, \ldots, k\}$,

$$\frac{\partial}{\partial x_i}(u_n v_n) = \frac{\partial u_n}{\partial x_i} \cdot v_n + u_n \cdot \frac{\partial v_n}{\partial x_i} \quad \forall n \in \mathbb{N},$$

hence

$$-\int_\Omega u_n v_n \frac{\partial \phi}{\partial x_i} = \int_\Omega \frac{\partial u_n}{\partial x_i} v \phi + \int_\Omega u_n \frac{\partial v_n}{\partial x_i} \phi, \qquad (*)$$

for all $\phi \in D(\Omega)$. We intend to pass to the limit in $(*)$. Pick an arbitrary $\phi \in D(\Omega)$. Denoting $C := \sup_\Omega |\partial \phi / \partial x_i| < \infty$, we can write

$$\left| \int_\Omega u_n v_n \frac{\partial \phi}{\partial x_i} - \int_\Omega uv \frac{\partial \phi}{\partial x_i} \right|$$

$$\leq C \int_\Omega |u_n v_n - uv|$$

$$\leq C \left(\int_\Omega |u_n(v_n - v)| + \int_\Omega |v(u_n - u)| \right)$$

$$\leq C \left(\|u_n\|_{L^2(\Omega)} \|v_n - v\|_{L^2(\Omega)} + \|v\|_{L^2(\Omega)} \|u_n - u\|_{L^2(\Omega)} \right)$$

$$\leq C^* \left(\|v_n - v\|_{L^2(\Omega)} + \|u_n - u\|_{L^2(\Omega)} \right) \quad \forall n \in \mathbb{N},$$

where C^* is a constant. So the left-hand side of $(*)$ converges to $-\int_\Omega uv \, (\partial \phi / \partial x_i)$ as $n \to \infty$. Similar arguments can be used for the two terms in the right-hand side of $(*)$. Thus we obtain by passing to the limit in $(*)$

$$-\int_\Omega uv \frac{\partial \phi}{\partial x_i} = \int_\Omega \frac{\partial u}{\partial x_i} \phi + \int_\Omega u \frac{\partial v_n}{\partial x_i} \phi,$$

for all $\phi \in D(\Omega)$ and $i = 1, 2, \ldots, k$, i.e.,

$$\frac{\partial}{\partial x_i}(uv) = \frac{\partial u}{\partial x_i} \cdot v + u \cdot \frac{\partial u}{\partial x_i}, \quad \text{in } D'(\Omega), \; i = 1, 2, \ldots, k.$$

Of course the above equalities are also satisfied in $L^1(\Omega)$, hence a.e. in Ω.

12.6   Answers to Exercises for Chap. 6

1. If $p = 2$ the corresponding norm, $\| \cdot \|_{L^2(\Omega)}$, is generated by the usual scalar product

$$(u, v)_{L^2(\Omega)} = \int_{\Omega} uv \, dx, \quad u, v \in L^2(\Omega),$$

so $\left(L^2(\Omega), \| \cdot \|_{L^2(\Omega)}\right)$ is a Hilbert space.

In order to conclude, it is sufficient to prove that, for $p \in (1, \infty) \setminus \{2\}$, $\| \cdot \|_{L^p(\Omega)}$ does not satisfy the parallelogram law (see Theorem 6.1 (Jordan–von Neumann)). To this end, we choose two disjoint open balls $B_1, B_2 \subset \Omega$ and two C^∞ functions ϕ_1, ϕ_2 with $\operatorname{supp} \phi_i \subset B_i$ and $\|\phi_i\|_{L^p(B_i)} = 1$, $i = 1, 2$. Obviously, ϕ_1 and ϕ_2 do not satisfy the parallelogram law.

2. Recall that for all $x, y \in H$ and $\alpha \in \mathbb{K}$,

$$\|x + \alpha y\|^2 = \|x\|^2 + 2 \operatorname{Re} \bar{\alpha}(x, y) + |\alpha|^2 \|y\|^2.$$

Assume that $|(x, y)| = \|x\| \cdot \|y\|$, $y \neq 0$. Choosing in the above identity $\alpha = -(x, y)/\|y\|^2$ we obtain

$$
\begin{aligned}
\|x + \alpha y\|^2 &= \|x\|^2 - \frac{|(x, y)|^2}{\|y\|^2} \\
&= \|x\|^2 - \frac{\|x\|^2 \|y\|^2}{\|y\|^2} = 0,
\end{aligned}
$$

so $x + \alpha y = 0$. Conversely, if x, y are linearly dependent, it follows easily that $|(x, y)| = \|x\| \cdot \|y\|$.

3. According to the Jordan–von Neumann theorem, it is enough to show that there are functions $u, v \in C[a, b]$ which do not satisfy the parallelogram law. Choose, for example, $u, v \in C[a, b]$ such that $0 \leq u \leq 1$, $0 \leq v \leq 1$, $\operatorname{supp} u \subset (a, (a + b)/2)$, $\operatorname{supp} v \subset ((a + b)/2, b)$, $\max u = \max v = 1$.

4. The space C is a finite dimensional subspace of $L^2(0, 1)$ (whose dimension is $n + 1$), hence C is a closed linear subspace. According to Theorem 6.4, for any $u \in L^2(0, 1)$, there exists a unique $p_u \in C$ which minimizes $\|u - p\|_{L^2(0,1)}$ over C, namely, $p_u = P_C u$.

5. (i) Observe that P is precisely the projection operator P_C, where C is the closed unit ball, so P is nonexpansive (see Sect. 6.3);

(ii) In this case we cannot use the previous argument (which is valid in Hilbert spaces). We distinguish three cases

(a) $u, v \in C$ is a trivial case;

(b) if $u \in C$, $v \in H \setminus C$, then

$$
\begin{aligned}
\|Pu - Pv\| &= \|u - \|v\|^{-1}v\| \\
&\leq \|u - v\| + \|v - \|v\|^{-1}v\| \\
&= \|u - v\| + \|v\| - 1 \\
&\leq \|u - v\| + \|v\| - \|u\| \\
&\leq 2\|u - v\|;
\end{aligned}
$$

(c) if $u, v \in H \setminus C$, then

$$
\begin{aligned}
\|Pu - Pv\| &\leq \|(1/\|u\|)u - (1/\|u\|)v\| \\
&\quad + \|(1/\|u\|)v - (1/\|v\|)v\| \\
&\leq \|u - v\| + (1/\|u\|) \cdot |\|v\| - \|u\|| \\
&\leq 2\|u - v\|.
\end{aligned}
$$

6. The space M is two-dimensional (representing a plane in $\mathbb{R}^3$), so it is closed. Clearly, the vector $v = (2, -1, -3)^T$ is orthogonal to M and $\text{Span}\{v\}$ is the orthogonal complement of M, i.e., $M^\perp = \text{Span}\{v\}$. The projection $P_M x$ of the given $x = (1, 2, -1)^T$ satisfies the conditions: $x - P_M x \in M^\perp$ (i.e., $x - P_M x = (2\alpha, -\alpha, -3\alpha)^T$) and $P_M x \in M$. Using these two conditions we can determine $\alpha = 3/14$, so $P_M x = (4/7, 31/14, -5/14)^T$, and $x = P_M x + (x - P_M x)$.

7. Obviously, M is a linear subspace of the Hilbert space $L^2(a, b)$. In fact, M is the nullspace of the linear continuous functional $\phi : L^2(a, b) \to \mathbb{R}$,

$$
\phi(u) = \int_a^b u(t)\, dt, \quad u \in L^2(a, b),
$$

so M is a closed linear space, with codim $M = 1$. We have $M^\perp =$ Span$\{1\}$, i.e., $M^\perp$ is the subspace of all constant functions. It is easily seen that any $u \in L^2(a,b)$ can be written as

$$u = \left(u - \frac{1}{b-a} \int_a^b u(t)\, dt \right) + \frac{1}{b-a} \int_a^b u(t)\, dt.$$

8. $M^\perp$ is the subspace of odd functions, i.e.,

$$M^\perp = \{ u \in L^2(-1,1);\ u(t) = -u(-t) \ \text{for a.a. } t \in (-1,1) \},$$

and for any $u \in L^2(-1,1)$ we have the decomposition

$$u(t) = \frac{u(t) + u(-t)}{2} + \frac{u(t) - u(-t)}{2} \quad \text{for a.a. } t \in (0,1).$$

9. Let us first prove that

$$Y^\perp = \left(\mathrm{Cl}\, Y \right)^\perp. \qquad (*)$$

Indeed, on the one hand,

$$Y \subset \mathrm{Cl}\, Y \implies \left(\mathrm{Cl}\, Y \right)^\perp \subset Y^\perp.$$

The converse inclusion is also true. Indeed, if $x \in Y^\perp$, then $(x, y) = 0$, $\forall y \in Y$, and this can be extended to all $y \in \mathrm{Cl}\, Y$, so $x \in \left(\mathrm{Cl}\, Y \right)^\perp$. Thus $Y^\perp \subset \left(\mathrm{Cl}\, Y \right)^\perp$, as claimed.

Now, taking into account $(*)$, we can write

$$\left(Y^\perp \right)^\perp = \left(\left(\mathrm{Cl}\, Y \right)^\perp \right)^\perp.$$

In order to conclude, it suffices to show that the right-hand side of the above equation equals $\mathrm{Cl}\, Y$. In fact, for any closed subspace $Z \subset H$, we have $\left(Z^\perp \right)^\perp = Z$. Indeed, $Z \subset \left(Z^\perp \right)^\perp$ and the converse inclusion follows easily: if $x \in \left(Z^\perp \right)^\perp$, then

$$x = x_1 + x_2, \quad x_1 \in Z,\ x_2 \in Z^\perp,$$

and since $0 = (x, x_2) = (x_1, x_2) + (x_2, x_2) = (x_2, x_2)$, it follows that $x_2 = 0$, so $x = x_1 \in Z$.

10. The subspace Y is not closed in $H = L^2(0, 1)$. In order to prove this, consider, e.g., the sequence (u_n) in H, defined by

$$
u_n(t) = \begin{cases} 0, & 0 < t < \frac{1}{n}, \\ (nt)^{-1/4}, & \frac{1}{n} < t < \frac{1}{2}, \\ -2\beta_n t, & \frac{1}{2} < t < 1, \end{cases}
$$

where β_n are constants, $n = 3, 4, \ldots$ We determine the β_n's such that $u_n \in Y$, i.e.,

$$
\begin{aligned}
0 &= \int_0^1 \frac{u(t)}{t}\, dt \\
&= n^{-1/4} \int_{1/n}^{1/2} t^{-1-1/4}\, dt - 2\beta_n \int_0^{1/2} dt.
\end{aligned}
$$

Hence,

$$
\beta_n = 4n^{-1/4}\left(n^{1/4} - 2^{1/4}\right) \;\to\; 4, \quad \text{as } n \to \infty.
$$

It is easily seen that $u_n \to u$ in H, where

$$
u(t) = \begin{cases} 0, & 0 < t < \frac{1}{2}, \\ -8t, & \frac{1}{2} < t < 1. \end{cases}
$$

Clearly,

$$
\int_0^1 \frac{u(t)}{t}\, dt = -4 \neq 0.
$$

11. We know that H^* is a Banach space, so it remains to prove that its norm is generated by a scalar product. Let x^*, y^* be two arbitrary elements of H^*. According to the Riesz Representation Theorem, there exist $x, y \in H$ such that $x^*(u) = (u, x)$, $y^*(u) = (u, y)$ $\forall u \in H$. Define $(x^*, y^*)_{H^*} = (y, x)$. It is easy to check that $(\cdot, \cdot)_{H^*}$ is a scalar product in H^* and $\|x^*\|_{H^*} = \sqrt{(x^*, x^*)_{H^*}} = \|x\|$ for all $x^* \in H^*$.

12. (i)   We have

$$\|v_n\|^2 = \frac{1}{n^2}\left(\sum_{i=1}^{n} a_i u_i, \sum_{j=1}^{n} a_j u_j\right)$$

$$= \frac{1}{n^2}\sum_{i=1}^{n} a_i^2\|u_i\|^2$$

$$= \frac{1}{n^2}\sum_{i=1}^{n} a_i^2$$

$$\leq \frac{C^2 n}{n^2} = \frac{C^2}{n} \to 0.$$

(ii)   We know from the above computation that $\sqrt{n}\|v_n\| \leq C$ for all $n \in \mathbb{N}$. Let $x \in H$ be arbitrary but fixed. Since $\{u_n\}_{n=1}^{\infty}$ is a basis in H, $x = \sum_{n=1}^{\infty}(x, u_n)u_n$. Denoting $x_N = \sum_{n=1}^{N}(x, u_n)u_n$, we have for $\varepsilon > 0$ small

$$\begin{aligned}|(\sqrt{n}v_n, x)| &\leq |(\sqrt{n}v_n, x - x_N)| + |(\sqrt{n}v_n, x_N)| \\ &\leq C\|x - x_N\| + \sqrt{n} \cdot |(v_n, x_N)| \\ &< \varepsilon + \sqrt{n} \cdot |(v_n, x_N)|, \quad N > N_\varepsilon.\end{aligned}$$

This estimate along with

$$\sqrt{n} \cdot |(v_n, x_N)| = \frac{\sqrt{n}}{n}|\sum_{i=1}^{N} a_i(x, u_i)|$$

$$\leq \frac{CN\|x\|}{\sqrt{n}}, \quad n \geq N$$

implies

$$\limsup_{n\to\infty}|(\sqrt{n}v_n, x)| < \varepsilon \;\; \forall \varepsilon > 0 \;\; \implies \;\; \lim_{n\to\infty}(\sqrt{n}v_n, x) = 0.$$

13. Assume (i) holds. It is easy to see that $R(A)$ is a closed subspace of H, so $H = R(A) \oplus R(A)^{\perp}$. We also infer from ($i$) that A is injective, so there exists $A^{-1} : R(A) \to H$ which is continuous. Define $B : H \to H$ by $By = A^{-1}P_{R(A)}y \;\; \forall y \in H$. Clearly, $B \in L(H)$ and $B \circ A = I$. Conversely, assuming (ii), we have

$$\|x\| = \|B(Ax)\| \leq \|B\| \cdot \|Ax\| \;\; \forall x \in H,$$

with $\|B\| \neq 0$ which is guaranteed by the relation $B \circ A = I$.

14. (i) It suffices to prove that $N(A) = R(A)^\perp$ (since $(\text{Cl}\,R(A))^\perp = R(A)^\perp$, see the solution of Exercise 6.9 above). Indeed, if $x \in N(A)$, then $(Av, v + x) \geq 0$ $\forall v \in H$, so replacing v by tv, $t \in \mathbb{R}$, we obtain $(Av, x) = 0$ $\forall v \in H$, i.e., $x \in R(A)^\perp$. Conversely, let $x \in R(A)^\perp$. We have $(A(v + x), v + x) \geq 0$ $\forall v \in H \Longrightarrow (Av + Ax, v) \geq 0$ $\forall v \in H$. Replacing v by tv we easily derive $(Ax, v) = 0$ $\forall v \in H \Longrightarrow Ax = 0$.

(ii) If $x + tAx = 0$, $t > 0$, then $\|x\|^2 + t(Ax, x) = 0 \Longrightarrow \|x\|^2 \leq 0 \Longrightarrow x = 0$, so $I + tA$ is injective. Let us prove that $I + tA$ ia also surjective. For an arbitrary $y \in H$ consider the equation $x + tAx = y$. Apply the Lax–Milgram Theorem with $a(u, v) = (u, v) + t(Au, v)$, $b(v) = (y, v)$ to deduce the existence of a unique $x \in H$ satisfying $a(x, v) = (y, v)$ $\forall v \in H \Longrightarrow x + tAx = y$.

Denote $J_t u = (I + tA)^{-1}u$, $u \in H$, $t > 0$. It is easily seen that J_t is a nonexpansive operator for all $t > 0$: we have just to multiply $J_t u + tA J_t u = u$ by $J_t u$ and use the positivity of A and the Bunyakovsky–Cauchy–Schwarz inequality.

Next, let $u \in H$ be arbitrary but fixed. According to (i), $u = u_1 + u_2$, $u_1 \in N(A)$, $u_2 \in \text{Cl}\,R(A)$. We have

$$J_t u_1 = u_1 = P_{N(A)}u \quad \forall t > 0. \tag{$*$}$$

Now, for $y \in R(A)$, i.e., $y = Av$ for some $v \in H$, and $t > 0$, we have $J_t y + A(t J_t y - v) = 0$. By the positivity of A we get

$$(J_t y, t J_t y - v) \leq 0 \quad \Longrightarrow \quad \|J_t y\| \leq \frac{\|v\|}{t}.$$

Thus, $\lim_{t \to \infty} J_t y = 0$. This property extends by density to all $y \in \text{Cl}\,R(A)$ since J_t is a nonexpansive operator. So we can write

$$\lim_{t \to \infty} J_t u_2 = 0. \tag{$**$}$$

From $(*)$ and $(**)$ we infer that $J_t u = J_t u_1 + J_t u_2 \to P_{N(A)}u$ as $t \to \infty$.

15. Since (u_n) converges weakly to u we have

$$\|u\| \leq \liminf_{n \to \infty} \|u_n\|. \tag{$*$}$$

In order to prove $(*)$ we can assume $u \neq 0$ (as the case $u = 0$ is trivial). We have

$$
\begin{aligned}
\|u\|^2 &= (u, u - u_n) + (u, u_n) \\
&\leq (u, u - u_n) + \|u\| \cdot \|u_n\|,
\end{aligned}
$$

which yields by passing to the limit

$$
\|u\|^2 \leq \|u\| \cdot \liminf_{n \to \infty} \|u_n\|,
$$

so $(*)$ holds true. Summarizing, we have

$$
\|u\| \leq \liminf_{n \to \infty} \|u_n\| \leq \limsup_{n \to \infty} \|u_n\| \leq \|u\|,
$$

hence $\|u_n\| \to \|u\|$. Now it is easy to conclude:

$$
\|u_n - u\|^2 = \|u_n\|^2 - 2\operatorname{Re}(u_n, u) + \|u\|^2 \to 0.
$$

16. Apply the Lax–Milgram Theorem (Theorem 6.17) with $H := H_0^1(0,1)$ (endowed with the H^1 norm) and

$$
a(u, v) = \int_0^1 u'v' \, dt + \int_0^1 uv \, dt, \quad b(v) = \int_0^1 fv \, dt.
$$

Since u satisfies the equation

$$
-u'' + u = f \quad \text{in } D'(0,1),
$$

it follows that $u'' \in L^1(0,1)$, hence $u \in W^{2,1}(0,1)$ and

$$
\begin{cases}
-u'' + u = f & \text{a.e. in } (0,1), \\
u(0) = 0, \ u(1) = 0.
\end{cases}
$$

17. (i) Assume u is a solution to problem (P). Let v be arbitrary in $H^1(0,1)$. Multiplication of the differential equation by $v(t)$ and integration over $(0,1)$ shows that u is a solution of $(\tilde{P})$. Now, assume that u is a solution of $(\tilde{P})$. Let v in $(\tilde{P})$ range $C_0^\infty(0,1)$. It follows that $u \in H^1(0,1)$ satisfies the equation

$$
-u'' + \alpha u = f \quad \text{in } D'(0,1).
$$

Since $\alpha u - f \in L^2(0,1)$, it follows that in fact $u \in H^2(0,1)$ and the above equation is satisfied for a.a. $t \in (0,1)$. Now, testing in

$(\tilde{P})$ with functions $v \in C^1[0,1]$ we readily infer that u satisfies the boundary conditions $u'(0) = 0$, $u'(1) = u(1)$.

(ii) In order to apply Lax–Milgram, consider $H = H^1(0,1)$ and define $a : H \times H \to \mathbb{R}$ and $b : H \to \mathbb{R}$,

$$a(u,v) = -u(1)v(1) + \int_0^1 u'v' + \alpha \int_0^1 uv, \quad b(v) = \int_0^1 fv.$$

Obviously, the functional a is bilinear and symmetric. It is also continuous on $H \times H$ (note that $(u,v) \longmapsto u(1)v(1)$ is continuous as $H^1(0,1)$ is compactly embedded in $C[0,1]$). We need to prove that a is coercive for large α. For $u \in H^1(0,1)$ we deduce from the obvious relation

$$u(1)^2 = u(t)^2 + 2\int_t^1 uu' \implies u(1)^2 \leq u(t)^2 + 2\int_0^1 |u| \cdot |u'|,$$

by integration over $[0,1]$,

$$\begin{aligned}
u(1)^2 &\leq \|u\|_{L^2(0,1)}^2 + 2\|u\|_{L^2(0,1)}\|u'\|_{L^2(0,1)} \\
&\leq \|u\|_{L^2(0,1)}^2 + \frac{1}{\varepsilon}\|u\|_{L^2(0,1)}^2 + \varepsilon\|u'\|_{L^2(0,1)}^2,
\end{aligned}$$

where ε is a positive number. If $\varepsilon \in (0,1)$, then for all $u \in H^1(0,1)$,

$$\begin{aligned}
a(u,u) &= -u(1)^2 + \int_0^1 (u')^2 + \alpha \int_0^1 u^2 \\
&\geq \left(\alpha - 1 - \frac{1}{\varepsilon}\right)\|u\|_{L^2(0,1)}^2 + (1 - \varepsilon)\|u'\|_{L^2(0,1)}^2,
\end{aligned}$$

which shows that a is coercive for α large enough.

It is also clear that b is linear and continuous, so all the conditions required by the Lax–Milgram Theorem are fulfilled.

(iii) Since a is symmetric, u is a minimizer of the functional

$$\begin{aligned}
v \longmapsto \frac{1}{2}a(v,v) - b(v) = \frac{1}{2}\left(-v(1)^2 + \int_0^1 (v')^2 + \alpha \int_0^1 v^2\right) \\
- \int_0^1 fv, \quad v \in H.
\end{aligned}$$

18. We are looking for $z := P_Y y$, i.e., z must satisfy two conditions: $z \in Y$ and $(y - z) \perp Y$. Note that Y itself is a Hilbert space, with the same scalar product and norm, having an orthonormal basis $\{u_n\}_{n=1}^{\infty}$, so the vector $z \in Y$ has the Fourier expansion $z = \sum_{n=1}^{\infty} (z, u_n) u_n$ (cf. Theorem 6.21 and Remark 6.22). The second condition is equivalent to

$$(y - z, u_n) = 0 \ \ \forall n \in \mathbb{N} \implies (z, u_n) = (y, u_n) \ \ \forall n \in \mathbb{N}.$$

Therefore, $z = \sum_{n=1}^{\infty} (z, u_n) u_n$.

19. According to Theorem 6.23, there exists a countable orthonormal basis of H, say $\{u_n\}_{n=1}^{\infty}$. We also know that $u_n \to 0$ weakly. On the other hand, if $\|x\| = 1$, then the constant sequence $x_n = x$, $n = 1, 2, \ldots$ satisfies the required properties. So we can assume $0 < \|x\| < 1$. Intuitively, we consider the sequence $x_n = \alpha_n u_n + x$, $n = 1, 2, \ldots$, where the α_n's are real numbers determined from the required condition

$$\|x_n\|^2 = 1 \iff \|\alpha_n u_n + x\|^2 = \alpha_n^2 + 2\alpha_n Re\,(u_n, x) + \|x\|^2$$
$$= 1, n = 1, 2, \ldots$$

Choose $\alpha_n = -Re(u_n, x) + \sqrt{|Re(u_n, x)|^2 + 1 - \|x\|^2} \longrightarrow \sqrt{1 - \|x\|^2}$ as $n \to \infty$. It follows that

$$(x_n, v) = \alpha_n (u_n, v) + (x, v) \to (x, v) \ \ \forall v \in H.$$

20. Left to the reader.

12.7   Answers to Exercises for Chap. 7

1. Left to the reader.

2. It suffices to show that $N(A^*) = \left(Cl\, R(A) \right)^{\perp}$, where

$$\left(Cl\, R(A) \right)^{\perp} = \{ y^* \in Y^*, \ y^*(y) = 0 \ \forall y \in Cl\, R(A) \}.$$

3. We need to prove that $D(A^*) \subset D(A)$. Take an arbitrary $y \in D(A^*)$. Since $R(A) = H$, there exists an $x \in D(A)$ such that $A^* y = Ax$. It is sufficient to prove that $y = x$. So, for any $w \in D(A)$ we have

$$(Aw, y) = (w, A^* y) = (w, Ax) = (Aw, x),$$

hence, as $R(A) = H$,

$$(u, y) = (u, x) \ \forall u \in H \implies y = x.$$

4. Left to the reader.

5. We have

$$\|A^* A\| \le \|A^*\| \cdot \|A\|^2. \tag{1}$$

On the other hand, for all $x \in H$,

$$\|Ax\|^2 = (Ax, Ax) = (x, A^* Ax) \le \|x\| \cdot \|A^* Ax\|$$
$$\le \|A^* A\| \cdot \|x\|^2,$$

which implies

$$\|A\|^2 \le \|A^* A\| \tag{2}$$

The claim follows from (1) and (2).

6. If A is symmetric, then

$$(Ax, x) = (x, A^* x) = (x, Ax) = \overline{(Ax, x)}$$
$$\implies (Ax, x) \in \mathbb{R} \ \forall x \in H.$$

Conversely, suppose that $(Ax, x) \in \mathbb{R} \ \forall x \in H$. We have

$$(Ax, y) = \frac{1}{4}[(A(x+y), x+y) - (A(x-y), x-y)$$
$$+ i(A(x+iy), x+iy) - i(A(x-iy), x-iy)].$$

Next,

$$(x, Ay) = \overline{(Ay, x)}$$
$$= \frac{1}{4}[(A(y+x), y+x) - (A(y-x), y-x)$$
$$- i(A(y+ix), y+ix) + i(A(y-ix), y-ix)]$$
$$= \frac{1}{4}[(A(x+y), x+y) - (A(x-y), x-y)$$
$$+ i(A(x+iy), x+iy) - i(A(x-iy), x-iy)]$$
$$= (Ax, y),$$

hence $A = A^*$.

7. We have for all $u \in H$

$$(Tu, u) = \|u\|^2 + a\|Au\|^2, \qquad (*)$$

where $\| \cdot \|$ is the norm induced by $(\cdot, \cdot)$. In particular, $(*)$ shows that $N(T) = \{0\}$, so T is injective. In order to prove that T is onto, one can apply the Lax–Milgram Theorem to

$$a(u, v) = b(v), \quad v \in H,$$

where

$$a(u, v) = (Tu, v) = (u, v) + a(Au, Av), \quad b(v) = (f, v), \quad f \in H.$$

So T is bijective, hence invertible, and obviously T^{-1} is linear. According to $(*)$, $\|Tu\| \geq \|u\|$ for all $u \in H$ and thus $T^{-1} \in L(H)$.

8. (a) Left to the reader.

(b) Using the fact that A is symmetric, we obtain

$$\|Tx\|^2 = (Ax + ix, Ax + ix) = \|Ax\|^2 + \|x\|^2 \ \forall x \in H. \ (*)$$

This shows that $N(T) = \{0\}$, hence T is injective.
Next, it follows from $(*)$ that

$$\|Tx\| \geq \|x\| \ \forall x \in H. \qquad (**)$$

This implies that $R(T)$ is closed in H. Indeed, if $y_n = Tx_n$ converges to some $y \in H$, then

$$\|x_n - x_m\| \leq \|y_n - y_m\| \ \forall m, n,$$

which shows that x_n converges to some $x \in H$. Hence, taking into account the continuity of T,

$$y = \lim_{n \to \infty} Tx_n = Tx \in R(T).$$

Thus, $H = R(T) \oplus R(T)^{\perp}$. Let us show that $R(T)^{\perp} = \{0\}$. Indeed, if $z \in R(T)^{\perp}$, we have

$$\begin{aligned}
0 &= (Tx, z) \\
&= (x, T^*z) \\
&= (x, Az - iz) \\
&= (x, Tz - 2iz) \ \forall x \in H.
\end{aligned}$$

It follows that $Tz = 2iz$, hence $z \in R(T) \implies z = 0 \implies R(T) = H$, so T is bijective. So T is invertible, T^{-1} is a linear operator, and, according to $(**)$, $T^{-1} \in L(H)$.

9. It is readily seen that for all $A \in L(H)$

$$P(A)^* = \bar{a}_0 I + \bar{a}_1 A^* + \bar{a}_2 (A^*)^2 + \cdots + \bar{a}_n (A^*)^n,$$

where the coefficients are the complex conjugates of the coefficients of $P(A)$.

(j) follows immediately from this identity;

(jj) Assume that $A^*A = AA^*$. Then

$$\begin{aligned} P(A^*)P(A) &= \sum_{i,j=1}^{n} \bar{a}_i a_j (A^*)^i A^j \\ &= \sum_{i,j=1}^{n} a_j \bar{a}_i A^j (A^*)^i \\ &= P(A)P(A^*). \end{aligned}$$

10. Left to the reader.

11. Left to the reader

12. One can assume $x \neq 0$. If $Ax = x$, then

$$\begin{aligned} \|x\|^2 &= (Ax, x) = (x, A^*x) \leq \|x\| \cdot \|A^*x\| \leq \|A^*\| \cdot \|x\|^2 \\ &= \|A\| \cdot \|x\|^2 \leq \|x\|^2, \end{aligned}$$

so we have equalities everywhere and in particular

$$(x, A^*x) = \|x\| \cdot \|A^*x\|.$$

From Exercise 6.2, we infer that x and A^*x are linearly dependent, i.e., there exists a scalar $\alpha \neq 0$ such that $A^*x = \alpha x$. Using the equality $\|x\|^2 = (x, A^*x)$ we see that $\alpha = 1$, hence $A^*x = x$. Conversely, let us assume that $A^*x = x$. Since $\|A^*\| = \|A\| \leq 1$, we infer by the previous argument that $(A^*)^*x = x \implies Ax = x$.

13. (a) We know that $C_0^\infty(0,1)$ is dense in H (see Theorem 5.8).
Since $C_0^\infty(0,1) \subset D(A)$, we infer that $D(A)$ is dense in H.
In order to prove that A is closed, let (u_n) be a sequence in
$D(A)$ such that $u_n \to u$ and $Au_n = u_n' \to v$ in H. Applying
the Arzelà–Ascoli criterion we infer that $u_n \to u$ in $C[0,1]$
and, in particular, $u(0) = 0$. Then $u_n' \to u'$ in $D'(0,1)$ and
in $H = L^2(0,1)$. Therefore, $u \in D(A)$ and $v = u'$.

(b) It is easy to see that $N(A) = \{0\}$ (hence A is injective),
and $R(A) = H$;

(c) If $v \in D(A^*)$, then the linear functional $f(u) := (Au, v) = \int_0^1 vu'$ is continuous on $D(A)$ with respect to the norm $\|\cdot\|$ of
$H = L^2(0,1)$. Since $\operatorname{Cl} D(A) = H$, the functional f can be
extended (by the Hahn–Banach Theorem or by continuity)
to the whole of H. This extension is again denoted by f.
So, according to the Riesz Representation Theorem, there
exists $w \in H$ such that

$$f(u) = (u, w) = \int_0^1 wu \quad \forall u \in H.$$

Interpreting v and w as distributions from $D'(0,1)$, we have
for all $\phi \in C_0^\infty(0,1)$

$$
\begin{aligned}
v'(\phi) &= -v(\phi') \\
&= -\int_0^1 v\phi' \\
&= -f(\phi) \\
&= -w(\phi),
\end{aligned}
$$

hence $v' = -w \in H \implies v \in H^1(0,1)$.
By the continuity of f on $D(A)$, there exists a constant
$k > 0$ such that

$$\left| \int_0^1 vu' \right| = \left| u(1)v(1) - \int_0^1 uv' \right| \le k\|u\| \quad \forall u \in D(A).$$

If, in addition, $u(1) = 1$, then we obtain

$$|v(1)| \le k\|u\| + \|v'\| \cdot \|u\| \quad \forall u \in D(A).$$

If we choose in this inequality $u = u_n$, where (u_n) is a
sequence in $D(A)$, with $u_n(1) = 1$ for all n, and such that

$u_n \to 0$ in H, we find $v(1) = 0$ by letting $n \to \infty$. It follows that $D(A^*) \subset \{v \in H^1(0,1); v(1) = 0\}$. In fact,

$$D(A^*) = \{v \in H^1(0,1); v(1) = 0\} \quad \text{and} \quad A^*v = -v'.$$

Clearly, $\operatorname{Cl} D(A^*) = H$.

14. Obviously, $D(A)$ is dense in H in both cases, hence it makes sense to define A^*. By arguments similar to those used for the previous exercise we obtain

   (a) $N(A) = \{0\}$, $R(A) = \{g \in H; \int_0^1 g = 0\}$, $D(A^*) = H^1(0,1)$, $A^*v = -v'$, $N(A^*) = \operatorname{Span}\{1\}$ (constant functions), $R(A^*) = H$.

   (b) We distinguish two cases.
   If $\alpha \neq 1$, then $N(A) = \{0\}$, $R(A) = H$, $D(A^*) = \{v \in H^1(0,1); v(1) = \alpha v(0)\}$, $A^*v = -v'$, $N(A^*) = \{0\}$, $R(A^*) = H^1(0,1)$.
   If $\alpha = 1$, then $N(A) = \operatorname{Span}\{1\}$, $R(A) = \{g \in H; \int_0^1 g = 0\}$, $D(A^*) = \{v \in H^1(0,1); v(1) = \alpha v(0)\}$, $A^*v = -v'$, $N(A^*) = \operatorname{Span}\{1\}$, $R(A^*) = \{g \in H; \int_0^1 g = 0\}$.

15. Obviously, in each of the above cases, $C_0^\infty(0,1) \subset D(A)$, so $D(A)$ is dense in H, hence A^* can be defined. Also, in each of the four cases, if $v \in D(A^*)$ then $f(u) = (Au, v)$ satisfies (for some constant C)
$$|f(u)| \leq C\|u\| \quad \forall u \in D(A).$$

Since $\operatorname{Cl} D(A) = H$, the functional f can be extended (by the Hahn–Banach Theorem or by continuity) to a functional from H^*, which is again denoted f. According to the Riesz Representation Theorem, there exists $w \in H$ such that

$$f(u) = (u, w) = \int_0^1 uw, \quad \forall u \in H.$$

On the other hand, interpreting v and w as elements of $D'(0,1)$, we can write for all $\phi \in C_0^\infty(0,1)$

$$
\begin{aligned}
v''(\phi) &= v(\phi'') \\
&= f(\phi) \\
&= (A\phi, v) \\
&= w(\phi).
\end{aligned}
$$

Hence $v'' = w \in H$. Of course, v' (as a primitive of v'') is absolutely continuous in $[0,1]$, so $v \in H^2(0,1)$ and $A^* v = v''$.

Now, all we have to do is to determine $D(A^*)$. Using the same idea as above (see the solution to Exercise 7.13), we find

(a) $D(A^*) = D(A)$, hence $A = A^*$;

(b) $D(A^*) = H^2(0,1)$, hence A is symmetric;

(c) $D(A^*) = D(A)$, hence $A = A^*$;

(d) $D(A^*) = \{u \in H^2(0,1);\ v(0) = v(1) = 0,\ v'(0) = v'(0) = v'(1)\}$.

16. (a) Obviously, A is linear and

$$\|Ax\|^2 = \sum_{j=1}^{\infty} |x_{p+j}|^2 \le \|x\|^2 \quad \forall x = (x_n) \in H,$$

which shows that A is continuous and $\|A\| \le 1$. In fact, $\|A\| = 1$ since for $\tilde{x} = (0,0,\ldots,0,1,0,0,\ldots)$, where 1 is placed on the $p+1$ position, we have $\|\tilde{x}\| = 1$ and $\|A\tilde{x}\| = 1$. In order to find A^* observe that

$$
\begin{aligned}
\langle Ax, y \rangle &= \sum_{m=1}^{\infty} x_{p+m} \bar{y}_m \\
&= \sum_{n=p+1}^{\infty} x_n \bar{y}_{n-p} \\
&= \langle x, A^* y \rangle \quad \forall x, y \in H,
\end{aligned}
$$

hence

$$A^* y = (0,0,\ldots,0,y_1,y_2,\ldots) \quad \forall y = (y_n) \in H,$$

where the zeroes occupy the first p positions.

(b) Clearly, $D(B) = H$, B is linear, and

$$\|Bx\|^2 = \sum_{n=1}^{\infty} \frac{n^{2\alpha}}{(1+n)^2} |x_n|^2 \le \|x\|^2 \quad \forall x = (x_n) \in H,$$

hence $B \in L(H)$ with $\|B\| \le 1$. It is easily seen that the sup in the definition of $\|B\|$ is reached, so $\|B\| = 1$;

(c) Obviously,

$$D(B) = \{x = (x_n) \in H; \sum_{n=1}^{\infty} \frac{n^{2\alpha}}{(1+n)^2} |x_n|^2 < \infty\},$$

hence $D(B)$ is a proper subset of H, which is dense in H: indeed, for any $x = (x_n) \in H$ and $\varepsilon > 0$ there exists

$$x^k = (x_1, x_2, \ldots, x_k, \frac{(k+1)^{\alpha}}{k+2} x_{k+1}, \frac{(k+2)^{\alpha}}{k+3} x_{k+2}, \ldots),$$

k large enough, such that $\|x^k - x\| < \varepsilon$.

(d) It is easy to see that

$$B^* x = \left(\frac{n^{\alpha}(-i)^n}{n+1} x_n\right) \quad \forall x = (x_n) \in D(B^*) = D(B).$$

(e) A is not normal, but B is normal (easy to check).

12.8 Answers to Exercises for Chap. 8

1. (a) $N(A) = \text{Span}\{1\}$ and $R(A) = \text{Span}\left(\{x, x^2, x^3\}\right)$;

    (b) By simple computations we find the following eigenvalues and corresponding eigenvalue sets:

    $$\lambda = 0, \ \text{Span}(\{1\}) \setminus \{0\} \ \text{ and } \lambda = i, \ \text{Span}(\{x^i\}) \setminus \{0\},$$
    $$i = 1, 2, 3.$$

2. (i) Left to the reader;

    (ii) If $a = 0$, then there is one eigenvalue of A, $\lambda = b$, and any $u \in X \setminus \{0\}$ is an eigenfunction corresponding to $\lambda = b$.

    If $a \neq 0$, then it is easily seen that A has no eigenvalue.

3. Assume λ is an eigenvalue of AB, i.e., there exists an $x \in X$, $x \neq 0$, such that $A(Bx) = \lambda x$. Note that, of necessity, $Bx \neq 0$. It follows that

    $$B(A(Bx)) = \lambda Bx \ \Rightarrow \ (BA)(Bx) = \lambda Bx,$$

    i.e., λ is also an eigenvalue of BA (Bx being a corresponding eigenvector).

    The converse implication is similar.

4. (a) Clearly, A maps X into itself and is a linear operator. De-
     note $K = \sup\{|k(t,s)|;\ (t,s) \in [0,1] \times [0,1]\} < \infty$ (since k
     is continuous on the compact set $[0,1] \times [0,1]$). We have

$$|(Au)(t)| \le K \int_0^1 |u(s)|\,ds, \quad t \in [0,1],$$

which implies

$$\|Au\|_X \le K\|u\|_X \quad \forall u \in X.$$

Thus $A \in L(X)$, as claimed.

(b) Consider in X the equation $Au = \lambda u$, $\lambda \in \mathbb{R}$.
     If $\lambda = 0$, this equation becomes $Au = 0$, which implies (by
     differentiation)

$$k(t,t)u(t) + \int_0^t \frac{\partial k}{\partial t}(t,s)u(s)\,ds = 0, \quad t \in [0,1],$$

hence

$$u(t) = -\frac{1}{k(t,t)} \int_0^t \frac{\partial k}{\partial t}(t,s)u(s)\,ds, \quad t \in [0,1].$$

This implies

$$|u(t)| \le K_1 \int_0^t |u(s)|\,ds, \quad t \in [0,1] \implies u \equiv 0,$$

so $\lambda = 0$ is not an eigenvalue of A.
If $\lambda \ne 0$, the equation $Au = \lambda u$ reads

$$\int_0^t k(t,s)u(s)\,ds = \lambda u(t), \quad t \in [0,1],$$

which leads to

$$|u(t)| \le K_2 \int_0^t |u(s)|\,ds, \quad t \in [0,1],$$

and thus we have again $u \equiv 0$.

The case $X = L^2(0,1)$ is similar.

5. (a) Obviously, A is linear, maps H into itself, and $A \in L(H)$. It is easily seen that $\|A\| = \sup_{n \in \mathbb{N}} |\lambda_n|$;

   (b) Easy to prove;

   (c) The set of eigenvalues consists of all distinct λ_n's.

6. (a) Left to the reader;

   (b) Apply the Arzelà–Ascoli criterion;

   (c) Let u, v be arbitrary elements of H. Taking into account the obvious equalities

   $$tv(t) = \frac{d}{dt}\left(\int_0^t sv(s)\,ds\right), \quad v(t) = -\frac{d}{dt}\left(v(s)\,ds\right)$$

   and integrating by parts, we obtain

   $$(Au, v) = \int_0^1 tv(t)\left(\int_t^1 u(s)\,ds\right)dt$$
   $$+ \int_0^1 v(t)\left(\int_0^t su(s)\,ds\right)dt = (u, Av).$$

   (d) Consider in H the equation $Au = \lambda u$, $\lambda \in \mathbb{R}$. Let us first examine the case $\lambda = 0$, i.e.,

   $$t\int_t^1 u(s)\,ds + \int_0^t su(s)\,ds = 0, \quad t \in [0, 1].$$

   By differentiation we obtain

   $$\int_t^1 u(s)\,ds = 0, \quad t \in [0, 1] \implies u \equiv 0,$$

   hence $\lambda = 0$ is not an eigenvalue of A, and $N(A) = \{0\}$. Now we are looking for nonzero eigenvalues of A. The equation $Au = \lambda u$ reads

   $$t\int_t^1 u(s)\,ds + \int_0^t su(s)\,ds = \lambda u(t), \quad t \in [0, 1].$$

   By differentiating this equation twice we find that u satisfies the equivalent problem

   $$\lambda u''(t) + u(t) = 0, \quad 0 \le t \le 1; \quad u(0) = 0, \quad u'(1) = 0.$$

Multiplication by $u(t)$ of this equation, followed by integration over $[0,1]$, shows that $\lambda > 0$. Solving the above problem we find

$$\lambda_n = \frac{1}{(1/2+n)^2\pi^2}, \quad u_n(t) = c_n \sin\left[(n+1/2)\pi t\right],$$
$$n = 0,1,2,\ldots$$

We determine the constants c_n by imposing $\|u_n\| = 1$, $n \in \mathbb{N}$, so

$$u_n(t) = \sqrt{2}\sin\left[(1/2+n)\pi t\right], \quad n = 0,1,2,\ldots$$

By Theorem 8.7 (Hilbert–Schmidt) we conclude that $B = \{\sqrt{2}\sin\left[(1/2+n)\pi t\right]\}_{n=0}^{\infty}$ is an orthonormal basis of H.

7. Assume that $Ax = \lambda x$ for some scalar λ. Then $\|Ax\| = |\lambda| \cdot \|x\|$ and so

$$\begin{aligned}|(Ax,x)| &= |\lambda| \cdot \|x\|^2 \\ &= \|Ax\| \cdot \|x\|.\end{aligned}$$

Conversely, let us assume that $|(Ax,x)| = \|Ax\| \cdot \|x\|$. For an arbitrary λ we have

$$\|Ax - \lambda x\|^2 = \|Ax\|^2 - 2\operatorname{Re}\bar{\lambda} \cdot (Ax,x) + |\lambda|^2\|x\|^2,$$

which (according to our assumption) equals zero for $\lambda = (Ax,x)/\|x\|^2$.

8. (a) Denote $e_1 = \|u\|^{-1}u$, $e_2 = \|v\|^{-1}v$, so $\{e_1,e_2\}$ is an orthonormal system. For all $x \in H$ we have

$$\begin{aligned}0 \leq \|(x,e_1)e_1 + (x,e_2)e_2 - x\|^2 &= -|(x,e_1)|^2 \\ &\quad - |(x,e_2)|^2 + \|x\|^2,\end{aligned}$$

which gives a particular case of the so-called Bessel inequality, i.e.,

$$|(x,e_1)|^2 + |(x,e_2)|^2 \leq \|x\|^2.$$

Now, for all $x \in H$

$$\begin{aligned}\|Ax\|^2 &= \|(x,v)u\|^2 + \|(x,u)v\|^2 \\ &= \|u\|^2\|v\|^2\left[|(x,\|u\|^{-1}u)|^2 + |(x,\|v\|^{-1}v)|^2\right] \\ &\leq \|u\|^2\|v\|^2\|x\|^2,\end{aligned}$$

which follows from the above Bessel inequality. Hence $\|A\| \leq \|u\| \cdot \|v\|$. In fact, $\|A\| = \|u\| \cdot \|v\|$, since

$$\|A(\|u\|^{-1}u)\| = \|(\|u\|^{-1}u, u)v\| = \|u\| \cdot \|v\|;$$

(b) Easy to check;

(c) Apply (a) with $H = L^2(-\pi, \pi)$, $u = \cos t$, $v = \sin t$ to find $\|A\| = \pi$;

(d) We first observe that any eigenvalue of A is a real number (since A is symmetric).

Denote $Y = \mathrm{Span}\{u, v\}$. Let us determine the nullspace $N(A)$. The equation $Ax = 0$ reads

$$(x, v)u + (x, u)v = 0 \implies |(x, v)|^2\|u\|^2 + |(x, u)|^2\|v\|^2 = 0,$$

which implies $x \perp Y$. Therefore, $N(A) = Y^\perp$. Note that $A(Y) = Y$.

In what follows we distinguish two cases:

(i) $\dim H > 2$. In this case, $N(A) = Y^\perp \neq \{0\}$, and $\lambda = 0$ is an eigenvalue of A, the corresponding eigenvectors being all nonzero vectors from $Y^\perp$. Next, consider the equation $Ax = \lambda x$, $\lambda \in \mathbb{R} \setminus \{0\}$, $x \in Y \setminus \{0\}$. By elementary computations we find two eigenvalues: $\lambda = \pm\|u\| \cdot \|v\|$, the corresponding eigenvectors being the nonzero multiples of $\|u\|^{-1}u \pm \|v\|^{-1}v$;

(ii) $H = Y$. In this case, $N(A) = \{0\}$ so $\lambda = 0$ is no longer an eigenvalue of A. As before, we find $\lambda = \pm\|u\| \cdot \|v\|$, the corresponding eigenvectors being the nonzero multiples of $\|u\|^{-1}u \pm \|v\|^{-1}v$.

9. (a) Clearly A is a linear operator. For all $x \in H$ we have

$$\|Ax\|^2 = \sum_{i=1}^{m} |c_i|^2 |(x, e_i)|^2$$

$$\leq \left(\max_{1 \leq i \leq m} |c_i| \right)^2 \sum_{i=1}^{m} |(x, e_i)|^2$$

$$\leq \left(\max_{1 \leq i \leq m} |c_i| \right)^2 \|x\|^2,$$

where we have used the Bessel inequality (see the solution of the previous exercise where the Bessel inequality is derived for $m = 2$). Hence $A \in \overline{L}(H)$ and $\|A\| \leq \beta :=$ $\max_{1 \leq i \leq m} |c_i|$. In fact, $\|A\| = \beta$, for if the maximum β is achieved for $i = i_0$, i.e., $\beta = |c_{i_0}|$, then observe that $\|Ae_{i_0}\| = |c_{i_0}|$, which confirms our claim.

It is readily seen that $R(A) = H_m := \mathrm{Span}\,(\{e_1, \ldots, e_m\})$ and $N(A) = H_m^\perp$;

(b) Easy to check;

(c) If $\dim H > m$, then $N(A) \neq \{0\}$, so $\lambda = 0$ is an eigenvalue of A, the corresponding eigenvectors being the nonzero vectors from $N(A) = H_m^\perp$. The other eigenvalues are determined from the equation $Ax = \lambda x$, $\lambda \in \mathbb{K} \setminus \{0\}$, $x = \sum_{i=1}^m \alpha_i e_i \in H_m \setminus \{0\}$, i.e., from the algebraic system

$$(\lambda - c_i)\alpha_i = 0, \quad i = 1, \ldots, m.$$

So, the eigenvalues we are looking for are the distinct c_i's, and the corresponding eigenvectors are the nonzero vectors $x = \sum_{i=1}^m \alpha_i e_i \in H_m$ with the α_i's satisfying the above system.

If $\dim H = m$, i.e., $H = H_m$, then we have only nonzero eigenvalues which can be determined as before.

10. (a) Obvious.

(b) Denote $u_0(t) = t/(1 + t)$, $t \in [0, 1]$. We have $R(A) = \mathrm{Span}\{u_0\}$ and $N(A) = (\mathrm{Span}\{u_0\})^\perp$.

(c) First, $\lambda = 0$ is an eigenvalue of A and the corresponding eigenfunctions are all nonzero functions of $N(A) = (\mathrm{Span}\{u_0\})^\perp$. Next, consider the equation $Au = \lambda u$, $u \in R(A) \setminus \{0\}$, $\lambda \neq 0$. Since $u(t) = Cu_0(t) = Ct/(1+t)$, $C \neq 0$, we obtain

$$\lambda = \int_0^1 \frac{s^2}{(1+s)^2}\, ds = \frac{3}{2} - 2\ln 2 \,.$$

The corresponding eigenfunctions are $u(t) = Cu_0(t)$, $C \in \mathbb{R} \setminus \{0\}$.

11. (a) It is easily seen that for $u \in H$, we have

$$(Au)(t) = v(t) = -\int_t^1 \left(\int_0^s u(\tau)d\tau \right) ds, \quad t \in [0,1].$$

Obviously, $A \in L(H)$ and $N(A) = \{0\}$.

(b) Integrating twice by parts shows that A is symmetric (hence self-adjoint). Also, A is compact (by the Arzelà–Ascoli criterion or the compact embedding of $H^2(0,1)$ in $H = L^2(0,1)$).

(c) The equation $Au = \lambda u$ reads

$$-\int_t^1 \left(\int_0^s u(\tau)d\tau \right) ds = \lambda u(t), \quad t \in [0,1].$$

Clearly, $\lambda = 0$ is not an eigenvalue of A, so consider $\lambda \in \mathbb{R} \setminus \{0\}$. We see that $u \in C^\infty[0,1]$ and satisfies the problem

$$\begin{cases} \lambda u''(t) = u(t), & t \in [0,1], \\ u(1) = 0, \ u'(0) = 0. \end{cases}$$

If we multiply the above equation by $u(t)$ and integrate over $[0,1]$, we obtain

$$-\lambda \int_0^1 (u')^2 dt = \int_0^1 u(t)^2 dt,$$

which shows that any eigenvalue $\lambda < 0$. Denote for convenience $\lambda = -1/\nu^2$, $\nu > 0$. Solving the above problem we find $u(t) = c\cos(\nu t)$, $\cos \nu = 0$, $c \in \mathbb{R} \setminus \{0\}$. Thus $\nu = n\pi + \pi/2$, $n = 0, 1, \ldots$ Therefore, A has eigenvalues

$$\lambda_n = \frac{1}{(n\pi + \pi/2)^2}, \quad n = 0, 1, \ldots$$

and the corresponding eigenfunctions are the nonzero multiples of the following normalized functions

$$u_n(t) = \sqrt{2}\cos\left((n\pi + \pi/2)t\right), \quad t \in [0,1], \ n = 0, 1, \ldots$$

According to Theorem 8.7 (Hilbert–Schmidt), the system $\{u_n\}_{n=0}^\infty$ is an orthonormal basis in H.

12. Search for u in the form $u(x) = u_1(x_1) \cdot u_2(x_2)$, with $u_1 \neq 0$, $u_2 \neq 0$. Thus the equation $-\Delta u = \lambda u$ reads

$$-\frac{u_1''}{u_1} = \frac{u_2''}{u_2} + \lambda.$$

Since the different sides of this equation depend on distinct variables, x_1 and x_2, they must be constant functions, so we obtain the following two eigenvalue problems:

$$\begin{cases} u_1'' + \nu u_1 = 0, & 0 < x_1 < a, \\ u_1(0) = 0, \ u_1(a) = 0, \end{cases}$$

$$\begin{cases} u_2'' + \mu u_2 = 0, & 0 < x_2 < b, \\ u_2(0) = 0, \ u_2(b) = 0, \end{cases}$$

with $\nu + \mu = \lambda$. If we multiply the equation $u_1'' + \nu u_1 = 0$ by u_1 and then integrate over $[0, a]$, we get

$$\int_0^a (u_1')^2 dx_1 = \nu \int_0^a u_1^2 \, dx_1,$$

which shows that $\nu > 0$. Similarly, $\mu > 0$, hence $\lambda > 0$ as well. Solving the above eigenvalue problems we find

$$\nu_n = \left(\frac{n\pi}{a}\right)^2, \quad u_{1,n}(x_1) = c_n \sin\left(\frac{n\pi}{a}x_1\right), \ n = 1, 2, \ldots$$

and

$$\mu_m = \left(\frac{m\pi}{b}\right)^2, \quad u_{2,m}(x_2) = \tilde{c}_n \sin\left(\frac{m\pi}{b}x_2\right), \ m = 1, 2, \ldots$$

Thus we have obtained the following eigenvalues of $-\Delta$

$$\lambda_{mn} = \left(\frac{n\pi}{a}\right)^2 + \left(\frac{m\pi}{b}\right)^2, \ m, n \in \mathbb{N},$$

the corresponding eigenfunctions being the nonzero multiples of

$$u_{mn}(x) = \frac{2}{\sqrt{ab}} \sin\left(\frac{n\pi}{a}x_1\right) \cdot \sin\left(\frac{m\pi}{b}x_2\right), \ m, n \in \mathbb{N}.$$

Note that the system $S = \{u_{mn}\}_{m,n=1}^\infty$ is an orthonormal basis of $H = L^2(\Omega)$, $\Omega = (0, a) \times (0, b)$. As S is an orthonormal system, it is enough to show that $\text{Span } S$ is dense in H (see Theorem 6.21).

Indeed, every function $u \in H$ can be approximated with respect to the L^2-norm by a function from $C_0^\infty(\Omega)$ (cf. Theorem 5.8), which in turn is close (even with respect to the uniform convergence topology) to a polynomial in x_1, x_2, i.e., a finite sum of product functions $u_1(x_1) \cdot u_2(x_2)$. Since the systems

$$\{\sqrt{2/a}\sin\left(\frac{n\pi}{a}x_1\right)\}_{n=1}^\infty, \quad \{\sqrt{2/b}\sin\left(\frac{m\pi}{b}x_2\right)\}_{m=1}^\infty$$

are bases in $L^2(0,a)$ and $L^2(0,b)$, respectively, it follows that every product function $u = u_1(x_1) \cdot u_2(x_2)$, with $u_1 \in L^2(0,a)$ and $u_2 \in L^2(0,b)$, is approximated in $H = L^2(\Omega)$ by functions from $\operatorname{Span} S$, hence $\operatorname{Span} S$ is dense in H, as claimed.

13. Proceed as for the previous exercise. Similarly, you can determine different orthonormal bases in $L^2(0,a) \times L^2(0,b)$.

12.9   Answers to Exercises for Chap. 9

1. (i) Apply the usual formula

$$e^{tA} = \sum_{k=0}^\infty \frac{t^k}{k!}A^k$$

and the observation that A^k is the null matrix for $k = 2, 3, \ldots$. Thus we find

$$e^{tA} = \begin{bmatrix} 1+t & t \\ -t & 1-t \end{bmatrix};$$

(ii) One can use the formula

$$e^{tA} = \sum_{k=0}^\infty \frac{t^k}{k!}A^k,$$

again, but we suggest another method. Recall that e^{tA} is the fundamental matrix of the differential linear system

$$\frac{d}{dt}\begin{pmatrix} x \\ y \end{pmatrix} = A\begin{pmatrix} x \\ y \end{pmatrix}$$

which equals the identity matrix for $t = 0$. We solve the above differential system with the initial conditions $x(0) = 1$, $y(0) = 0$ and $x(0) = 0$, $y(0) = 1$, and find

$$e^{tA} = \begin{bmatrix} \cos t & \sin t \\ -\sin t & \cos t \end{bmatrix};$$

(*iii*)

$$e^{tA} = \begin{bmatrix} 2e^{-2t} - e^{-3t} & -e^{-2t} + e{-3t} \\ 2e^{-2t} - 2e^{-3t} & -e^{-2t} + 2e^{-3t} \end{bmatrix}.$$

2. By the classic Jordan decomposition theorem we have $A = B^{-1}JB$, where B is a nonsingular $n \times n$ matrix and J has Jordan blocks $J_0, J_1, \ldots, J_m$ on its diagonal and 0 in the rest. Here, denoting the simple eigenvalues of A by $\lambda_1, \lambda_2, \ldots, \lambda_p$ and the other eigenvalues of A by $\lambda_{p+1}, \ldots, \lambda_{p+m}$, we have $J_0 = \text{diag}\,(\lambda_1, \lambda_2, \ldots, \lambda_p)$, $J_i = \lambda_{p+i} I_{p_i} + B_{p_i}$, $i = 1, \ldots, m$, where I_{p_i} is the $p_i \times p_i$ identity matrix and B_{p_i} is the $p_i \times p_i$ matrix having all entries situated above the principal diagonal equal to 1 and 0 otherwise. Note that

$$e^{tJ_0} = \text{diag}\,(e^{t\lambda_1}, \ldots, e^{t\lambda_p}), \ \ e^{tJ_i} = e^{t\lambda_{p+i}} e^{tB_{p_i}},$$

where $e^{tB_{p_i}}$ has a special form involving $\{1, t, t^2, \ldots, t^{p_i-1}\}$, $i = 1, \ldots, m$. Thus, since

$$e^{tA} = B^{-1} e^{tJ} B = B^{-1} \cdot \text{diag}\{e^{tJ_0}, e^{tJ_1}, \ldots, e^{tJ_m}\} \cdot B,$$

then both (*a*) and (*b*) follow easily.

3. Left to the reader.

4. For a given pair $(t_0, x_0) \in [0, \infty) \times X$ and for all $(t, x) \in (0, \infty) \times X$, $t > t_0$, we have

$$\begin{aligned} \|T(t)x - T(t_0)x_0\| &\le \|T(t)x - T(t)x_0\| \\ &\quad + \|T(t)x_0 - T(t_0)x_0\| \\ &= \|T(t)(x - x_0)\| \\ &\quad + \|T(t_0)\big[T(t - t_0)x_0 - x_0\big]\| \\ &\le M e^{\omega t} \|x - x_0\| \\ &\quad + M e^{\omega t_0} \cdot \|T(t - t_0)x_0 - x_0\|. \end{aligned}$$

On the other hand, if $t_0 > 0$ and $t \in (0, t_0)$, we have

$$\begin{aligned} \|T(t)x - T(t_0)x_0\| &\le \|T(t)x - T(t)x_0\| + \|T(t)x_0 - T(t_0)x_0\| \\ &= \|T(t)(x - x_0)\| + \|T(t)\big(x_0 - T(t_0 - t)x_0\big)\| \\ &\le M e^{\omega t} \big[\|x - x_0\| + \|T(t_0 - t)x_0 - x_0\|\big]. \end{aligned}$$

The claim follows from the above estimates.

5. (a) It is easy to check that $\{G(t) : X \to X; t \in \mathbb{R}\}$ is a uniformly continuous group. Its generator is $A \in L(X)$ given by

$$(Af)(x) = \lambda[f(x - \delta) - f(x)], \quad f \in X, \ x \in \mathbb{R}.$$

(b) If $t \geq 0$, we have $\|G(t)f\|_X \leq 1$ for all $f \in X$ satisfying $\|f\|_X \leq 1$, and for $\hat{f} \equiv 1$ we have $\|G(t)\hat{f}\|_X = 1$. So $\|G(t)\| = 1 \ \forall t \geq 0$.

For $t < 0$ we easily deduce that $\|G(t)\| \leq e^{-2\lambda t}$, and this upper bound is reached for $\tilde{f}(x) = \cos(\pi x/\delta)$ (indeed, $(G(t)\tilde{f})$ $(0) = e^{-2\lambda t}$).

6. This is a translation group and its generator $A : D(A) \subset X \to X$ is defined by

$$D(A) = \{f \in X; \ f \text{ is differentiable on } \mathbb{R} \text{ and } f' \in X\},$$
$$Af = f'.$$

Use arguments similar to those in Sect. 9.5. Obviously, $\|T(t)\| = 1$ for all $t \in \mathbb{R}$.

7. (a) It is easy to see that $\{G(t) : X \to X; t \in \mathbb{R}\}$ is a C_0-group and its infinitesimal generator is $A : D(A) \subset X \to X$ is given by

$$D(A) = W^{1,1}(\mathbb{R}^k), \quad (Af)(x) = -\nabla f(x) \cdot Mx,$$

for all $f \in D(A)$ and a.a. $x \in \mathbb{R}^k$.

(b) Assume that $\sum_{i=1}^{k} m_{ii} = 0$. Denote by $W(t)$ the determinant of e^{tM} whose columns are solutions of the differential linear system $u'(t) = Mu(t)$. Recall that $W(t)$ is known as the Wronski determinant of the system of solutions of $u'(t) = Mu(t)$ that are here given by the columns of $X(t) = e^{tM}$. Using the definition of a determinant, we can see that the derivative of $W(t)$ is the sum of k determinants that are obtained by differentiating one by one the rows of $W(t)$. Noting that the derivative of each row contains linear combinations of the other rows, we derive

$$W'(t) = \Big(\sum_{i=1}^{k} m_{ii} \Big) \cdot W(t) = 0 \ \forall t \in \mathbb{R}.$$

So W is a constant function, hence $W(t) = W(0) = 1 \ \forall t \in \mathbb{R}$.

Next, by using the change $x = e^{tM}y$, we obtain

$$\int_{\mathbb{R}^k} |(G(t)f)(x)|^p dx = \int_{\mathbb{R}^k} |f(e^{-tM}x)|^p dx$$

$$= \int_{\mathbb{R}^k} |(f)(y)|^p W(t)\, dy,$$

$$= \int_{\mathbb{R}^k} |(f)(x)|^p dx,$$

for all $f \in X$ and $t \in \mathbb{R}$, hence $\|G(t)\| = 1$ for all $t \in \mathbb{R}$.

8. It is easy to check that all the C_0-semigroup properties are fulfilled in this case. The infinitesimal generator is given by

$$D(A) = \{f \in X;\ f' \text{ exists and belongs to } X\}, \quad Af = -f'.$$

9. (a) Let us only check the continuity at $t = 0$, the other properties being trivially satisfied. Let $(x_n)_{n \in \mathbb{N}} \in X$ be arbitrary but fixed. For every $\varepsilon > 0$ there exists an $m \in \mathbb{N}$ such that $\sum_{j=m+1}^{\infty} |x_j|^p < \varepsilon$.

So we have

$$\left(\sum_{j=m+1}^{\infty} |e^{-c_j t}x_j - x_j|^p \right)^{1/p}$$

$$\leq \left(\sum_{j=m+1}^{\infty} |e^{-c_j t}x_j|^p \right)^{1/p} + \left(\sum_{j=m+1}^{\infty} |x_j|^p \right)^{1/p}$$

$$\leq 2 \left(\sum_{j=m+1}^{\infty} |x_j|^p \right)^{1/p}$$

$$\leq 2\varepsilon^{1/p}.$$

This implies

$$\|T(t)(x_n) - (x_n)\|_p^p \leq \sum_{j=1}^{m} |e^{-c_j t}x_j - x_j|^p + 2^p \varepsilon.$$

It follows that

$$\limsup_{t \to 0^+} \|T(t)(x_n) - (x_n)\|_p^p \leq 2^p \varepsilon \quad \forall \varepsilon > 0,$$

which proves the claim.

(b) If A denotes the infinitesimal generator of the semigroup and $\lim_{h\to 0^+} h^{-1}[T(h)(x_n) - (x_n)]$ exists, then $A(x_n) = -(c_n x_n)$ with $(c_n x_n) \in X$. In fact,

$$D(A) = \{(x_n) \in X;\ (c_n x_n) \in X\}.$$

Indeed, noting that (by the Mean Value Theorem)

$$\frac{e^{-c_j h} - 1}{h} + c_j = -c_j e^{-c_j \theta_j} + c_j,\ \ 0 < \theta_j < h,$$

we can write for $h,\ \varepsilon > 0$ and $(x_n) \in D(A)$ (defined above)

$$\left\| h^{-1}\big[T(h)(x_n) - (x_n)\big] + (c_n x_n) \right\|^p$$

$$\leq \sum_{j=1}^{N} \Big| \frac{e^{-c_j h} x_j - x_j}{h} + c_j x_j \Big|^p$$

$$+ \sum_{j=N+1}^{\infty} |1 - e^{-c_j \theta_j}|^p \cdot |c_j|^p \cdot |x_j|^p$$

$$\leq \sum_{j=1}^{N} \Big| \frac{e^{-c_j h} x_j - x_j}{h} + c_n x_n \Big|^p + \sum_{j=N+1}^{\infty} |c_j|^p \cdot |x_j|^p$$

$$\leq \sum_{j=1}^{N} \Big| \frac{e^{-c_j h} x_j - x_j}{h} + c_j x_j \Big|^p + \varepsilon,$$

where $N = N_\varepsilon$ comes from $(c_n x_n) \in X$, i.e., $\sum_{n=1}^{\infty} |c_n|^p \cdot |x_n|^p < \infty$. Therefore,

$$\limsup_{h\to 0^+} \Big\| \frac{T(h)(x_n) - (x_n)}{h} + (c_n x_n) \Big\|^p$$

$$\leq \sum_{j=1}^{N} \Big| \frac{e^{-c_j h} x_j - x_j}{h} + c_j x_j \Big|^p + \varepsilon,$$

which implies

$$\limsup_{h\to 0^+} \Big\| \frac{T(h)(x_n) - (x_n)}{h} + (c_n x_n) \Big\|^p \leq \varepsilon.$$

This concludes the proof.

    (c) The semigroup is uniformly continuous $\Longleftrightarrow D(A) = X$ and $A \in L(X)$ (see Theorems 9.5 and 9.14). In our case, $D(A) = X \Longleftrightarrow (c_n)$ is bounded. If (c_n) is bounded, then obviously $A \in L(X)$.

10. It is easy to see that A satisfies all the conditions of the Hille–Yosida Generation Theorem (Theorem 9.22), so A generates a C_0-semigroup of contractions $\{T(t) : X \to X; t \geq 0\}$. We know that, for $u_0 \in D(A)$, $u(t, \cdot) = T(t)u_0$ satisfies

$$\frac{d}{dt} u(t, \cdot) = Au(t, \cdot), \ t \geq 0,$$

in X, i.e., $u_t + u_x = 0$ in $\Omega = (0, \infty) \times (0, 1)$. Noting that this equation has the characteristic $x - t = C$, we can derive

$$(T(t)u_0)(x) = \begin{cases} u_0(x - t), & 0 \leq t \leq x, \\ 0, & t > x. \end{cases}$$

This formula extends by density to all $u_0 \in X$, hence the semigroup is completely determined. We saw that, for $u_0 \in D(A)$, the function $u(t, x) = (T(t)u_0)(x)$ satisfies in the classical sense the transport equation $u_t + u_x = 0$ in Ω. Obviously, this u also satisfies the transport equation in the sense of distributions, i.e.,

$$\int_\Omega u(\phi_t + \phi_x) \, dt \, dx = 0 \ \ \forall \phi \in D(\Omega).$$

This relation extends by density to all $u_0 \in X = L^2(0, 1)$, hence $u_t + u_x = 0$ in $D'(\Omega)$.

11. First of all, observe that the substitution $v(t, x) = e^{at}u(t, x)$ leads to a similar initial-boundary value problem for the equation

$$v_t - v_{xx} = e^{at} f(t, x),$$

so one can assume $a = 0$. As in Sect. 9.12.1, one can express this problem as a Cauchy problem for an evolution equation in $X = L^2(0, 1)$ associated with the operator $A : D(A) \subset X \to X$ defined by

$$D(A) = \{v \in H^2(0, 1); \ v(0) = 0, \ v'(1) + \alpha v(1) = 0\},$$
$$Av = v'', \ v \in D(A).$$

This operator is linear, densely defined (since $C_0^\infty(0,1) \subset D(A)$ and $C_0^\infty(0,1)$ is dense in X), closed, self-adjoint, and dissipative. The reader can easily check that all these properties of A hold true. According to Theorem 9.29, A is the generator of a C_0-semigroup of contractions. So, applying the theory developed in Sect. 9.11, we conclude that, for an arbitrary $r > 0$, the Cauchy problem in X

$$\begin{cases} u'(t) = Au(t) + f(t), & 0 \le t \le r, \\ u(0) = u_0, \end{cases}$$

where $u(t) := u(t, \cdot)$, $f(t) := f(t, \cdot)$, has a unique mild solution u on every interval $[0, r]$, $r > 0$, hence $u \in C([0, \infty); L^2(0, 1))$. If $u_0 \in D(A)$ and $f \in C^1([0, \infty); L^2(0, 1))$, then $u \in C([0, \infty); L^2(0, 1)) \cap C^1((0, \infty); L^2(0, 1))$ (cf. Theorem 9.47), i.e., u is a classical solution.

If the term au is replaced by $h(u)$ with h a Lipschitz function, then we can also prove the existence of a unique mild solution on every interval $[0, r]$, $r > 0$ (see Remark 9.50).

12. This problem is similar to the initial-boundary value problem discussed in Sect. 9.12.2. However, since the boundary conditions are different, separate analysis is needed. Denote

$$X = \{p \in H^1(0, 1); \ p(0) = 0\} \times L^2(0, 1)$$

and endow X with the scalar product

$$\langle [p_1, q_1], [p_2, q_2] \rangle = \int_0^1 p_1' p_2' \, dx + \int_0^1 q_1 q_2 \, dx$$

and the corresponding induced norm. It is easily seen that X is a real Hilbert space. Define $A : D(A) \subset X \to X$ by

$$D(A) = \{[p, q] \in H^2(0, 1) \times H^1(0, 1); \ p(0) = p'(1) = 0,$$
$$q(0) = 0\},$$

and $A[p, q] = [q, p'']$. The given problem can be expressed as the following Cauchy problem in X

$$\frac{d}{dt}[u(t, \cdot), v(t, \cdot)] = A[u(t, \cdot), v(t, \cdot)] + [0, f(t, \cdot)], \quad t \ge 0,$$
$$[u(0, \cdot), v(0, \cdot)] = [u_0, v_0] \in X.$$

Denote by (CP) this Cauchy problem. In order to derive existence results for (CP), we are going to show that A is the generator of a C_0-group of isometries. For this purpose, we can use Corollary 9.34.

First of all, note that $D(A)$ is dense in $X = Y \times L^2(0,1)$, where $Y = \{p \in H^1(0,1); p(0) = 0\}$. Indeed, Y is dense in $L^2(0,1)$ (in fact $C_0^\infty(0,1)$ is dense in $L^2(0,1)$). Next, let u be arbitrary in Y. Then, $y(x) = u(x) + x(x-2)u(1)$ belongs to $H_0^1(0,1)$, so there exists a sequence (ϕ_n) in $C_0^\infty(0,1)$ which converges in $H^1(0,1)$ (hence in $C[0,1]$) to y. Construct a sequence (u_n) by $u_n(x) = \phi_n(x) - x(x-2)u(1)$. Clearly, $u_n(0) = 0$, $u_n'(1) = 0$ for all n, and $u_n \to u$ in $H^1(0,1)$, which concludes the proof of the claim (that $D(A)$ is dense in X). The condition $(kk)^*$ of Corollary 9.34 is also satisfied, so A generates a C_0-group of isometries, say $\{G(t) : X \to X; t \in \mathbb{R}\}$. In order to finish we can follow the discussion in Sect. 9.12.2 (where the case $u(t,0) = 0 = u(t,1)$ was addressed).

13. Note that the second boundary condition is a dynamic one (as it involves the derivative v_t) so this initial-boundary value problem needs special analysis. The main idea towards solving this problem is to consider an appropriate framework. Specifically, we shall consider the real Hilbert space $X = L^2(0,1) \times L^2(0,1) \times \mathbb{R}$ with the scalar product

$$\langle [f_1, g_1, \xi_1], [f_2, g_2, \xi_2] \rangle = L \int_0^1 f_1 f_2 \, dx + C \int_0^1 g_1 g_2 \, dx + C_1 \xi_1 \xi_2,$$

and the induced norm.

(a) Define $A : D(A) \subset X \to X$ and $B : X \to X$ by

$$D(A) = \{[f, g, \xi] \in H^1(0,1)^2 \times \mathbb{R}; \xi = g(1), -g(0) = R_0 f(0)\},$$

$$A[f, g, \xi] = [-L^{-1}g', -C^{-1}f', C_1^{-1}f(1)] \; \forall [f, g, \xi] \in D(A),$$

$$B[f, g, \xi] = -[L^{-1}Rf, C^{-1}Gg, C_1^{-1}D_1\xi] \; \forall [f, g, \xi] \in X.$$

Note that the given initial-boundary value problem can be expressed as the following Cauchy problem in X, denoted

(CP),

$$\begin{cases} \frac{d}{dt}[i(t,\cdot),v(t,\cdot),\xi(t)] = (A+B)[i(t,\cdot),v(t,\cdot),\xi(t)]+ \\ [L^{-1}e(t,\cdot),0,C_1^{-1}e_1(t)], \ t>0, \\ [i(0,\cdot),v(0,\cdot),\xi(0)] = [i_0(\cdot),v_0(\cdot),\xi_0], \end{cases}$$

where $\xi(t) = v(t,1)$ and $\xi_0 = v_0(1)$.

Let us show that A is the generator of a C_0-semigroup. First of all, observe that $D(A)$ is dense in X. Indeed, for any $[f,g,\xi] \in X$, there exist two sequences (f_n), (ϕ_n) in $C_0^\infty(0,1)$ such that $f_n \to f$ and $\phi_n \to h = h(x) := g(x)-\xi x$ in $L^2(0,1)$. Hence, denoting $g_n(x) = \phi_n(x)+\xi x$, we can see that $[f_n,g_n,\xi] \in D(A)$ for all n and $[f_n,g_n,\xi] \to [f,g,\xi]$ in X.

By a straightforward computation it follows that

$$\langle A[f,g,\xi],[f,g,\xi] \rangle \leq 0 \ \ \forall [f,g,\xi] \in D(A),$$

so A is dissipative. Let us prove that $R(\lambda I - A) = X$ for all $\lambda > 0$, i.e., A is m-dissipative, hence A is the infinitesimal generator of a C_0-semigroup of contractions (see Theorem 9.25 (Lumer–Phillips)). In fact, it is enough to show that $R(\lambda I - A) = X$ for some $\lambda > 0$, i.e., in other words, for every $[p,q,\eta] \in X$ there exists $[f,g,g(1)] \in D(A)$ such that

$$\begin{cases} \lambda L f + g' = Lp, \\ \lambda C g = f' = Cq, \end{cases}$$

with

$$g(0) + R_0 f(0) = 0, \ \lambda C_1 g(1) - f(1) = C_1\eta.$$

The above differential system can be solved for f,g by using the substitutions

$$\sqrt{\lambda C} g = z_1 - z_2, \ \sqrt{\lambda L} f = z_1 + z_2,$$

which give (by addition and subtraction of the two equations) an equivalent system of two differential equations in z_1 and z_2. Requiring the general solution $[f,g]$ of the above differential system to satisfy the boundary conditions we see that (for $\lambda > 0$ large enough) there exists a unique solution $[f, g]$ of the above boundary value problem, with $[f, g, g(1)] \in D(A)$, as claimed.

Therefore, A generates a C_0-semigroup of contractions, say $\{S(t) : X \to X; t \geq 0\}$. Note that the density of $D(A)$ in X also follows from Theorem 9.27, since X is a Hilbert space and therefore a reflexive Banach space.

Since $B \in L(X)$, it follows by Theorem 9.35, that $A + B$ is the generator of a C_0-semigroup $\{T(t) : X \to X; t \geq 0\}$. In fact, this semigroup is also a semigroup of contractions since B is m-dissipative. Having this semigroup, one can express the solution of the above Cauchy problem (CP) by using the variation of constants formula.

For $[i_0, v_0, v_0(1)] \in D(A)$, $e \in C^1([0, \infty); L^2(0, 1))$, and $e_1 \in C^1[0, \infty)$, there exists a unique strong solution $[u(t, \cdot), v(t, \cdot), v(t, 1)]$, hence $[u, v]$ can be regarded as a classical solution of the original problem.

On the other hand, since $D(A)$ is dense in X, we have that for $[i_0, v_0, \xi] \in X$, $e \in L^1_{\text{loc}}([0, \infty); L^2(0, 1))$ and $e_1 \in L^1_{\text{loc}}[0, \infty)$, the Cauchy problem (CP) only has a mild solution given by the formula of variation of constants. In this case, the third component of the initial datum, ξ_0, can be chosen independently of v_0, and the third component of the solution, $\xi(t)$, may not satisfy the identity $\xi(t) = v(t, 1)$, $t > 0$, as in the classical case. This means that the evolution at the boundary point $x = 1$ is weakly dependent on the evolution in $(0, 1)$.

(b) In this case, we observe that B is a Lipschitz operator from X into itself, so one can prove existence of a mild solution under appropriate conditions (see Remark 9.50).

12.10 Answers to Exercises for Chap. 10

1. (i) We find that $D(Q)$ is dense in $H = L^2(0, 1)$ since $C_0^\infty(0, 1) \subset D(Q)$ and is dense in H.

Now, we easily obtain by integration by parts

$$(Qv, w) = -\int_0^1 v''w\,dx = -\int_0^1 vw''\,dx$$
$$= (v, Qw) \forall v, w \in D(Q),$$

so $D(Q) \subset D(Q^*)$ and $Q^*w = Qw$ for all $w \in D(Q)$. Let us prove the converse inclusion: $D(Q^*) \subset D(Q)$. Choose a $w \in D(Q^*)$, i.e., $w \in H$ and $f(v) = (Qv, w)$ satisfies (for a constant C)

$$|f(v)| \leq C\|v\| \forall v \in D(Q).$$

By arguments already used before (see the solution to Exercise 7.15) we can deduce that $w \in H^2(0, 1)$. On the other hand, from

$$
\begin{aligned}
|f(v)| &= |(Qv, w)| \\
&= |\int_0^1 v''w\,dx| \\
&= |-v'(0)w(0) - v(1)w'(1) + \int_0^1 vw''\,dx| \\
&\leq C\|v\| \forall v \in D(Q),
\end{aligned}
$$

we get

$$|v'(0)w(0) + v(1)w'(1)| \leq C\|v\| \forall v \in D(Q).$$

Choosing $v := v_n$ in the last inequality, where (v_n) is a sequence in $D(Q)$ satisfying $v_n'(0) = 1$, $v_n(1) = 0$ $\forall n$, and $v_n \to 0$ in H, we obtain by letting $n \to \infty$ that $w(0) = 0$. By a similar argument, we also get $w'(1) = 0$. Summarizing, $w \in D(Q)$, hence $D(Q^*) \subset D(Q)$, as claimed.

The fact that $Q^* = Q$ also follows by the arguments used in the proof of Proposition 7.14.

Moreover, Q is strongly positive, since the Poincaré inequality still holds for functions $v \in H^1(0, 1)$ with $v(0) = 0$. Indeed, by using the Hölder inequality, we have

$$v(x) = \int_0^x v'(s)\,ds, x \in [0, 1] \implies \int_0^1 [v(x)]^2 dx$$
$$\leq \int_0^1 [v'(x)]^2 dx.$$

(ii)   By easy computations we find

$$\lambda_n = \left(n + \frac{1}{2}\right)^2 \pi^2, \quad e_n(x) = \sqrt{2}\sin\left((n + \frac{1}{2})\pi x\right), \quad n = 0, 1, 2, \ldots$$

(iii) $H_E = \{v \in H^1(0,1); \ v(0) = 0\}$, with the scalar product $(p,q)_E = \int_0^1 p'q'\,dx$. Obviously, this space is compactly embedded in H. An orthonormal basis of H_E is $\{\frac{2\sqrt{2}}{(2n+1)\pi}\sin\left((n + \frac{1}{2})\pi x\right)\}_{n=0}^{\infty}$.

(iv) $u(t,x) = \sum_{n=0}^{\infty} u_n(t)e_n(x)$, where

$$\begin{cases} u_n'(t) + (n + \frac{1}{2})^2\pi^2 u_n(t) = f_n(t), & 0 < t < T, \\ u_n(0) = u_{0n}, & n = 0, 1, 2, \ldots, \end{cases}$$

where

$$f_n(t) = (f(t,\cdot), e_n) = \sqrt{2}\int_0^1 f(t,x)\sin\left((n + \frac{1}{2})\pi x\right)dx,$$

$$u_{0n} = (u_0, e_n) = \sqrt{2}\int_0^1 u_0(x)\sin\left((n + \frac{1}{2})\pi x\right), \quad n = 0, 1, 2, \ldots$$

The rest is left to the reader.

2. The temperature $u = u(t,x)$ satisfies the initial-boundary value problem

$$\begin{cases} u_t - \alpha u_{xx} = 0, & t \in (0,T), \ x \in (0,l), \\ u(t,0) = u_1, \ u(t,l) = u_2, & t \in [0,T], \\ u(0,x) = u_0, & x \in [0,l], \end{cases}$$

where $\alpha > 0$ is the diffusivity of the rod material. We want to convert this problem into a similar one with homogeneous boundary conditions. It is easy to see that y defined by

$$y(t,x) = u(t,x) - \left[\left(1 - \frac{x}{l}\right)u_1 + \frac{x}{l}u_2\right], \quad (t,x) \in [0,T] \times [0,l],$$

satisfies

$$\begin{cases} y_t - \alpha y_{xx} = 0, & t \in (0,T), \ x \in (0,l), \\ y(t,0) = 0, \ y(t,l) = 0, & t \in [0,T], \\ y(0,x) = u_0 - \left[\left(1 - \frac{x}{l}\right)u_1 + \frac{x}{l}u_2\right], & x \in [0,l]. \end{cases}$$

This problem can be expressed as a Cauchy problem in $H = L^2(0, l)$ associated with the operator $Q : D(Q) \subset H \to H$, defined by

$$D(Q) = H^2(0, l) \cap H^1_0(0, l), \quad Qv = -\alpha v'' \ \forall v \in D(Q).$$

This Q satisfies both conditions **(a)** and **(b)** of Theorem 10.1 (with $H_E = H^1_0(0, l)$). Its eigenvalues are

$$\lambda_n = \alpha \frac{n^2 \pi^2}{l^2}, \quad n = 1, 2, \ldots$$

The corresponding orthonormal basis consists of the functions $e_n(x) = \sqrt{2} \sin \frac{n\pi x}{l}$, $n = 1, 2, \ldots$ The solution y is given by

$$y(t, x) = \sum_{n=1}^{\infty} y_n(t) e_n(x),$$

where the y_n's satisfy

$$\begin{cases} y'_n(t) + \alpha \frac{n^2 \pi^2}{l^2} y_n(t) = 0, & 0 < t < T, \\ y_n(0) = y_{0n}, & n = 1, 2, \ldots, \end{cases}$$

with $y_{0n} = (y(0, \cdot), e_n)_H$, $n = 1, 2, \ldots$

The rest is left to the reader.

3. In this case we define $Q : D(Q) \subset H \to H$ by

$$\begin{aligned} D(Q) &= \{v \in H^2(0, 1); \ -v'(0) + \alpha v(0) = 0, \ v'(1) = 0\}, \\ Qv &= -v'' \ \forall v \in D(Q). \end{aligned}$$

(a) Obviously, $D(Q)$ is dense in H. It is also easy to check that

$$(Qv, w) = (v, Qw) \ \forall v, w \in D(Q),$$

so $D(Q) \subset D(Q^*)$ and $Q^*v = Qv \ \forall v \in D(Q)$. Let us prove that $D(Q^*) \subset D(Q)$ to conclude that $Q^* = Q$. Choose an arbitrary $w \in D(Q^*)$. By arguments similar to those used for

Exercise 7.15, we infer that $w \in H^2(0,1)$. Next, since

$$
\begin{aligned}
f(v) := (Qv, w) &= - \int_0^1 v'' w \, dx \\
&= v'(0)w(0) + v(1)w'(1) \\
&\quad - v(0)w'(0) - \int_0^1 vw'' \, dx \\
&= v(0)[\alpha w(0) - w'(0)] \\
&\quad + v(1)w'(1) - \int_0^1 vw'' \, dx
\end{aligned}
$$

for all $v \in D(Q)$ and $|f(v)| \le C\|v\|$ for all $v \in D(Q)$ and a constant C, we deduce

$$
|v(0)[\alpha w(0) - w'(0)] + v(1)w'(1)| \le C\|v\| \quad \forall v \in D(Q).
$$

Choosing in this inequality $v := v_n \in D(Q)$ such that $v_n(0) = 1$, $v_n(1) = 0$, $v_n \to 0$ in H, we obtain by letting $n \to \infty$

$$
-w'(0) + \alpha w(0) = 0.
$$

Similarly, $w'(1) = 0$, hence $w \in D(Q)$, so $D(Q^*) \subset D(Q)$, as claimed, i.e., $Q^* = Q$.

Let us prove that Q is also strongly monotone. First of all, for all $v \in H^1(0,1)$ we have

$$
\begin{aligned}
v(x) = v(0) + \int_0^x v'(s) \, ds, \ x \in [0,1] \implies |v(x)| &\le |v(0)| \\
+ \|v'\|, \ x \in [0,1],
\end{aligned}
$$

so

$$
\|v\|^2 \le 2\big(|v(0)|^2 + \|v'\|^2\big).
$$

On the other hand, for all $v \in D(Q)$ we have

$$
\begin{aligned}
(Qv, v) &= - \int_0^1 v'' v \, dx \\
&= v'(0)v(0) + \|v'\|^2 \\
&= \alpha |v(0)|^2 + \|v'\|^2,
\end{aligned}
$$

Therefore, Q is indeed strongly positive.

(b) It is easily seen that $H_E = H^1(0,1)$ with the scalar product

$$(v_1, v_2)_E = \alpha v_1(0)v_2(0) + \int_0^1 v_1' v_2' \, dx \quad \forall v_1, v_2 \in H_E,$$

and the induced norm, which is equivalent to the usual norm of $H^1(0,1)$ (easy to check). Thus H_E is compactly embedded in H.

4. Denoting again $H = L^2(0,1)$, one can define $Q : D(Q) \subset H \to H$ by $D(Q) = \{v \in H^2(0,1); \, v'(0) = v'(1) = 0\}$, $Q(v) = -v''$. But this operator is not strongly positive. In order to remedy this, let us consider a perturbation of Q, again denoted Q, defined on the same $D(Q)$ by $Qv = -v'' + v$, $v \in D(Q)$. With the substitutions

$$y(t,x) = e^{-t}u(t,x), \quad \tilde{f}(t,x) = e^{-t}f(t,x),$$

the original problem becomes

$$\begin{cases} y_t - y_{xx} + y = \tilde{f}(t,x), & t \in (0,T), \ x \in (0,1), \\ y_x(t,0) = 0, \ y_x(t,1) = 0, & t \in [0,T], \\ y_0(x) = u_0(x), & x \in (0,1), \end{cases}$$

which can be expressed as the Cauchy problem in H

$$\begin{cases} y'(t) + Qy(t) = \tilde{f}(t), & 0 < t < T, \\ y(0) = u_0, \end{cases}$$

where $y(t) := y(t, \cdot) \in H$.

(i) By arguments already used before we can see that Q satisfies **(a)**.

(ii) $\lambda_n = 1 + n^2\pi^2$, $n = 0, 1, 2, \ldots$; $e_0(x) = 1$, $e_n(x) = \sqrt{2}\sin(n\pi x)$, $n = 1, 2, \ldots$

(iii) $H_E = H^1(0,1)$ with the usual scalar product and norm, and the corresponding orthonormal basis in H_E is

$$\hat{e}_0(x) = 1, \quad \hat{e}_n(x) = \frac{\sqrt{2}}{\sqrt{1 + n^2\pi^2}}\sin(n\pi x), \quad n = 1, 2, \ldots$$

(iv) Left to the reader.

5. First of all, observe that this exercise is an abstract extension of the previous exercise.

  (a) Since A is not strongly positive, we cannot apply Theorem 10.1 directly to the given Cauchy problem (CP). Fortunately, by the changes

$$y(t) = e^{-\alpha t} u(t), \quad \tilde{f}(t) = e^{-\alpha t} f(t),$$

  the given (CP) becomes

$$\begin{cases} y'(t) + Qy(t) = \tilde{f}(t), \ 0 < t < T, \\ y(0) = u_0, \end{cases}$$

  for which Theorem 10.1 is applicable. The reader is encouraged to discuss (using Theorem 10.1) the existence, uniqueness, and regularity of the solution u with respect to the regularity of u_0 and f.

  (b) According to Theorem 10.3, the Cauchy problem above governed by Q has a unique periodic solution y: $y(0) = y(T)$. In terms of u and α this is equivalent to $e^{\alpha T} u_0 = u(T)$, which proves the claim.

  It should be noted that, in fact, u_0 determined here belongs to $D(A)$, hence the corresponding solution u is more regular and $\|u(T) - u_0\|_E$ is small. This follows from [36, Theorem 2.4, p. 56 and Theorem 2.1, p. 48], since Q is a maximal monotone, self-adjoint operator. We encourage the reader to read Chapter I in [36] to understand why $u_0 \in D(A)$.

6. The eigenvalues of $Q = -\Delta$ with Dirichlet conditions on $\partial\Omega$ have been already found (see the solution to Exercise 8.12), namely,

$$\lambda_{mn} = \left(\frac{n\pi}{a}\right)^2 + \left(\frac{m\pi}{b}\right)^2, \quad m, n = 1, 2, \ldots$$

Correspondingly, we have the following orthonormal basis in $H = L^2(\Omega)$,

$$e_{mn}(x) = \frac{2}{\sqrt{ab}} \sin\left(\frac{n\pi}{a} x_1\right) \cdot \sin\left(\frac{m\pi}{b} x_2\right), \quad m, n = 1, 2, \ldots$$

Thus the Fourier expansion solution is $u(t,x) = \sum_{m,n=1}^{\infty} u_{mn}(t)$ $\times e_{mn}(x)$, where the u_{mn}'s are determined from

$$\begin{cases} u'_{mn}(t) + \left[\left(\frac{n\pi}{a}\right)^2 + \left(\frac{m\pi}{b}\right)^2\right] u_{mn}(t) = f_{mn}(t), \\ u_{mn}(0) = u_{0mn}, \quad m,n = 1,2,\ldots, \end{cases}$$

where

$$u_{0mn} = \frac{2}{\sqrt{ab}} \int_{\Omega} u_0(\xi) \sin\left(\frac{n\pi}{a}\xi_1\right) \cdot \sin\left(\frac{m\pi}{b}\xi_2\right) d\xi_1 d\xi_2,$$

$$m,n = 1,2,\ldots,$$

and

$$f_{mn}(t) = \frac{2}{\sqrt{ab}} \int_{\Omega} f(t,\xi) \sin\left(\frac{n\pi}{a}\xi_1\right) \cdot \sin\left(\frac{m\pi}{b}\xi_2\right) d\xi_1 d\xi_2,$$

$$m,n = 1,2,\ldots$$

The rest is left to the reader.

7. The solution is similar to that of the preceding exercise, being based on Exercise 8.13).

8. The given problem is governed by the operator $Q : D(Q) \subset H = L^2(0,3) \to H$ defined by $D(Q) = H^2(0,3) \cap H_0^1(0,3)$, $Qv = -v'' \; \forall v \in D(Q)$. Its energetic extension maps $H_E = H_0^1(0,3)$ into $(H_E)^*$. So the given problem can be regarded as a Cauchy problem for an evolution equation in $(H_E)^*$ and solved by the Fourier method (see [22, Chapter 7]).

On the other hand, in this case we can convert the given problem into a usual one. First of all, observe that the second order distributional derivative of the function

$$\tilde{u}(x) = \alpha(x-1)H(x-1) + \beta(x-2)H(x-2)$$

is precisely $\alpha\delta_1 + \beta\delta_2$, where H denotes the usual Heaviside function. In fact, you just need to observe that the second derivative in the sense of distributions of $x \mapsto (x - x_0)H(x - x_0)$ is δ_{x_0}. Thus, the substitution $y(t,x) = u(t,x) + \tilde{u}(x)$ leads us to the problem

$$\begin{cases} y_t - y_{xx} = 0, & (t,x) \in (0,\infty) \times (0,3), \\ y(t,0) = 0, \; y(t,3) = 2\alpha + \beta, & t \geq 0, \\ y(0,x) = \tilde{u}(x), & x \in [0,3]. \end{cases}$$

A new substitution, namely $z(t,x) = y(t,x) - (2\alpha + \beta)x/3$, leads us to the problem

$$\begin{cases} z_t - z_{xx} = 0, & (t,x) \in (0,\infty) \times (0,3), \\ z(t,0) = 0, \; z(t,3) = 0, & t \geq 0, \\ z(0,x) = \tilde{u}(x) - \frac{2\alpha+\beta}{3}x, & x \in [0,3], \end{cases}$$

which can easily be solved by the Fourier method (please do it!).

9. Here is the mathematical model (see, e.g., [5, p. 460]).

$$\begin{cases} u_{tt} - a^2 u_{xx} = 0, & t \geq 0, \; 0 \leq x \leq l, \\ u(t,0) = 0, \; u(t,l) = 0, & t \geq 0, \\ u(0,x) = 0, \; u_t(0,x) = v_0(x), & 0 \leq x \leq l. \end{cases}$$

Here $a^2 = H/\rho$, where H is the tension in the string, and ρ is the mass per unit length of the string material.

The governing operator is $Q : D(Q) \subset H = L^2(0,l) \to H$, with $D(Q) = H^2(0,l) \cap H^1_0(0,l)$, $Qv = -a^2 v'' \;\; \forall v \in D(Q)$. Its eigenvalues are $\lambda_n = (an\pi/l)^2$, $n = 1,2,\ldots$ and the corresponding orthonormal basis consists of $e_n(x) = \sqrt{2/l}\sin(n\pi x/l)$, $n = 1,2,\ldots$

Using the Fourier method, one can find the solution u of the initial-boundary value problem above as $u(t,x) = \sum_{n=1}^{\infty} u_n(t) e_n(x)$, where the u_n's satisfy the problems

$$\begin{cases} u_n''(t) + (an\pi/l)^2 u_n(t) = 0, & t \geq 0, \\ u_n(0) = u_{0n}, \; u_n'(0) = v_{0n}, & n = 1,2,\ldots, \end{cases}$$

with

$$u_{0n} = 0, \quad v_{0n} = \sqrt{2/l} \int_0^l v_0(\xi)\sin(n\pi\xi/l)\,d\xi, \quad n = 1,2,\ldots,$$

and so on.

10. The mathematical model is similar to that in the previous exercise, with some changes corresponding to the new situation, namely,

$$\begin{cases} u_{tt} - a^2 u_{xx} = 0, & t \geq 0, \; 0 \leq x \leq l, \\ u(t,0) = 0, \; u_x(t,l) = 0, & t \geq 0, \\ u(0,x) = u_0(x), \; u_t(0,x) = 0, & 0 \leq x \leq l. \end{cases}$$

The solution can be determined as a Fourier expansion. In order to find the eigenvalues and eigenfunctions of the governing operator Q, look at the solution to Exercise 10.1 above. For the regularity of the solution, see Theorem 10.4. You can also use the semigroup approach to deduce the existence of a mild solution for $u_0 \in H^1(0, l)$ with $u_0(0) = 0$.

11. Left to the reader.

12. The mathematical model is the following:

$$\begin{cases} u_{tt} - a^2 u_{xx} = \delta(x - l/2), & (t, x) \in (0, \infty) \times (0, l), \\ u(t, 0) = 0, \ u_x(t, l) = 0, & t \geq 0, \\ u(0, x) = 0, \ u_t(0, x) = 0, & 0 \leq x \leq l. \end{cases}$$

As in the case of Exercise 10.8 above, we use a change of the form $y(t, x) = u(t, x) + \tilde{u}(x)$, where

$$\tilde{u}(x) = \frac{c}{a^2}(x - l/2)H(x - l/2),$$

to convert the above initial-boundary value problem into

$$\begin{cases} y_t - y_{xx} = 0, & (t, x) \in (0, \infty) \times (0, l), \\ y(t, 0) = 0, \ y(t, l) = \frac{cl}{2a^2}, & t \geq 0, \\ y(0, x) = \tilde{u}(x), \ y_t(0, x) = 0, & x \in [0, l]. \end{cases}$$

In order to homogenize the boundary condition we use a new change, $z(t, x) = y(t, x) - cx/(2a^2)$, which leads to

$$\begin{cases} z_t - z_{xx} = 0, & (t, x) \in (0, \infty) \times (0, l), \\ z(t, 0) = 0, \ z(t, l) = 0, & t \geq 0, \\ z(0, x) = \tilde{u}(x) - cx/(2a^2), \ z_t(0, x) = 0, & x \in [0, l]. \end{cases}$$

This problem can be easily solved by using the Fourier method and this task is left to the reader. Of course, we could have used a single change, namely,

$$z(t, x) = u(t, x) + \tilde{u}(t, x) - \frac{cx}{2a^2},$$

but we wanted to follow a more transparent procedure.

13. Recall that in this case (see the solution to Exercise 8.12) the eigenvalues of $-\Delta$ with Dirichlet boundary conditions are

$$\lambda_{mn} = \left(\frac{n\pi}{a}\right)^2 + \left(\frac{m\pi}{b}\right)^2, \quad m, n = 1, 2, \ldots$$

and the corresponding orthonormal basis in $H = L^2(\Omega)$ consists of

$$e_{mn}(x) = \frac{2}{\sqrt{ab}} \sin\left(\frac{n\pi}{a}x_1\right) \cdot \sin\left(\frac{m\pi}{b}x_2\right), \quad m, n = 1, 2, \ldots$$

To complete the task see the solution to Exercise 10.6 above.

12.11 Answers to Exercises for Chap. 11

1. Recall that for a given kernel $k = k(t, s) \in C(\Delta)$, $\Delta = \{(t, s) \in \mathbb{R}; a \le s \le t \le b\}$, the resolvent kernel $R(t, s)$ is defined by

$$R(t, s) = \sum_{n=1}^{\infty} k_n(t, s), \quad (t, s) \in \Delta,$$

where

$$k_1(t, s) = k(t, s),$$
$$k_n(t, s) = \int_s^t k(t, \tau)k_{n-1}(\tau, s)\, d\tau, \quad (t, s) \in \Delta, \ n = 2, 3, \ldots$$

Note that the interval $[a, b]$ could be replaced by $[a, +\infty)$ if the corresponding Volterra equation is considered on $[a, +\infty)$.

(a) By easy computations we find

$$R(t, s) = e^{t^2 - s^2 + t - s}, \quad x(t) = e^{t(t+1)}, \quad t \ge 0.$$

Alternatively, denoting $y(t) = e^{-t^2}x(t)$, the given equation can be written as

$$y(t) = 1 + \int_0^t y(s)\, ds, \quad t \ge 0,$$

which is equivalent to the problem

$$\begin{cases} y'(t) = y(t), \quad t \ge 0, \\ y(0) = 1, \end{cases}$$

so we obtain again the solution x.

(b) $R(t, s) = \dfrac{2 + \cos t}{2 + \cos s} e^{t-s}$,

   $x(t) = e^t \sin t + e^t (2 + \cos t) \ln \dfrac{3}{2t + \cos t}$.

(c) $R(t, s) = \sinh(t - s)$, $x(t) = \sinh t$.

2. (a) If x is a solution of the given integral equation, then $x(0) = 0$ and

$$x'(t) = 1 - \frac{t^2}{2} + x(t) + \int_0^t x(s)\, ds,\ \ t \ge 0.$$

Thus $x'(0) = 1$ and

$$x'' = x'(t) + x(t) - t.$$

Thus we have obtained the Cauchy problem

$$\begin{cases} x'' - x'(t) - x(t) = -t,\ \ t \ge 0, \\ x(0) = 0,\ x'(0) = 1. \end{cases}$$

Conversely, if x is a solution to this problem, then x satisfies the given integral equation.

By easy computations we find

$$x(t) = c_1 e^{(1+\sqrt{5})t/2} + c_2 e^{(1-\sqrt{5})t/2} + t - 1,$$

with

$$c_1 = \frac{5 - \sqrt{5}}{2}, \quad c_2 = \frac{5 + \sqrt{5}}{2}.$$

(b) The equivalent Cauchy problem is

$$\begin{cases} x''(t) + x(t) = 6t, \qquad t \ge 0, \\ x(0) = 1,\ x'(0) = 0. \end{cases}$$

By easy computations we find

$$x(t) = \cos t - 6 \sin t + 6t,\ \ t \ge 0.$$

(c) If x is a solution, then $x(0) = 0$. By differentiation we obtain from the given integral equation

$$x'(t) = 3 - x(t) - \underbrace{\int_0^t e^{t-s} x(s)\, ds}_{3t - x(t)},\ \ t \ge 0,$$

hence x satisfies the problem

$$\begin{cases} x'(t) = 3 - 3t, & t \geq 0, \\ x(0) = 0, \end{cases}$$

which is equivalent to the given integral equation and has the solution

$$x(t) = \frac{3}{2}t(2 - t), \quad t \geq 0.$$

3. (a) From the given integral equation we obtain by differentiation

$$x(t) - 2t \int_0^t x(s)\,ds = t, \quad t \geq 0.$$

Then $y(t) = \int_0^t x(s)\,ds$ satisfies the Cauchy problem

$$\begin{cases} y'(t) - 2ty(t) = t, & t \geq 0, \\ y(0) = 0 \end{cases}$$

which has the solution

$$y(t) = \frac{1}{2}(e^{t^2} - 1), \quad t \geq 0 \Longrightarrow x(t) = te^{t^2}, \quad t \geq 0.$$

(b) From the given integral equation we obtain by differentiation

$$x(t) - \int_0^t \sin(t - s) \cdot x(s)\,ds = 2(2t + 1), \quad t \geq 0,$$

and so $x(0) = 2$. Another differentiation leads to

$$x'(t) - \underbrace{\int_0^t \cos(t - s) \cdot x(s)\,ds}_{2t(t+1)} = 4, \quad t \geq 0.$$

So we have obtained the problem

$$\begin{cases} x'(t) = 2(t^2 + t + 2), & t \geq 0, \\ x(0) = 2, \end{cases}$$

which is equivalent to the given integral equation and gives the solution

$$x(t) = \frac{2}{3}t^3 + t^2 + 4t + 2, \quad t \geq 0.$$

(c) $x(t) = (\cos t - t\cos t - t\sin t)e^{-2t}, \quad t \geq 0.$

4. $R(t,s)$ is a continuous function on the triangle $\Delta_0 = \{(t,s); 0 \le s \le t \le b\}$, being defined by

$$R(t,s) = \sum_{n=1}^{\infty} k_n(t,s), \quad (t,s) \in \Delta_0,$$

where

$$k_1(t,s) = k(t,s) = h(t-s),$$

$$k_n(t,s) = \int_s^t k(t,\tau)k_{n-1}(\tau,s)\,d\tau$$

$$= \int_s^t h(t-\tau)k_{n-1}(\tau,s)\,d\tau, \quad (t,s) \in \Delta_0, \ n = 2, 3, \ldots$$

Since k_1 depends on $t-s$ only, we can easily observe (by a change of variable) that so is k_2. It follows by induction that all the k_n's depend on $t-s$ only $\implies$ so is R.

5. We can easily show by induction that $R(t,s) \ge 0$, $0 \le s \le t \le b$. Next, denote

$$\phi(t) = f(t) + \int_a^t k(t,s)x(s)\,ds - x(t) \ge 0, \quad t \in [a,b].$$

Hence,

$$x(t) = f(t) - \phi(t) + \int_a^t k(t,s)x(s)\,ds, \quad t \in [a,b],$$

which implies

$$x(t) = f(t) - \phi(t) + \int_a^t R(t,s)[f(s) - \phi(s)]\,ds$$

$$= f(t) + \int_a^t R(t,s)f(s)\,ds$$

$$\underbrace{- \left[\phi(t) + \int_a^t R(t,s)\phi(s)\,ds\right]}_{\ge 0}, \quad t \in [a,b].$$

So the conclusion is obvious.

6. Left to the reader.

7. We prefer to use the following Bielecki-like norm in $X = C(D)$:

$$\| g \|_B = \sup_{(t,s) \in Q} e^{-M(t+s)} \mid g(t,s) \mid, \quad g \in X,$$

where M is a large positive constant. Define an operator P on X by

$$(Pg)(t,s) = f(t,s) + \int_0^t \int_0^s k(t,s,\xi,\eta) g(\xi,\eta) \, d\eta d\xi,$$

$(t,s) \in D, \ g \in X.$

Clearly, P maps X into itself, and for $g_1, g_2 \in X$ and $(t,s) \in D$ we have

$$\mid (Pg_1)(t,s) - (Pg_2)(t,s) \mid$$

$$\leq C \int_0^t \int_0^s \mid g_1(\xi,\eta) - g_2(\xi,\eta) \mid \, d\eta d\xi$$

$$= C \int_0^t \int_0^s e^{+M(\eta+\xi)} e^{-M(\eta+\xi)} \mid g_1(\xi,\eta) - g_2(\xi,\eta) \mid \, d\eta d\xi$$

$$\leq C \| g_1 - g_2 \|_B \int_0^t \int_0^s e^{M(\eta+\xi)} \, d\eta d\xi$$

$$= \frac{C}{M^2} \| g_1 - g_2 \|_B (e^{Mt} - 1)(e^{Ms} - 1),$$

where $C = \sup_{(t,s,\xi,\eta) \in Q} \mid k(t,s,\xi,\eta) \mid < \infty$. It follows that

$$e^{-M(t+s)} \mid (Pg_1)(t,s) - (Pg_2)(t,s) \mid$$

$$\leq \frac{C}{M^2} \| g_1 - g_2 \|_B (1 - e^{-Mt})(1 - e^{-Ms})$$

$$\leq \frac{C}{M^2} \| g_1 - g_2 \|_B,$$

for all $(t,s) \in D, \ g_1, g_2 \in X$. Hence

$$\| Pg_1 - Pg_2 \|_B \leq \frac{C}{M^2} \| g_1 - g_2 \|_B, \quad g_1, g_2 \in X,$$

so P is a contraction on $(X, \| \cdot \|_B)$ for $M^2 > C$. Therefore, according to the Banach Contraction Principle, P has a unique fixed point $x = x(t, s) \in X$ which is the unique solution of equation (E).

In order to prove the result the reader may also use other methods, similar to those discussed in Sect. 11.1.

8. The given problem is equivalent to the following integral equation in $X = C[0, T]$

$$x(t) = x_0 + \int_0^t f(s)\,ds + \int_0^t \left(\int_0^s k(s, \tau) x(\tau)\,d\tau \right) ds, \ \ t \in [0, T].$$
$$(*)$$

Define $P : X \to X$ by

$$(Pg)(t) = x_0 + \int_0^t f(s)\,ds + \int_0^t \left(\int_0^s k(s, \tau) g(\tau)\,d\tau \right) ds,$$
$$t \in [0, T], \ g \in X.$$

One can show by a fixed point approach that P has a unique fixed point $x \in X$, which is the unique solution of equation $(*)$, and hence of the given problem.

9. (a) The equation can be written as

$$x(t) = \cos t + \lambda c_1 \sin t + \lambda c_2 \cos t, \qquad (*)$$

with

$$c_1 = \int_0^\pi \cos s \cdot x(s)\,ds, \quad c_2 = -\int_0^\pi \sin s \cdot x(s)\,ds. \quad (**)$$

If we substitute $(*)$ into $(**)$, we obtain the following algebraic system in c_1, c_2 :

$$\begin{cases} c_1 - \frac{\lambda\pi}{2} c_2 = \frac{\pi}{2}, \\ \frac{\lambda\pi}{2} c_1 + c_2 = 0. \end{cases}$$

Note that the determinant of this system is positive for all $\lambda \in \mathbb{R}$, so there exists a unique solution (c_1, c_2) which gives the solution of the given integral equation (see $(*)$)

$$x(t) = \frac{2(2\cos t + \lambda\pi \sin t)}{4 + \lambda^2\pi^2}.$$

(b) We have $x(t) = t + \lambda c \sin t$, where $c = \int_0^{2\pi} |\pi - s| \cdot x(s)\, ds$. Substituting $x(t)$ given by the first relation into the second yields

$$c = \int_0^{2\pi} |\pi - s| \cdot (s + \lambda c \sin s)\, ds \iff c(1 - 2\lambda \pi) = \pi^3.$$

Therefore, if $\lambda = 1/(2\pi)$ the given integral equation has no solution, otherwise (i.e., for $\lambda \in \mathbb{R} \setminus \{1/(2\pi)\}$) the equation has the solution

$$x(t) = t + \frac{\lambda \pi^3}{1 - 2\lambda \pi} \sin t.$$

(c) We have

$$x(t) = f(t) + \lambda c_1 - 3\lambda c_2 t, \qquad (*)$$

where

$$c_1 = \int_0^t x(s)\, ds, \quad c_2 = \int_0^t s x(s)\, ds.$$

Thus we get the system

$$\begin{cases} c_1 = \int_0^1 [f(s) + \lambda c_1 - 3\lambda c_2 s]\, ds, \\ c_2 = \int_0^1 s[f(s) + \lambda c_1 - 3\lambda c_2 s]\, ds, \end{cases}$$

or

$$\begin{cases} (1 - \lambda)c_1 + \frac{3}{2}\lambda c_2 = \int_0^1 f(s)\, ds, \\ -\frac{1}{2}c_1 + (1 + \lambda)c_2 = \int_0^1 s f(s)\, ds. \end{cases}$$

The determinant of this algebraic system is $\Delta = (4 - \lambda^2)/4$. So, for each $\lambda \in \mathbb{R} \setminus \{-2, +2\}, c_1, c_2$ can be uniquely determined and the solution of the given integral equation can be explicitly expressed by using formula $(*)$.

If $\lambda = -2$ the above algebraic system has solutions if and only if

$$\int_0^1 f(s)\, ds = 3 \int_0^1 s f(s)\, ds \qquad (**)$$

and in this case there are infinitely many solutions of the given integral equation, namely (see $(*)$),

$$x(t) = f(t) + 2c_1(3t - 1) - 2t \int_0^1 f(s)\, ds, \quad c_1 \in \mathbb{R}.$$

An example of a function satisfying condition $(**)$ above is $f(t) = t - 1$.

If condition $(**)$ is not satisfied, then the given integral equation has no solution.

If $\lambda = +2$ the compatibility condition for the above algebraic system is

$$\int_0^1 f(s)\, ds = \int_0^1 sf(s)\, ds$$

and, if this condition is satisfied (e.g., $f(t) = 3t - 1$), we have again infinitely many solutions for the given integral equation,

$$x(t) = f(t) + 2c_1(1 - t) - 2t \int_0^1 f(s)\, ds, \quad c_1 \in \mathbb{R}.$$

Otherwise, the given integral equation has no solution.

10. Denote

$$c = \begin{bmatrix} c_1 \\ \vdots \\ c_n \end{bmatrix}, \quad g = \begin{bmatrix} f_1 \\ \vdots \\ f_n \end{bmatrix}, \quad K = (k_{ij})_{1 \le i, j \le n}.$$

So (3) can be written as follows:

$$(I - \lambda K)c = g. \tag{3'}$$

There is a bijective correspondence between the solution sets of (F) and $(3')$.

The following alternative for equation $(3')$ is well known:

(j) if $\det(I - \lambda K) \ne 0$, then there is a unique solution of $(3')$ given by

$$c = (I - \lambda K)^{-1}g,$$

which gives the solution of (F) by means of (2);

(jj) otherwise, $\det(I - \lambda K) = 0$ and equation $(3')$ has solutions $\iff g$ is orthogonal to $N(I - \bar{\lambda}K^*) = N(I - \bar{\lambda}\bar{K}^T)$, so equation (F) has infinitely many solutions,

$$x(t) = x_p(t) + \sum_{i=1}^{m} \alpha_i x_i(t),$$

where x_p is a particular solution of (F), $\alpha_1, \ldots, \alpha_m \in \mathbb{K}$. and $x_1, \ldots, x_m$ are independent solutions of the homogeneous integral equation (which can be calculated explicitly).

11. If $\lambda = 0$, then there is a unique solution $x = f$. From now on we consider $\lambda \in \mathbb{K} \setminus \{0\}$. Denote $H_m = \text{Span}(\{e_1, \ldots, e_m\})$. From the solution to Exercise 8.9, we know that A is symmetric (hence its eigenvalues are real numbers), $R(A) = H_m$ and $N(A) = H_m^\perp$. In fact, it is easy to see that the eigenvalues of A are $\mu_k = k$, $k = 1, \ldots, m$, with $e_1, \ldots, e_m$ as corresponding eigenvectors.

We distinguish two cases:

Case 1. $\dim H = m$, i.e., $H = H_m$. Then the given Fredholm equation is a simple algebraic system,

$$(I - \lambda A)x = f. \tag{1}$$

If $\lambda \in \mathbb{K} \setminus \{1, 1/2, \ldots, 1/m\}$, then (1) has the unique solution

$$x = \sum_{k=1}^m \frac{(f, e_k)}{1 - \lambda k} e_k.$$

If $\lambda = 1/j$ for some $j \in \{1, \ldots, m\}$, then system (1) is solvable if and only if $(f, e_j) = 0$. In this situation, there are infinitely many solutions x with coordinates $x_k = j(f, e_k)/(j - k), k \neq j$, and $x_j \in \mathbb{K}$ being arbitrary.

Case 2. $\dim H > m$. Of course, $H = H_m \oplus H_m^\perp$ with $H_m^\perp \neq \{0\}$. We look for x of the form $x = x_1 + x_2, x_1 \in H_m, x_2 \in H_m^\perp$. Using a similar decomposition for f, i.e., $f = f_1 + f_2, f_1 \in H_m, f_2 \in H_m^\perp$, we derive from (1) that $x_2 = f_2$, and $(I - \lambda A)x_1 = f_1$. Using the same discussion as before, we can find x_1, when it exists, so we conclude that $x = x_1 + f_2$.

12. By the Weierstrass M−test, we have

$$k \in C(\bar{Q}) \subset L^2(Q), \quad Q = (0, 1) \times (0, 1).$$

Obviously, A is self-adjoint and compact.

Case $m = 0$. In this case, $N(A) = \{0\}$. Indeed, $Ag = 0$ implies

$$0 = (Ag, g)_{L^2} = \sum_{n=1}^\infty \frac{1}{(n+1)^2} (g, u_n)_{L^2}^2,$$

hence

$$(g, u_n)_{L^2} = 0 \quad \forall n \in \{0, 1, 2, \ldots\} \quad \Longrightarrow \quad g = 0,$$

since the system $\{u_n\}_{n=0}^{\infty}$ is a basis in $H = L^2(0, 1)$.

In order to determine the eigenpairs of A consider the equation

$$Ag = \mu g,$$

which can be written as

$$\sum_{n=0}^{\infty} \frac{(g, u_n)_{L^2}}{(n+1)^2} u_n = \mu \sum_{n=0}^{\infty} (g, u_n)_{L^2} u_n,$$

where we have used the Fourier expansion of g. As $\{u_n\}_{n=0}^{\infty}$ is a basis in H, we have

$$\left(\mu - \frac{1}{(n+1)^2}\right) (g, u_n)_{L^2} = 0, \quad n = 0, 1, 2, \ldots \qquad (*)$$

If $\mu \neq \frac{1}{(n+1)^2}$ for all $n \in \{0, 1, 2, \ldots\}$ then

$$(g, u_n)_{L^2} = 0 \quad \forall n \in \{0, 1, 2, \ldots\} \quad \Longrightarrow \quad g = 0,$$

hence such μ's are not eigenvalues of A. For $\mu = \mu_n = \frac{1}{(n+1)^2}$ we have from $(*)$

$$(g, u_k)_{L^2} = 0, \quad \forall k \in \mathbb{N}, \quad k \neq n,$$

so the eigenfunctions corresponding to $\mu_n = \frac{1}{(n+1)^2}$ are nonzero multiples of u_n.

According to the Schmidt formula we have for $\lambda \in \mathbb{R} \setminus \{1, 2^2, 3^2, \ldots\}$ and a.a. $t \in (0, 1)$,

$$x(t) = f(t) + 2\lambda \sum_{k=0}^{\infty} \frac{\int_0^1 f(s) \cos\left((k + 1/2)\pi s\right) ds}{(k+1)^2 - \lambda}$$

$$\times \cos\left((k + 1/2)\pi t\right) + \alpha \cos\left((n + 1/2)\pi t\right), \quad \alpha \in \mathbb{R}.$$

Case $m \geq 1$. In this case $Y_0 := N(A) = \text{Span}(\{u_0, u_1, \ldots, u_{m-1}\})$ and $H = Y_0 \oplus Y_1$, where $Y_1 = N(A)^{\perp} = \text{Span}(\{u_m, u_{m+1}, \ldots\})$.

Denote by A_1 the restriction of A to Y_1 which is a Hilbert space with respect to the scalar product and norm of $H = L^2(0, 1)$. Obviously, A_1 maps Y_1 to itself, being compact, self-adjoint, with

$N(A_1) = \{0\}$, and with eigenvalues $\mu_n = 1/(n+1)^2$ and eigen-functions $u_n, n \geq m+1$. In fact, Y_1 and A_1 play the roles of H and A we had before.

The equation

$$x = f + \lambda A x$$

can be written as

$$x_0 + x_1 = f_0 + f_1 + \lambda A x_1,$$

where $x_0, f_0 \in Y_0$ and $x_1, f_1 \in Y_1$, so $x_0 = f_0$ and

$$x_1 = f_1 + \lambda A_1 x_1. \qquad (**)$$

Based on the above arguments, we have

- if $\lambda \neq (n+1)^2$ for all $n \geq m$ then

$$x(t) = f_0(t) + x_1(t)$$
$$= f(t) + 2\lambda \sum_{k=m}^{\infty} \frac{\int_0^1 f_1(s) \cos\left((k+1/2)\pi s\right) ds}{(k+1)^2 - \lambda}$$
$$\times \cos\left((k+1/2)\pi t\right),$$

and

- if $\lambda = (n+1)^2$ for some $n \geq m$, then the Fredholm equation $(**)$ has solutions if and only if $f_1 \perp u_n \iff f \perp u_n$, and in this case

$$x(t) = f(t) + (2n+1)^2$$
$$\times \sum_{k \geq m, k \neq n} \frac{\int_0^1 f_1(s) \cos\left((k+1/2)\pi s\right) ds}{(k+1)^2 - \lambda}$$
$$\times \cos\left((k+1/2)\pi t\right) + \alpha \cos\left((n+1/2)\pi t\right), \alpha \in \mathbb{R}.$$

Bibliography

[1] Adams, R. A., *Sobolev Spaces*, Academic Press, New York–San Francisco–London, 1975.

[2] Ambrosetti, A. and Arcoya, D., *An Introduction to Nonlinear Functional Analysis and Elliptic Problems*, Springer, 2011.

[3] Banţă, V., *Partial Differential Equations. Collection of Problems*, University of Bucharest, Bucharest, 1989 (in Romanian).

[4] Barbu, V., *Partial Differential Equations and Boundary Value Problems*. Mathematics and its Applications, 44, Kluwer, Dordrecht, 1988.

[5] Boyce, W. E. and DiPrima, R. C., *Elementary Differential Equations and Boundary Value Problems*, Second Edition, John Wiley, New York–London–Sydney–Toronto, 1969.

[6] Brezis, H., *Functional Analysis, Sobolev Spaces and Partial Differential Equations*, Springer, 2011.

[7] Butzer, P. L. and Berens, H., *Semi-Groups of Operators and Approximation*, Springer, 1967.

[8] Corduneanu, C., *Principles of Differential and Integral Equations*, Second Edition, Chelsea Publishing Co., Bronx, N.Y., 1977.

[9] Corduneanu, C., *Integral Equations and Applications*, Cambridge University Press, Cambridge, 1991.

[10] Costara, C. and Popa, D., *Exercises in Functional Analysis*, Kluwer, 2003.

© Springer Nature Switzerland AG 2019

G. Moroşanu, *Functional Analysis for the Applied Sciences*,
Universitext, https://doi.org/10.1007/978-3-030-27153-4

[11] Cronin, J., *Differential Equations. Introduction and Qualitative Theory*, Third edition, Chapman & Hall/CRC, 2008.

[12] Engel, K.-J. and Nagel, R., *One-Parameter Semigroups for Linear Evolution Equations*, Graduate Texts in Math., Vol. 194, Springer-Verlag, 2000.

[13] Engel, K.-J. and Nagel, R., *A Short Course on Operator Semigroups*, Springer-Verlag, 2010.

[14] Evans, L. C., *Partial Differential Equations*, Graduate Studies in Math. 19, Amer. Math. Soc., Providence, Rhode Island, 1998.

[15] Friedman, A., *Foundation of Modern Analysis*, Dover, New York, 1982.

[16] Gel'fand, I. M., *Lectures on Linear Algebra*, Dover, 1989.

[17] Gel'fand, I. M. and Shilov, G. E., *Generalized Functions. Vol. 2. Spaces of Fundamental and Generalized Functions*, Academic Press, New York–London, 1968.

[18] Godunov, A. N., The Peano theorem in Banach spaces, *Funct. Anal. Appl* **9** (1975), no. 1, 53–55.

[19] Goldstein, J. A., *Semigroups of Operators and Applications*, Oxford University Press, 1985.

[20] Hammerstein, A., Nichtlineare integralgleichungen nebst anwendungen, *Acta Math.* **54** (1930), No. 1, 117–176.

[21] Hille, E. and Phillips, R. S., *Functional Analysis and Semigroups*, Amer. Math. Soc. Coll. Publ., Vol. 31, Amer. Math. Soc., 1957.

[22] Hokkanen, V.-M. and Moroşanu, G., *Functional Methods in Differential Equations*, Chapman & Hall/CRC, 2002.

[23] Iftimie, V., *Partial Differential Equations*, University of Bucharest, 1980 (in Romanian).

[24] Kato, T., *Remarks on pseudo-resolvents and infinitesimal generators of semi-groups*, Proc. Japan Acad. **35** (1959), 467–468.

[25] Kōmura, Y., Nonlinear semi-groups in Hilbert space, *J. Math. Soc. Japan* **19** (1967), No. 4, 493–507.

[26] Krasnosel'skii, M. A., *Topological Methods in the Theory of Non-linear Integral Equations*, Pergamon, 1964.

[27] Krasnov, M., Kissélev, A. and Makarenko, G., *Équations intégrales*, Mir, Moscou, 1976.

[28] Kurosh, A., *Cours d'algèbre supérieure*, Mir, Moscou, 1973.

[29] Lang, S., *Real and Functional Analysis*, Third Edition, Springer, New York, 1993.

[30] Lebedev, N. N., *Special Functions and Their Applications*, Revised English edition, translated and edited by R.A. Silverman, Prentice-Hall, Inc., Englewood Cliffs, N.J., 1965.

[31] Lions, J. L. and Magenes, E., *Problèmes aux limites non homogènes et applications*, Vol. 1, Dunod, Paris, 1968.

[32] Mărcuş, A., *Introduction to Mathematical Logic and Set Theory*, Course Notes, 2017 (in Romanian).

[33] Marsden, J. E. and Hoffman, M. J., *Elementary Classical Analysis*, Second edition, W. H. Freedman & Co., New York, 1993.

[34] Micu, S. and Zuazua, E., An introduction to the controllability of partial differential equations, in "Quelques questions de théorie du contrôl", T. Sari, ed., Collection Travaux en Cours Hermann, 2004, pp. 69–157.

[35] Milman, V., *An Introduction to Functional Analysis*, World, 1999.

[36] Moroşanu, G., *Nonlinear Evolution Equations and Applications*, D. Reidel, Dordrecht–Boston–Lancaster–Tokyo, 1988.

[37] Moroşanu, G., *Elements of Linear Algebra and Analytic Geometry*, Matrix Rom, Bucharest, 2000 (in Romanian).

[38] Natanson, I. P., *Theory of Functions of a Real Variable*, Editura tehnică, Bucharest, 1957 (in Romanian).

[39] Pazy, A., *Semigroups of Linear Operators and Applications to Partial Differential Equations*, Springer-Verlag, 1983.

[40] Popa, E., *Collection of Functional Analysis Problems*, Editura didactică şi pedagogică, Bucharest, 1981 (in Romanian).

[41] Rosenlicht, M., *Introduction to Analysis*, Dover, New York, 1968.

[42] Rudin, W., *Principles of Mathematical Analysis*, Third Edition, McGraw-Hill, 1976.

[43] Schechter, M., *Principles of Functional Analysis*, Academic Press, New York–London, 1971.

[44] Shilov, G. Ye., *Mathematical Analysis. A Special Course*, Pergamon Press, Oxford–New York–Paris, 1965.

[45] Showalter, R. E., *Monotone Operators in Banach Space and Nonlinear Partial Differential Equations*, Math. Surveys and Monographs, Vol. 49, Amer. Math. Soc., 1997.

[46] Stein, E. M. and Shakarchi, R., *Real Analysis. Measure Theory, Integration, and Hilbert Spaces*, Princeton University Press, Princeton and Oxford, 2005.

[47] Stroock, D. W., Weyl's lemma, one of many, *Groups and Analysis*, 164–173, *London Math. Soc. Lecture Notes Ser.*, 354, Cambridge Univ. Press, Cambridge, 2008.

[48] Trotter, H. F., *Approximation of semi-groups of operators*, Pacific J. Math. **8** (1958), 887–919.

[49] Vrabie, I. I., *Semigroups of Linear Operators and Applications*, Editura Universităţii "Alexandru Ioan Cuza", Iaşi, 2001 (in Romanian).

[50] Wheeden, R. L. and Zygmund, A., *Measure and Integral. An Introduction to Real Analysis*, Marcel Dekker, Inc., 1977.

[51] Yosida, K., *Functional Analysis*, Third Edition, Springer, 1971.

[52] Zeidler, E., *Applied Functional Analysis. Applications to Mathematical Physics*, Appl. Math. Sci. 108, Springer-Verlag, 1995.

[53] Zeidler, E., *Applied Functional Analysis. Main Principles and Their Applications*, Appl. Math. Sci. 109, Springer-Verlag, 1995.

Printed in the United States
By Bookmasters